Moses among the Moderns

# Scientific and Learned Cultures and Their Institutions

*Edited by*

Mordechai Feingold (*California Institute of Technology*)

**VOLUME 36**

# Moses among the Moderns

*German Constructions of Biblical Law, 1750–1930*

*Edited by*

Paul Michael Kurtz

BRILL

LEIDEN | BOSTON

Cover illustration: Postcard, ca. 1906; illustration by D. Bernstein of Brussels; object in the private holding of Paul Kurtz; image in the public domain.

Library of Congress Cataloging-in-Publication Data

Names: Kurtz, Paul Michael, editor.
Title: Moses among the moderns : German constructions of biblical law, 1750–1930 / edited by Paul Michael Kurtz.
Description: Leiden ; Boston : Brill, [2024] | Series: Scientific and learned cultures and their institutions, 2352–1325 ; volume 36 | Includes bibliographical references and index.
Identifiers: LCCN 2024002793 (print) | LCCN 2024002794 (ebook) | ISBN 9789004691766 (hardback) | ISBN 9789004691780 (ebook)
Subjects: LCSH: Moses (Biblical leader)–Influence. | Leadership–Religious aspects. | Jewish law. | Bible. Old Testament–Criticism, interpretation, etc.
Classification: LCC BS580.M6 M665 2024  (print) | LCC BS580.M6  (ebook) | DDC 296.1/8094309034–dc23/eng/20240213
LC record available at https://lccn.loc.gov/2024002793
LC ebook record available at https://lccn.loc.gov/2024002794


Typeface for the Latin, Greek, and Cyrillic scripts: "Brill". See and download: brill.com/brill-typeface.

ISSN 2352-1325
ISBN 978-90-04-69176-6 (hardback)
ISBN 978-90-04-69178-0 (e-book)
DOI 10.1163/9789004691780

# Contents

# Acknowledgements

The foundation of this volume is a workshop convened at St John's College, Cambridge, on 22–24 July 2019. Its bedrock is 'Writing Jewish History: Ancient Judaism as a Political Problem in Central Europe at the Rise of the Nation State', a project hosted at the University of Cambridge in 2017–2019 and funded by the European Union (Horizon 2020 research and innovation program, Marie Skłodowska-Curie grant agreement No. 749628). In addition to that Marie Curie fellowship, generous support for that workshop came from the DAAD–Cambridge Research Hub in German Studies as well as the Cambridge Inter-faith Programme. The Cambridge Faculty of Divinity and the DAAD–Cambridge Research Hub also lent assistance for organization and promotion.

The subsequent stages of editorial work received financial backing through a postdoctoral fellowship from the Flemish Research Council (FWO) for a project entitled 'Modern Germans, Ancient Jews: Political Representations of Early Judaism in the German-Speaking Lands of the 19th Century'. Financing for open-access publication flowed from a different one, sponsored by the Ghent University Special Research Fund and called 'Philology, Queen of Science: The Making of a Knowledge-System in the Nineteenth Century (PhiQoS)'.

Intellectual, advisory, and moral support came quickly, kindly, and ceaselessly from Nathan MacDonald. So too Suzanne Marchand offered valuable guidance in the early conceptual stages, during the workshop in the middle, and on the final manuscript in the end. As ever, Daniel Pioske, Emiliano Urciuoli, and Rebecca Van Hove provided insight and encouragement from start to finish. Christian Hoekema generously helped with compiling the index. My gratitude also extends to Mordechai Feingold, who showed much enthusiasm for the volume and great patience all along the way.

*Paul Kurtz*
Groningen, July 2023

# Figures

# Contributors

*Judith Frishman*
Leiden University, The Netherlands (J.Frishman@hum.leidenuniv.nl)

*Ofri Ilany*
Van Leer Jerusalem Institute, Israel (ofrilany@gmail.com)

*Carolin Kosuch*
Georgia Augusta University of Göttingen, Germany (carolin.kosuch@uni-goettingen.de)

*Paul Michael Kurtz*
Ghent University, Belgium (paulmichael.kurtz@ugent.be)

*Michael Ledger-Lomas*
King's College London, United Kingdom (michael.ledger-lomas@kcl.ac.uk)

*Suzanne Marchand*
Louisiana State University, United States (smarch1@lsu.edu)

*Carlotta Santini*
École normale supérieure de Paris – Pays Germaniques (UMR 8547), France (carlottasantini@hotmail.it)

*Felix Wiedemann*
Freie Universität Berlin, Germany (felix.wiedemann@fu-berlin.de)

*Irene Zwiep*
University of Amsterdam, The Netherlands (I.E.Zwiep@uva.nl)

# Introduction

*Moses in Modernity*

*Paul Michael Kurtz*

One summer's day in 1898, an uncle wrote his nephew about the latter's upcoming trip abroad. It was furniture he wanted. The episode might be otherwise unremarkable were that uncle not the grand duke of Baden, that nephew the king of Prussia, and that furniture the Ark of the Covenant. This relic from Hebrew antiquity, a wooden chest covered in gold, reportedly contained the work of Moses – and with it a key to preserving bible, church, and crown.

When the Baden duke wrote the Prussian king, the impetus had come from a Swedish surveyor and archivist by the name of Henning Melander (1858–1933). Earlier that spring, the Swede had published a series of articles in the Zionist weekly *Die Welt*, beneath the title 'Could the Ark of the Covenant Be Found Again?'.[1] Melander not only answered yes but even cast himself as just the man for the job. He was convinced this sacred artifact still lay there for the taking, buried long ago in Jerusalem by the prophet Jeremiah. More than serve a practical purpose or as an aesthetic decoration, the chest encapsulated the cultural, or rather religious, heritage of Mosaic Israel. Evoking J.W. Goethe's (1749–1832) sense of morphology, from his letter to J.G. Herder (1744–1802), Melander asserted, 'The whole of Israel is concentrated in the Ark of the Covenant, as the plant contains the whole tree in itself', which included 'the archive kept safe with it'.[2]

But he convinced more than himself. The article by Melander entered the ducal hands together with Theodor Herzl's 1896 *Judenstaat* (*Jewish State*), a foundational document of Zionism. It was an eccentric priest who put them there: William Henry Hechler (1845–1931), chaplain to the British embassy in Vienna and champion for a Jewish homeland in Palestine. In fact, this German-born Anglican had helped put Melander's article in *Die Welt* in the first place, a journal founded by Herzl (1860–1904). The reverend had also placed his own work in the pages of its very first issue, wherein he expressed his conviction that 'a "*Jewish state*" must rise again in Palestine', all the more 'with the agreement

---

1  Henning Melander, 'Könnte man die Bundeslade wiederfinden?', *Die Zeit* 2/16 (22 April 1898), 3–4; 2/17 (29 April 1898), 2–4; 2/18 (6 May 1898), 7–8; 2/19 (13 May 1898), 5–6; 2/20 (20 May 1898), 9. He ended the article with a plea for financing, citing lack of interest in his homeland and lack of personal funds.

2  Ibid., 2/17, 2, 3.

and kind help of the European princes'.[3] Hechler formed one half of a Zionist odd couple with Herzl. The committed Christian showed no less dedication to the goal than the secular Jew – but far more devotion to the precise geography. Certainly well-connected, if not always well-respected, the Christian Zionist helped leverage his political connections for the cause.

Herzl and Hechler worked partly in tandem, partly at cross purposes as they chased those princes around Europe in search of political backing.[4] They found it in Friedrich I of Baden (1826–1907), at least for a time. In July of 1898, after a couple years' imploring and cajoling, the Zionist duo prevailed upon Friedrich I to write to Wilhelm II (1859–1941), the Prussian king and German emperor. Broaching Jewish settlements in Palestine, Herzl's plea for an audience, and Hechler's theo-archaeological ambitions, the nobleman reported on the clergyman, 'He also revealed to me the secret of the "Ark of the Covenant" and said: all the hopes for recovering it would turn on you.'[5] In this missive to his imperial nephew, the royal uncle included the 'scientific, thorough work' printed in *Die Zeit* – checked by 'German and British theologians' and championed by Hechler – which discussed 'theologically, historically' the quest for the holy grail that was the holy ark. He opined, 'The historical interest in the course of the peregrination of the Ark of the Covenant up to the place it now lies is, in fact, very meaningful – but of course the act of discovery would be full of meaning as a historical moment in the entire world'.[6] The duke also forwarded a suggestion for the German emperor to obtain from the Ottoman sultan, Abdul Hamid II (1842–1918), the area in question for 'antiquarian research' – though, crucially, without divulging 'the goal of the research' since he 'would otherwise probably not be inclined to comply'.

3  William Henry Hechler, 'Christen über die Judenfrage', *Die Welt* 1/2 (1897), 7–9, at 7, 8. He continued, 'But if the Zionist movement progresses so eagerly and actively as it now does in the entire world, then this wonderful nineteenth century of electricity and the railroad – where everything moves fast and which has seen the formation of the new German Empire and of other empires – can finally still witness the foundation of the new Jewish state' (ibid., 8).

4  For more on this story, see Isaiah Friedman, *Germany, Turkey, and Zionism, 1897–1918* (Oxford, 1977); Paul Charles Merkley, *The Politics of Christian Zionism, 1891–1948* (New York, 1998), 3–34; Victoria Clark, *Allies for Armageddon: The Rise of Christian Zionism* (New Haven, 2007).

5  Friedrich I to Wilhelm II, 28 July 1898, in Hermann Ellern and Bessi Ellern, *Herzl, Hechler, the Grand Duke of Baden and the German Emperor, 1896–1904, documents found ... reproduced in facsimile* (Tel Aviv, 1961), letter 12, pp. 32–35, at p. 34; transcribed and reprinted in Walther Peter Fuchs, *Großherzog Friedrich I. von Baden und die Reichspolitik, 1871–1907*, vol. 4, *1898–1907*, Veröffentlichungen der Kommission für geschichtliche Landeskunde in Baden-Württemberg, Series A/32 (Stuttgart, 1980), letter 1879, pp. 68–69, at 68.

6  Ibid., 35 / 69.

Friedrich knew how to zig and zag between religious and political aspirations – or to differentiate, as Herzl recalled him saying, 'world-historical' and 'theological' perspectives.[7] (Such a distinction also featured between Hechler and Herzl themselves.) He well understood Wilhelm's ambitions of empire as well as the potential expediency of Jews. Amidst the global jostling and jockeying of European powers, the idea emerged that Jewish emigration could warrant declaration of a protectorate for German Jews in Palestine at the expected fall of the Ottoman empire.[8] Yet finding Moses' chest might also bring a badge to the emperor's. And Wilhelm was known to like shiny things. In this century of imperial rivalry and national display, museums in Europe were quickly filling up, often through depredation from colonial pursuits. No doubt, the Ark of the Covenant would look handsome in Berlin, next to the recently plundered metopes from the Pergamon Altar.

That autumn of 1898, Friedrich received a long letter from the chaplain. Anticipating the emperor's imminent journey to Jerusalem, the frenetic Hechler was hoping to advance his agenda via the German ambassador:

> I purpose [sic] telling him all about mount Nebo and try to persuade him to have that whole district of East Jordan, near the Dead Sea, given to the Emperor of Germany by the Sultan, so that, when the Ark of the Covenant is found, his Majesty will possess it with the two tables of stone with the 10 Commandments written by God on mount Sinai, and probably the original MS. of the 5 books of Moses, written by Moses, which were hid in the Ark and which will prove how foolishly so called 'Higher Criticism' tries to make out that Moses could not have written this and that, etc. etc.[9]

---

7   Theodor Herzl, 3 September 1898, in Raphael Patai, ed., *The Complete Diaries of Theodor Herzl*, trans. Harry Zohn, 5 vols (New York, 1960), 2:659.

8   For more on the potential alliance of imperialist and Zionist agendas, see Friedman, *Germany, Turkey, and Zionism*; Walther Peter Fuchs, *Studien zu Großherzog Friedrich I. von Baden*, Veröffentlichungen der Kommission für geschichtliche Landeskunde in Baden-Württemberg, Series B/100 (Stuttgart, 1995), esp. 185–220; John C.G. Röhl, *Wilhelm II: The Kaiser's Personal Monarchy, 1888–1900*, trans. Sheila de Bellaigue (Cambridge, 2004), esp. 924–965; cf. also Röhl, 'Herzl and Kaiser Wilhelm II: A German Protectorate in Palestine?', in *Theodor Herzl and the Origins of Zionism*, eds Ritchie Robertson and Edward Timms, Austrian Studies 8 (Edinburgh, 1997), 27–38.

9   William Henry Hechler to the Grand Duke Friedrich I of Baden, 26 September 1898, in Ellern and Ellern, *Herzl, Hechler, the Grand Duke of Baden and the German Emperor, 1896–1904*, letter 15, pp. 38–47, at p. 41; partially republished in Fuchs, *Großherzog Friedrich I. von Baden und die Reichspolitik, 1871–1907*, vol. 4, *1898–1907*, letter 1890, pp. 76–77. Though placed in good narrative form, the passage sees misattribution – to the duke himself, rather than Hechler – and therefore misinterpretation in Shalom Goldman, *Zeal for Zion: Christians,*

By finding Moses' box, Hechler hoped to close Pandora's, to contain the demons of historical criticism. Just as the divine cloud had surrounded Moses on Mount Sinai in the biblical account, so now clouds of doubt had been encircling him again in the minds of modern thinkers. Like other presumably historical figures – Homer, Plato, Jesus – the arrows of philology had been assailing the balloon of this Hebrew prophet over the century, both the unity of his writings and the historicity of the man.[10] In the case of Moses, it was well over a century, at least as far back as Baruch Spinoza (1632–1677). Whether as figment or as fraud, he had become suspect, together with his legacy: from his religious laws to his political constitution to his historical deeds. Discovering the ark, with the tablets of stone and manuscripts of Moses, could, hoped Hechler, bring certainty to the Christian faith – and glory to the German empire. Herzl and Hechler, however, would only find disappointment: in spade and crown alike.

If this motley crew hoped to find the corpus of Moses buried away in Southwest Asia, its spirit was on full display in Central Europe already, or rather still. Images of Moses and/or his Ten Commandments adorned public buildings across the Germanies. Some had stood for centuries, like the Decalogue scenes painted by Lucas Cranach the Elder (1472–1553) on the townhall of Wittenberg, Saxony, in the early 1500s, the statue of Moses erected on the Bremen *Rathaus* about a century later, or the figure carved into wooden paneling of the *Friedenssaal* in the city hall of Münster sometime in between. Others were of more recent vintage, such as on a ceiling in the guildhall of Monheim, Bavaria: the residence of a 'court Jew' built in the 1730s yet sold to the city upon expulsion of Jewish people not too long thereafter. Still other likenesses had been recently restored, including the *Rathaus* façade in Lindau, Swabia, during the 1880s. Yet Moses had made more recent appearances as well. A mural in the Neues Museum of Berlin, by Wilhelm von Kaulbach (1804–1874), starred him along with other lawgivers ancient and modern: Solon (ca. 630–560 BCE), Charlemagne (ca. 747–814), and Frederick the Great (1712–1786) (Figure 1).

*Jews, & the Idea of the Promised Land* (Chapel Hill, 2009), 112–17. The same (ostensibly direct) quotation looks like an almost different one altogether in Merkley, *The Politics of Christian Zionism*, 31, which renders it back into English from the French translation of Claude Duvernoy, *Le prince et le prophète* (Jerusalem, 1966) but without confusing the parties concerned.

10    See Paul Michael Kurtz, 'A Historical, Critical Retrospective on Historical Criticism', in *The New Cambridge Companion to Biblical Interpretation*, eds Ian Boxall and Bradley C. Gregory (Cambridge, 2022), 15–36.

FIGURE 1     Photographs by Gustav Schauer of sketches for mural paintings of Moses, Solon, Charlemagne, and Frederick the Great in the Neues Museum of Berlin Published in Karl Frenzel, *Die Wandgemälde W. von Kaulbach's Treppenhause des Neuen Museums zu Berlin* (Berlin, 1870). DIGITAL IMAGE COURTESY OF THE RIJKSSTUDIO OF THE RIJKSMUSEUM AMSTERDAM, THE NETHERLANDS.

In 1904, he would also ornament the renovated *Grossratssaal* of Basel, Switzerland. Alongside such civic sightings, Mosaic manifestations could long no less be spotted on sacred sites as well, from the *Mosesbrunnen* beside the Minster in Bern – built in 1544 and rebuilt 250 years later – to the graven image of him sculpted for the Berlin cathedral in the 1890s (Figure 2).[11]

---

11     For more on Moses in graphic representation, see the digest in Elisabeth L. Flynne, 'Moses in the Visual Arts', *Interpretation* 44/3 (1990), 265–76, and the array of figures in Friedrich Wilhelm Graf, *Moses Vermächtnis. Über göttliche und menschliche Gesetze*, 3rd ed. (Munich, 2006).

Just as Moses featured as the giver of law on physical structures – sacred and civic alike – so too he figured in conceptual ones. His portfolio, however, had shrunk over the years, or better millennia, since Philo (ca. 20 BCE–50 CE) once listed his functions as king, lawgiver, priest, and prophet.[12] Most important for Philo, in *On the Life of Moses*, was his charge as philosopher-sage. Though it was one of the last to go, he was increasingly relieved of this capacity as well.[13] As Dmitri Levitin has argued,

> By the end of the [seventeenth] century, the narrative of Judaic primacy and of Moses as pioneering philosopher-sage was almost dead. This happened not under the aegis of heterodoxy or 'early enlightenment', but from the slow dissemination of new sources, from new approaches to the existing sources, and from the theological pressures that shaped these scholarly developments.[14]

Yet the narrative of Moses as pioneering lawgiver lived on – be it bruised, maimed, or reincarnated – even as the nature of that law (civil or religious), its significance (historical or normative), and its scope (universal or particular) remained very much contested. Although Spinoza killed the author Moses, and Voltaire (1694–1778) assassinated his moral character, Rousseau (1712–1778) could still laud the effects of his legislation.[15] Many followed Montesquieu (1689–1755), whose *On the Spirit of Law* relativized such historic legislation by appealing to the words of Solon and God themselves that conceded to having

---

12     Philo, *Mos.*, 2.292. See further Maren R. Niehoff, *Philo of Alexandria: An Intellectual Biography*, Anchor Yale Bible Reference Library (New Haven, 2018), 110–20; Louis H. Feldman, *Philo's Portrayal of Moses in the Context of Ancient Judaism*, Christianity and Judaism in Antiquity 15 (Notre Dame, 2007). For a longer, wider reception history of Moses' roles, starting from scripture itself, see Jane Beal, ed., *Illuminating Moses: A History of Reception from Exodus to the Renaissance*, Commentaria 4 (Leiden, 2014).

13     Cf. Hywel Clifford, 'Moses as Philosopher-Sage in Philo', in *Moses in Biblical and Extra-Biblical Traditions*, eds Axel Graupner and Michael Wolter, Beihefte zur Zeitschrift für die alttestamentliche Wissenschaft 372 (Berlin, 2007), 151–67.

14     Dmitri Levitin, *Ancient Wisdom in the Age of the New Science: Histories of Philosophy in England, c. 1640–1700*, Ideas in Context (Cambridge, 2015), 114. For ancient debates among Jews, Christians, and pagans over the priority and primacy of Moses and Homer, see Arthur J. Droge, *Homer or Moses? Early Christian Interpretations of the History of Culture*, Hermeneutische Untersuchungen zur Theologie 26 (Tübingen, 1989).

15     Cf. Ronald Schechter, *Obstinate Hebrews: Representations of Jews in France, 1715–1815*, Studies on the History of Society and Culture (Berkeley, 2003).

FIGURE 2    Statue of Moses and the Ten Commandments on the
            Berlin Cathedral
            PHOTOGRAPH BY VOLLWERTBIT; IMAGE COURTESY
            OF WIKIMEDIACOMMONS, CC BY-SA 3.0. HTTPS:
            //COMMONS.WIKIMEDIA.ORG/WIKI/FILE:BERLINER
            _DOM_MOSE_BERLIN2007.JPG

given imperfect laws, which then furnished a 'sponge that wipes out all the difficulties that are to be found in the law of Moses'.[16] Nonetheless, this title role
of historic legislator remained a point of departure and return.

Despite these conceptual shifts, older – even ancient – practices perdured
in processing Mosaic law. In his monumental *Mosaisches Recht*, or *Mosaic*

---

16    [Montesquieu], *De l'esprit des loix* ..., new ed., 3 vols (Geneva, 1749), 2:140, bk 19, ch. 21,
      alluding to Solon's response to the Athenians as told by Plutarch (Plut. *Sol.* 15.2) and God
      to a Hebrew prophet (Ezek 20:25).

*Jurisprudence*, Johann David Michaelis (1717–1791) declared, 'whoever wants to look at the laws [*Gesetze*] with the eye of a Montesquieu, for him it is indispensable to know the legal systems [*Rechte*] of other peoples: the further in time and place, the better'.[17] For Michaelis and many more, Athens had much to do with Jerusalem: 'Everything that can move us to devote our hard work to Greek law will also recommend the Mosaic law to us as remarkable'.[18] Or was it Jerusalem and Cairo? Already the *Geography* of Strabo (ca. 62–24 BCE) portrayed Moses as an Egyptian priest: this antique sketch associating law in Bible and religion in Egypt was transformed into a bright mural in the Enlightenment.[19]

The study of comparative lawgiving represented a time-honored tradition. Ever since antiquity, Moses was made to stand shoulder to shoulder with – or head and shoulders above – other peoples' lawgivers. In his *Library of History*, Diodorus of Sicily (1st cent. BCE) placed Moses, with his god 'Iao' and his people the Jews, in a lineup of divine lawgivers: alongside Hermes through Menes in Egypt, Zeus through Minos in Crete, Apollo through Lycurgus for the Spartans, the Good Spirit through Zarathustra for the Aryans, and Hestia through Zalmoxis for the Getae.[20] The obscure late-antique work known as *Lex Dei* or *Mosaicarum et Romanarum legum collatio* (*The Law of God* or *Compilation of the Mosaic and Roman Laws*), a more technical, less topical comparison, even juxtaposed extracts from the Hebrew scriptures with some from Roman legal writing.[21] Attracting savants since the sixteenth century, this collection underwent its foundational editing by Theodor Mommsen (1817–1903) in 1890. Albert

---

17      Johann David Michaelis, *Mosaisches Recht*, 6 vols (Frankfurt am Main, 1770–75), 1:2. Michaelis has attracted considerable interest in the last twenty years: cf. Jonathan Sheehan, *The Enlightenment Bible: Translation, Scholarship, Culture* (Princeton, 2005); Michael M. Carhart, *The Science of Culture in Enlightenment Germany*, Harvard Historical Studies 159 (Cambridge, MA, 2007); Suzanne L. Marchand, *German Orientalism in the Age of Empire: Religion, Race, and Scholarship*, Publications of the German Historical Institute, Washington D.C. (Cambridge, 2009); Michael Legaspi, *The Death of Scripture and the Rise of Biblical Studies*, Oxford Studies in Historical Theology (Oxford, 2010); Avi Lifschitz, *Language & Enlightenment: The Berlin Debates of the Eighteenth Century*, Oxford Historical Monographs (Oxford, 2012); Ofri Ilany, *In Search of the Hebrew People: Bible and Nation in the German Enlightenment*, trans. Ishai Mishroy, German Jewish Cultures (Bloomington, 2018); Yael Almog, *Secularism and Hermeneutics*, Intellectual History of the Modern Age (Philadelphia, 2019).

18      Michaelis, *Mosaisches Recht*, 1:5.

19      Strab. 16.2.34–41. See further Jan Assmann, *Moses the Egyptian: The Memory of Egypt in Western Monotheism* (Cambridge, MA, 1998), 91–143.

20      Diod. Sic. 1.94.1–2.

21      See further Robert M. Frakes, *Compiling the* Collatio Legum Mosaicarum et Romanarum *in Late Antiquity*, Oxford Studies in Roman Society & Law (Oxford, 2011).

Montefiore Hyamson (1875–1954), a British civil servant and Zionist leader, commenced his English translation and revision of that edition because of 'the prospect of an interesting comparison between two great [legal] systems' but had to confess 'this promise was illusory'.[22] He was not the last to find interest – or illusions – in comparing the code of Moses to more worldly systems of law.

A historic lawgiver and the founder of a nation, of an ancient Hebrew people, Moses therefore occupied a central place as a both a father of Judaism and a framer of European civilization, two of his major roles – by turns in tension, in harmony, or in parallel – within much wider, longer cultural history.[23] Throughout the long nineteenth century, Mosaic patrimony came to the fore especially in debates over the legacy of Judaism in the West, over what the Christian nations of modern Europe owed to an ancient Semitic people of the Middle East. Moses loomed equally large as Jewish Europeans contemplated questions of tradition, identity, and assimilation. From the poet Heinrich Heine (1797–1856) to novelist Franz Kafka (1883–1924) to composer Arnold Schoenberg (1874–1951), Germanophone Jews in particular often framed their reflections on Jewishness and its relationship to modern European cultures by exploring the figure of Moses.[24] Psychoanalyst Sigmund Freud (1856–1939) devoted no small amount of attention to him too, which itself has attracted much analysis.[25] In like manner, the philosopher Hermann Cohen (1842–1918)

---

22    Albert Montefiore Hyamson, *Mosaicarum et romanarum legum collatio: With introduction, facsimile and transcription of the Berlin codex, translation, notes and appendices* (London, 1913), vii.

23    Cf., e.g., Wolf-Daniel Hartwich, *Die Sendung Moses. Von der Aufklärung bis Thomas Mann* (Munich, 1997); Assmann, *Moses the Egyptian*; Melanie Jane Wright, *Moses in America: The Cultural Uses of Biblical Narrative* (Oxford, 2002); Brian Britt, *Rewriting Moses: The Narrative Eclipse of the Text*, Journal for the Study of the Old Testament Supplement Series 402 (London, 2004); Barbara Johnson, *Moses and Multiculturalism* (Berkeley, 2010); Theodore Ziolkowski, *Uses and Abuses of Moses: Literary Representations since the Enlightenment* (Notre Dame, 2016).

24    See Bluma Goldstein, *Reinscribing Moses: Heine, Kafka, Freud, and Schoenberg in a European Wilderness* (Cambridge, MA, 1992). Good bibliography appears in Pamela Cooper-White, 'Freud's Moses, Schoenberg's Moses, and the Tragic Quest for Purity,' *American Imago* 79/1 (2022), 89–122.

25    Richard J. Bernstein, *Freud and the Legacy of Moses*, Cambridge Studies in Religion and Critical Thought 4 (Cambridge, 1998); Ruth Ginsburg and Ilana Pardes, eds, *New Perspectives on Freud's* Moses and Monotheism, Conditio Judaica 60 (Tübingen, 2006); Gilad Sharvit and Karen S. Feldman, eds, *Freud and Monotheism: Moses and the Violent Origins of Religion*, Berkeley Forum in the Humanities (New York, 2018); Lawrence J. Brown, ed., *On Freud's 'Moses and Monotheism'*, The International Psychoanalytical

reflected on the authority of Mosaic law as guarantor of human morality and defender of pure monotheism.[26]

However, Moses was not just built into the architecture of modern Europe, both real and ideal. His law was even built into its legal structures. Michael Carhart writes, accordingly, 'As law codes were revised in the seventeenth and eighteenth centuries, frequently parts or even all of the law of Moses were incorporated into local legal systems, and most regions of Europe at least rendered the Ten Commandments as a basic part of the regional law'.[27] However, revision could also turn into rejection, as was the case from at least the later seventeenth century onward. Transforming age-old theological questions about not only the particularity and universality of Jewish law but also natural *vis-à-vis* revealed or 'positive' religion, a wide array of discussions both between and among Jews and Christians reassessed the normativity of ancient Mosaic law in modern European society. Many among the Christians converged with a larger movement, in the words of Nils H. Roemer, 'to decenter Judaism's elevated role in world history'.[28] Between the Christians, not a few Protestants deployed certain stereotypes of Judaism – as hidebound legalism – to polemicize against Catholicism: availing themselves of ancient Jewish law to assail contemporary Catholics as legalistic, irrational, and unmodern. Long disputes ensued as to which laws, if any, were still binding and why.

---

Association Series Contemporary Freud: Turning Points and Critical Issues (London, 2022); Yosef Hayim Yerushalmi, *Freud's Moses: Judaism Terminable and Interminable*, Franz Rosenzweig Lecture Series (New Haven, 1991).

26    See George Y. Kohler, 'Finding God's Purpose: Hermann Cohen's Use of Maimonides to Establish the Authority of Mosaic Law', *Journal of Jewish Thought and Philosophy* 18/1 (2010), 75–105.

27    Carhart, *The Science of Culture in Enlightenment Germany*, 45. For an overview of nineteenth-century codification in Germany, see Susan Gaylord Gale's classic article 'A Very German Legal Science: Savigny and the Historical School', *Stanford Journal of International Law* 18/1 (1982), 123–46; Michael John, *Politics and the Law in Late Nineteenth-Century Germany: The Origins of the Civil Code* (Oxford, 1989). For the incorporation of Mosaic law in Lutheran lands – alongside Roman, custom, and canon – during codification processes since the eighteenth century, see Dieter Strauch, 'Quellen, Aufbau und Inhalt des Gesetzbuches', in *Das schwedische Reichsgesetzbuch (Sveriges Rikes Lag) von 1734. Beiträge zur Entstehungs- und Entwicklungsgeschichte einer vollständigen Kodifikation*, ed. Wolfgang Wagner, Ius Commune: Veröffentlichungen des Max-Planck-Instituts für Europäische Rechtsgeschichte, Sonderhefte: Studien zur europäischen Rechtsgeschichte 29 (Frankfurt, 1986), 61–106, esp. 93–98.

28    Nils H. Roemer, *Jewish Scholarship and Culture in Nineteenth-Century Germany: Between History and Faith*, Studies in German Jewish Cultural History and Literature (Madison, 2005), 18; cf. also Bernd Witte, *Moses und Homer. Griechen, Juden Deutsche: Eine andere Geschichte der deutschen Kultur* (Berlin, 2018).

Claims and convictions, like their consequences, could cut across confessions. Some of those discussions were theoretical in nature, ranging from reason and revelation to ethics and the nature of religion. As criticism of the Old Testament grew at the end of the eighteenth century to match that of the New, much Enlightenment thought was dispensing with *adiaphora* in order to save the faith.[29] New frames thus formed around an older question, one Christine Hayes articulates in the title of her astute study on antiquity as 'what's divine about divine law?'[30] In his inaugural lecture at Jena of 1789 – the year before a title change in his position from history to philosophy – Friedrich Schiller (1759–1805) attributed much enlightened thought to 'Mosaic religion', insofar as the teaching of a single god undergirded the concept of reason. Founded on truth and founding a state, argued Schiller, the Mosaic legislation fostered happiness among the Hebrews and laid a solid foundation for future enlightenment.[31] Around the same time, G.W.F. Hegel (1770–1831) became convinced that the moral theory of Immanuel Kant (1724–1804) had essentially interiorized the 'positive' law of Moses.[32] So too the *Haskalah*, or Jewish Enlightenment, supported a comparable move to internalize and privatize traditional practice. As Leora Batnitzky has shown, this trajectory imagined 'that Judaism, and especially Jewish law, is not political but instead concerns the intellectual and spiritual dimensions of human experience' – which allowed for its construction as a 'religion' structurally analogous to Christianity.[33] As the reform movement in Judaism gained steam, the medieval philosopher

---

29    Cf. Sheehan, *The Enlightenment Bible*, esp. 151–52.

30    Christine Hayes, *What's Divine about Divine Law? Early Perspectives* (Princeton, 2015). See now also the ERC project 'How God Became a Lawgiver', under the direction of Konrad Schmid at the University of Zurich.

31    On representations of Jews and Judaism in Schiller's address 'Die Sendung Moses' ('The Legation of Moses'), see Martha B. Helfer, *The Word Unheard: Legacies of Anti-Semitism in German Literature and Culture* (Chicago, 2011), 23–55.

32    Henry Silton Harris, *Hegel's Ladder: A Commentary on Hegel's* Phenomenology of Spirit, 2 vols (Indianapolis, 1997), esp. 1:377 n.25, 1:613 n.69, 2:70 n.53. For more on Hegel's understanding of freedom, rationality, and morality in ancient Judaism – especially with regard to Mosaic law and Roman rule – see Shlomo Avineri, *Hegel's Theory of the Modern State*, Cambridge Studies in the History and Theory of Politics (Cambridge, 1972), 15–24.

33    Leora Batnitzky, *How Judaism Became a Religion: An Introduction to Modern Jewish Thought* (Princeton, 2011), 59. As a contemporaneous counterpoint to such 'Talmudophobia', Jews in the Russian Empire looked to Catholic tradition: see the critical intervention of Eliyahu Stern, 'Catholic Judaism: The Political Theology of the Nineteenth-Century Russian Jewish Enlightenment', *Harvard Theological Review* 109/4 (2016), 483–511, esp. 495, 501.

Maimonides (ca. 1138–1204) became a chief source in debates on the binding nature of Jewish law.[34]

Other discussions were historical, driven by recasting, as the old pun went, the Mosaic Bible into a biblical mosaic. While critical analysis of biblical literature narrowed the textual inventory ascribable to Moses (or even to deep antiquity), the move to historicize and/or particularize Mosaic law expanded, whether the intention was to salvage or to spurn it. The lion's share of literature on biblical learning in this period tends to focus on Protestant scholarship, in the matrix of theology, philology, and history. Yet Jews, no less than Christians, were confronted by the concomitant conundrums. Jewish writers generated an equally broad set of solutions when faced with such historicist thinking.[35] In his voluminous work on Jewish antiquity, Isaak Markus Jost (1793–1860), for instance, accepted a distinction between 'Mosaism' and 'Judaism', between biblical and postbiblical Jewish history: a theologically freighted, often pernicious periodization long deployed in Protestant polemic. However, he argued against the implementation of Mosaic law in that earlier era, calling it more theory than practice, and argued for the realization of its monotheistic and moral dimensions only in the latter epoch, which amounted to a rise – not a fall – in the legacy of Moses.[36] Later, Benno Jacob (1862–1945) conceded critical arguments against Mosaic authorship, in his popular book *Die Thora Moses* (*The Torah of Moses*), but still claimed 'the spirit of Moses' gave unity to the Pentateuch and thus warranted continued reference to 'the *Torah of Moses*', even more if than two hands wrote under the name.[37]

---

34    George Y. Kohler, *Reading Maimonides' Philosophy in 19th Century Germany: The Guide to Religious Reform*, Amsterdam Studies in Jewish Philosophy 15 (Dordrecht, 2012).

35    See Michael A. Meyer, *Response to Modernity: A History of the Reform Movement in Judaism*, Studies in Jewish History (Oxford, 1988); Ismar Schorsch, *From Text to Context: The Turn to History in Modern Judaism*, Tauber Institute for the Study of European Jewry Series (Hanover, NH, 1994); Susannah Heschel, *Abraham Geiger and the Jewish Jesus*, Chicago Studies in the History of Judaism (Chicago, 1998); David N. Myers, *Resisting History: Historicism and Its Discontents in German-Jewish Thought*, Jews, Christians, and Muslims from the Ancient to the Modern World (Princeton, 2003); and Yaacov Shavit and Mordechai Eran, *The Hebrew Bible Reborn: From Holy Scripture to the Book of Books. A History of Biblical Culture and the Battles over the Bible in Modern Judaism*, trans. Chaya Naor, Studia Judaica 38 (Berlin, 2007).

36    Ran HaCohen, *Reclaiming the Hebrew Bible: German-Jewish Reception of Biblical Criticism*, trans. Michelle Engel, Studia Judaica 56 (Berlin, 2010), esp. 54–62.

37    Benno Jacob, *Die Thora Moses*, Volksschriften über die jüdische Religion 1/3–4 (Frankfurt, 1912/13), 94; see further Christian Wiese, *Challenging Colonial Discourse: Jewish Studies and Protestant Theology in Wilhelmine Germany*, trans. Barbara Harshav and Christian Wiese, Studies in European Judaism 10 (Leiden, 2005), 220–30.

Still other discussions were political, or rather overtly so. Concerns with the normativity of biblical legislation did not, of course, arise for the first time in this period. Perennial questions had long looked to the Mosaic law and an imagined Hebrew republic for matters juridical and legal, governmental and national. As Adam Sutcliffe has written of the early modern period, 'The most politically contested field of Hebraism, and the most useful area for supporters of republicanism, was Jewish law.'[38] Eric Nelson has further emphasized how rabbinic texts, not the biblical alone, served as sources for such political thought.[39] Yet the long nineteenth century, as an age of reform, revolution, and emancipation, brought new possibilities and challenges – and with them new questions and answers. In 1788, Abbé Grégoire (1750–1831) transferred ideas – of Moses Mendelssohn (1729–1786) most of all – from Germany to France, and from Jewish to Christian thought, when he distinguished 'in Mosaic law, between what lies essentially in the religious realm and what is only the object of civil and criminal jurisprudence, for the two are separable'.[40] But not everyone considered politics so separable from religion within the Mosaic writings. In a venomous series of lectures, Heinrich Leo (1799–1878) moved seamlessly from textual and historical arguments against the 'Mosaic' part of Mosaic law – what he deemed a devious project of later Jewish priestcraft, hiding behind the authority of Moses – into a diatribe against hierarchy (as in rule by priests), theocracy, and democracy, even pairing Robespierre (1748–1794) with Pharisees and portraying him as 'the consequence' of such priestly rule: a rule always empty of feeling and full of fanaticism.[41] Just as Pharisaism enabled a superior Christianity to emerge, he argued, so also Catholicism allowed a higher

---

38     Adam Sutcliffe, *What are Jews For? History, Peoplehood, and Purpose* (Princeton, 2020), 39. See further idem, *Judaism and Enlightenment*, Ideas in Context (Cambridge, 2005); Graham Hammill, *The Mosaic Constitution: Political Theology and Imagination from Machiavelli to Milton* (Chicago, 2012); Markus M. Totzeck, *Die politischen Gesetze des Mose. Entstehung und Einflüsse der* politica-judaica *Literatur in der Frühen Neuzeit*, Refo500 Academic Studies 49 (Göttingen, 2019).

39     Eric Nelson, *The Hebrew Republic: Jewish Sources and the Transformation of European Political Thought* (Cambridge, MA, 2010).

40     Cited in Michael Graetz, *The Jews in Nineteenth-Century France: From the French Revolution to the Alliance Israélite Universelle*, trans. Jane Marie Todd, Stanford Studies in Jewish History (Stanford, 1996), 177 – the original source being *Essai sur la régénération physique, morale et politique des juifs. Ouvrage couronné par la Société Royale des Sciences et des Arts de Metz, le 23 Août 1788* (Metz, 1789), 155.

41     Heinrich Leo, *Vorlesungen über die Geschichte des Jüdischen Staates; gehalten an der Universität zu Berlin* (Berlin, 1828), 57; for more on Leo, see Christhard Hoffmann, *Juden und Judentum im Werk deutscher Althistoriker des 19. und 20. Jahrhunderts*, Studies in Judaism in Modern Times 9 (Leiden, 1988), 42–73. Leo's lectures drew a sustained

Protestantism to form. If politics and religion were inseparable in the law of Moses itself, they were no less indivisible in discussions of that inseparability. Long after the idea of Moses the philosopher-sage had faded, Moses the lawgiver remained alive and well enough. But here too, new sources, new approaches, and new pressures led to new conceptual formations. In 1903, a few years after meeting Herzl thrice that autumn of 1898, in Constantinople, Mikve-Israel, and Jerusalem (thanks to efforts by Hechler and Friedrich I), Wilhelm II may not have dismantled but nonetheless did diminish the pedestal of Mosaic law. Responding to clouds of doubt gathering amidst all the inscriptions and monuments that had recently come to light in the Middle East, he released a statement that distinguished two kinds of revelation: a universal line for the development of humanity (Hammurabi, Moses, Abraham, Homer, Charlemagne, Luther, Shakespeare, Goethe, Kant, Wilhelm I) and a particular one 'more religious' in nature (Abraham, Moses, prophets, psalmists, Jesus). The statement, widely circulated, read in part:

> When Moses had to reburnish the well[-]known paragraphs of the law, perhaps derived from the code of Hammurabi ... here the historian can perhaps construe from the sense or wording a connection with the laws of Hammurabi, the friend of Abraham. [...] But that will never disguise the fact that God incited Moses thereto and in so far revealed himself to the people of Israel.[42]

Although the name of the Hebrew legislator appeared on both lists, his 'legislative act on Sinai' underwent reinterpretation: 'only regarded as symbolically inspired by God'. But if the law of Moses had become a dead letter to some, for

---

response by Henry John Rose of St John's, Cambridge, who sought to confront 'a view of Jewish History founded on the modern German School of Philology': Rose, *The Law of Moses Viewed in Connexion with the History and Character of the Jews, with a Defence of the Book of Joshua against Professor Leo of Berlin: Being the Hulsean Lectures for 1833. To Which is Added An Appendix Containing Remarks on the Arrangement of the Historical Scriptures Adopted by Gesenius, de Wette, and Others* (Cambridge, 1834), viii.

42    Dated 15 February 1903, the letter was published in Chicago as 'Kaiser Wilhelm on "Babel and Bible". (Letter from His Majesty Emperor William II. To Admiral Hollman, President of the Oriental Society)', *The Open Court* 7 (1903), 432–36. The publisher reprinted it, among other appendices, to a translation of Friedrich Delitzsch's first two lectures: Delitzsch, *Babel and Bible: Two Lectures on the Significance of Assyriological Research for Religion, Embodying the Most Important Criticisms and the Author's Replies, Profusely Illustrated*, trans. Thomas J. McCormack and W.H. Carruth (Chicago, 1903). On the political dimensions to this confession of faith, see John C.G. Röhl, *Wilhelm II: Into the Abyss of War and Exile, 1900–1941*, trans. Sheila de Bellaigue and Roy Bridge (Cambridge, 2014), 498–521.

many his legacy lived on – so many, in fact, that a German emperor felt compelled to intervene in a perilous theological, and political, debate.

With its citation of prophets, Jesus, and Luther, the kaiser's intervention suggests two other aspects of Moses in modernity.[43] On the one hand, the Pentateuch was often read through the Prophets, especially by Christians. Some such interpreters latched onto a selection of prooftexts – like the passage in Isaiah 1 where God asks, 'What need have I of all your sacrifices?' (JPS) – to claim the law had already been rejected, by God or the prophets themselves, even before Christianity superseded Judaism. The commentary by Ferdinand Wilhelm Weber (1836–1879), a Protestant pastor and active missionary to Jews, went in this direction: 'So little delight does the LOrd have in the nature of this people's worship, then, that he wants to put an end to it himself.'[44] The hermeneutical navigation here, of course, came from a particular reading of Paul (ca. 5–64 CE) and, further still, of Paul as read by Martin Luther (1483–1546): with a theological animosity to law.[45] On the other hand, the framing of Mosaic law was largely limited to the Bible, not the Talmud. Few Christians engaged with rabbinic materials, and those who did usually depended on translations and anthologies like that by Weber, who sought to define a 'theology' for ancient Judaism.[46] Many more simply disliked, even disparaged, the rabbis and their battery of discussions on halacha. Hermann L. Strack (1848–1922), the theologian and orientalist, was thus a rare Christian expert in rabbinics – enough to merit the respect of Jewish scholars – though not a rarity in his dedication to converting Jews.[47] In consequence, the question of Mosaic law was mostly confined to the (Christian) Bible.

---

43     This paragraph draws on insights kindly offered by an anonymous reviewer of the volume.

44     Ferdinand Wilhelm Weber, *Der Profet Jesaja, in Bibelstunden ausgelegt*, 2 vols (Nördlingen, 1875, 1876), 1:11, orthography original; cf. also the preface, which discusses Christian interpretation of Hebrew prophecy.

45     For more compelling understandings of Paul on the law, see Daniel Boyarin, *A Radical Jew: Paul and the Politics of Identity*, Contraversions: Critical Studies in Jewish Literature, Culture, and Society (Berkeley, 1994); Paula Fredriksen, *Paul: The Pagans' Apostle* (New Haven, 2017).

46     Ferdinand Wilhelm Weber, *System der altsynagogalen palästinischen Theologie aus Targum, Midrasch und Talmud dargestellt*, eds Franz Delitzsch and Georg Schnedermann, 1st ed. (Leipzig, 1880), repr. with a new title as *Die Lehren des Talmud, quellenmässig, systematisch und gemeinverständlich dargestellt*, Schriften des Institutum Judaicum 2/1 (Leipzig, 1886) and then in its 2nd ed. as *Jüdische Theologie auf Grund des Talmud und verwandter Schriften gemeinfasslich dargestellt* (Leipzig, 1897). For similar efforts in English, see the work of Paul Isaac Hershon (1817–1888), a Jewish convert and missionary to Jews.

47     For an introduction to Christian work on the Talmud in this period – especially at the nexus of proselytism and antisemitism – see Christian Wiese, 'Ein "aufrichtiger Freund

True to its title, this volume focuses on the figure of Moses to examine diverse representations and appropriations of biblical law across the German lands from the late eighteenth to early twentieth centuries. Its chapters chart how reflections on that ancient law shaped debates on modern structures – legal, political, religious, scholarly – and how contemporary concerns impacted, conversely, the apprehension of Mosaic law. Did Moses copy Hammurabi? Why would sacred legislation have any place in a modern state? Must Jews be forever bound by ancient law? How radical were the politics of Moses? In doing so, these chapters foreground the entanglements of secular and religious, of past and present, and of biblical, classical, and orientalist traditions – all against such background as legal reformation, administrative integration, colonial extraction, and civil emancipation. The selection of sources, sites, and speakers further shows how 'German' discourse remained inextricable from durable, if at times disavowed, connections across and beyond Europe.

Part 1, 'Representations of the Past', focuses on historical constructions of Moses and biblical law. In 'The Early Speech of Nations: Biblical Poetry and the Emergence of Germanic Myth', Ofri Ilany investigates the juxtaposition of things Hebraic and things Germanic: from language, literature, and law to religion, people, and custom. Positioning figures like Johann Gottfried Herder, Friedrich Gottlieb Klopstock (1724–1803), and Wilhelm Martin Leberecht de Wette (1780–1849) within a wider, deeper galaxy of intellectual culture, he documents the promotion (or demotion) of sacred scripture into the national heritage of an ancient, primitive people: no longer a legal code, moral guide, history book, or doctrinal system but now primarily a cultural artifact. Herder did much to recast Moses as a people's poet, a nation's founder. 'Which lawgiver could claim to have an effect *deeper on his people's customs* [*Sitten*] than Moses?', Herder once asked, in a source this chapter cites. 'Not even Lycurgus can be compared to him; and now if he summed up the effect of his existence into words, it became – a *song*.'[48] Ilany argues for entanglement between the study of Hebrew national epics and the creation of German national

---

des Judentums"? "Judenmission", christliche Judaistik, und Wissenschaft des Judentums im deutschen Kaiserreich am Beispiel Hermann L. Stracks', in *Gottes Sprache in der philologischen Werkstatt. Hebraistik von 15. bis zum 19. Jahrhundert*, eds Giuseppe Veltri and Gerold Necker, Studies in European Judaism 11 (Leiden, 2004), 277–316; cf. also Hans-Günther Waubke, *Die Pharisäer in der protestantischen Bibelwissenschaft des 19. Jahrhunderts*, Beiträge zur historischen Theologie 107 (Tübingen, 1998).

48   Johann Gottfried Herder, 'Ueber die Wirkung der Dichtkunst auf die Sitten der Völker in alten und neuen Zeiten. Eine Preißschrift. (1778)', repr. in *Johann Gottfried von Herder's sämmtliche Werke*, Section 2, *Zur schönen Literatur und Kunst*, vol. 9, ed. Johann von Müller (Tübingen, 1807), 378.

mythology: both were deeply concerned with naturalism, poesy, and original-ity. His analysis uncovers how ideas of poetic nations governed understand-ings of the Bible and how conceptions of the ancient Hebrews, in turn, shaped notions of other peoples' pasts. But if sacred scripture morphed into a cultural inscription, reading it proved no easier – and no less contentious.

'I am speaking here not of the historical but of the mythical Moses,' Heymann Steinthal (1823–1899) specified in 1862, 'and hope the reader will be inclined to distinguish these two just as much as with the historical and legendary Charlemagne.' The Jewish philosopher, psychologist, and philologist contin-ued: 'Now the mythical Moses is, in essence, comparable with Prometheus.'[49] Like Steinthal in this passage, whom she herself discusses in her essay, Carlotta Santini moves from things Germanic to things Greek, examining how Bible and Homer underwent analysis in the new science of myth, circa 1800. With 'The Rise of Jewish Mythology: Biblical Exegesis and the Scientific Study of Myth', Santini follows the cross-pollination across fields now seen as fenced off. Alongside Ilany, Santini traces a transformation from the Bible as holy writ of God to wholly written by humans: be it composed or collected by Moses, com-piled in his name or created by the people as collective. In this trajectory from sacred scripture to national monument, she also offers readings, from a dif-ferent perspective, on Michaelis, Robert Lowth (1710–1787), Christian Gottlob Heyne (1729–1812), and Herder – the same Herder who once declared, 'What Homer is to the Greeks, Moses is to the Hebrews'.[50]

However, Santini's study of biblical and classical studies centers on mythol-ogy as both subject and object, as discipline and material, following the fate of this approach as it found advocates and adversaries. She presents Lowth treating Moses and Homer together to consider myth in both and, conversely, Heyne moving from the authority of the Bible to validate the inquiry into myth, which in turn validated it as a hermeneutical lens. Along the way, her chapter discerns tensions in collaboration, demonstrating the difficulty of interpreting ancient myths as well as modern mythologists. She exposes differ-ence in rhetorical or generic parameters between Heyne and Martin Gottfried Hermann (1755–1822) and disagreement over interpretation of the first book

---

49 Heymann Steinthal, 'Die ursprüngliche Form der Sage von Prometheus (Mit Bezug auf: Kuhn, *Die Herabkunft des Feuers und des Göttertranks*)', *Zeitschrift für Völkerpsychologie und Sprachwissenschaft* 2 (1862), 1–29.

50 Johann Gottfried Herder, *Vom Geist der Ebräischen Poesie. Eine Anleitung für die Liebhaber derselben, und der ältesten Geschichte des menschlichen Geistes*, 2 vols (Dessau, 1782–83), 2:83.

of Moses between the editor Johann Philipp Gabler (1753–1826) and author Johann Gottfried Eichhorn (1752–1827). Eichhorn's ideas, she argues, then entered *Völkerpsychologie* ('folk psychology') through Friedrich August Carus (1770–1807). Half a century before Steinthal invoked Prometheus, Carus had produced a psychology of ancient Hebrews, which appraised the Mosaic writings as sources and further assessed the 'psychological formation' (*psychologische Bildung*) of Moses and his contemporaries.[51] The chapter sketches other contours, including attempts to discover laws that governed myth, laws of human cultural development, and laws of interpretation. Identifying where writers were not willing to go and what they were unwilling to relinquish, she advances her core argument: scholarship on mythic material still carries within it a theology that molded the study of myth as a discipline. According to Santini, the rise of mythology as a science meant the Bible had to undergo such analysis or risk losing its status as a legitimate object of study, with 'religion' then deposed as a superior form of consciousness. Therefore, confessional convictions that once held faith and science together still exercise a bonding power in mythology today.

Moses for the Hebrews, Confucius for the Chinese, Jesus for the Christians, Muhammad for the Muslims: these pairings appeared in the entry for 'legislator' in a major French lexicon of the period. 'Religion is the first law of any society which begins; therefore, those who founded religions have been called legislators', the author, an attorney general, registered before adding political and civil laws: and with them the exemplary lawgivers Lycurgus and Solon for Greece (i.e. for Sparta and Athens, respectively) and Napoleon (1769–1821) for France.[52] If the dictionary had gone through a third edition, in the final third of the century, that entry could not have omitted Hammurabi for the Babylonians. In his essay 'Moses or Hammurabi? Law, Morality & Modernity in Ancient Near Eastern Studies,' Felix Wiedemann shifts from the pairing of Jerusalem with Athens or Berlin to that of Jerusalem with Babylon. Like Ilany, Wiedemann pursues constructions of the ancient Hebrews as a more natural,

---

51    Cf. Friedrich August Carus, *Psychologie der Hebräer* (Leipzig, 1809), 95, 97.
52    A. Gastambide, 'Législateur', *Dictionnaire de la conversation et de la lecture*, 1st ed., vol. 34 (Paris, 1837), 486; 2nd ed., vol. 12 (Paris, 1864), 212. This legislator lineup was widespread across the theological and political spectrum across the century: from Hugh James Rose's *Notices of the Mosaic Law: With Some Account of the Opinions of Recent French Writers Concerning It* (London, 1831) to Louis Jacolliot's intended yet incomplete *Les législateurs religieux: Manou – Moïse – Mahomet. Traditions religieuses comparés des lois de Manou, de la Bible, du Coran, du ritual égyptien, du Zend-Avesta des Parses et des traditions finnoises* (Paris, 1876), the latter both fused with antisemitism and formative for Friedrich Nietzsche.

more primitive people – at a lower stage of 'civilisation' – as well as conceptions of Jewish heritage in European culture. So too his chapter complements Santini's in addressing the obstacles that emerged when the books of Moses were set alongside texts from other ancient peoples: but here Moses sits alongside Hammurabi, instead of Homer. This essay depicts several ways in which comparison opened fault lines within fields and fissures within confessions. As with myth, so too law could shake up or shore up the faith.

Wiedemann shows how the code of Hammurabi could be too close for comfort. Both biblical and Babylonian legislation entailed comparable content, commensurate claims of divine origins, and corresponding human lawgivers. Concentrating less on new ways of reading than new sets of data, Wiedemann traces an international arms race – or rather race of hands – in the deciphering, transcribing, editing, and processing of sources, and he tracks a series of interpretative contests that not only included German, Austrian, and Swiss or even British, French, and American scholars but also incorporated Jews, Catholics, and Protestants, involved rabbis and pastors, and encompassed ivory tower and public square alike. His central argument insists the controversies over the Mosaic and Hammurabic codes represented no mere chapter in the saga of the *Babel-Bibel-Streit* but constituted a story in its own right. Wiedemann's essay situates this discourse within wider debates on the origins of civilisation and foundations of law as well as discussions over which ancient people proved to be the greatest – or at least the greatest contributor to 'modern European culture'. With an eye on the labile relationship between archaeology and exegesis as well as the anxieties about autonomy (independent innovation, not diffusion), antiquity (older being better), and authenticity (original over copy), Wiedemann trains his sights on *Sittlichkeit*, which he considers a 'key concept' in this period, one that blurred morality, culture, and law. His analysis further reveals that the debate took as its starting point the confrontation between a sacred Moses and a secular Hammurabi, with the former conflating morality and law and the latter distinguishing the two. In the end, Wiedemann explains how much of these debates about the past proved to be, in fact, about the present: from law and religion to ethics and even capitalism.

Had he written several decades later, Moritz Duschak (1815–1890) would have had to tackle those Babylonians. But in 1869, the Moravian rabbi, like the attorney's entry in the *Dictionnaire de la conversation et de la lecture* cited above, could still ignore them as he sought to systematise Jewish law. In *Das mosaisch-talmudische Strafrecht. Ein Beitrag zur historischen Rechtswissenschaft* (*Mosaic-Talmudic Criminal Law: A Contribution to Historical Legal Studies*), Duschak argued that the rabbis had not only sought to harmonise, in a distinctive way, the few Mosaic legal provisions with the expanded legal ideas and experiences among the Jews themselves but also borrowed a great deal from the legislations of other peoples, Persians, Greeks, and Romans – albeit with wise concealment of the origin – and merged it with Jewish law.[53]

Not yet excavated, much less codified into the inventory of inescapable cultural comparisons, the Code of Hammurabi did not make his list.

Thirty years onward, a different eye looked at criminal law in comparative perspective. That eye belonged to Theodor Mommsen, and it looked to specialists in other fields of legal history – jurists and philologists – during work on the Roman case, for a title in the 'Systematic Handbook of German Legal Studies'.[54] While it did not make contact with the ancient empires of Mesopotamia either, that line of sight had become more difficult to maintain. After Mommsen's death, contributors to the 1905 *Zum ältesten Strafrecht der Kulturvölker* (*On the Oldest Criminal Law of the Civilized Peoples*) wanted to add other 'peoples of culture' for publication – the Babylonians, Parsees, Slavs – but the decision rested with preserving Mommsen's own parameters.[55] Two aspects stand out. First, the volume included dedicated discussions of Greek, Roman, Germanic, Indian, Arabic, and Islamic law – but not Jewish. (Jews as well as Christians authored entries for the book.) As clarified in the foreword, an initial print run had credited, inter alia, Theodor Nöldeke (1836–1930) for 'Arabic', Julius Wellhausen (1844–1918) for 'Arabic-Jewish', and Ignaz Goldziher (1850–1921) for 'Muslim' criminal law.[56] The final publication of Wellhausen's entry, however, bore the title 'Arabic-Israelite'. It therefore bound ancient Israel to pre-Islamic Arabs: on the apparent assumption of a common substrate to peoples, in this case, both primitive and Semitic.[57] Second, this construction divided Torah and Talmud, thereby fracturing the Jewish legal tradition. In specifying 'Israelite', the title highlighted one part of Jewish history while

53    Moritz Duschak, *Das mosaisch-talmudische Strafrecht. Ein Beitrag zur historischen Rechtswissenschaft* (Vienna, 1869), vi–vii.

54    Theodor Mommsen, *Römisches Strafrecht*, Systematisches Handbuch der Deutschen Rechtswissenschaft (Leipzig, 1899).

55    Karl Binding, foreword, *Zum ältesten Strafrecht der Kulturvölker. Fragen zur Rechtsvergleichung gestellt von Theodor Mommsen, beantwortet von H. Brunner, B. Freudenthal, J. Goldziher, H.F. Hitzig, Th. Noeldeke, H. Oldenberg, G. Roethe, J. Wellhausen, U. von Wilamowitz-Moellendorff* (Leipzig, 1905), viii.

56    Ibid., vii.

57    On such reasoning according to 'primitivity' and 'Semiticity', especially in Wellhausen, see Paul Michael Kurtz, *Kaiser, Christ, and Canaan: The Religion of Israel in Protestant*

blurring later (and especially for Christians, theologically freighted and polemically charged) periods of ancient Judaism. Meanwhile, the content of the chapter, insofar as it concerned Jewish antiquity, referred squarely to the Old Testament. In this way, Jewish law was compressed into the biblical and conflated with pre-Islamic Arabic.

But what some would separate, others joined together. Duschak's book was a sort of sequel. In 1864, he had undertaken a similar effort for civil law, hoping to ensure that guidelines on marriage did not conflict with moral, religious ones. There too, he treated the body of Jewish law, exegesis, and commentary as a single corpus, reflected in the very title *Das mosaisch-talmudische Eherecht* (*Mosaic-Talmudic Marital Law*), and here again arose questions of history and its consequences. Duschak was less concerned with the autonomy or authenticity of Jewish tradition than with its stability and integrity and thus posed two contentious queries for Jews: 'Is the Mosaic-Talmudic marriage law a doctrine that was once established, concluded, not to be altered or modified? Who would claim this?'.[58]

Part 2 of *Moses among the Moderns*, under the heading 'Transformations in the Present', considers some who asked and answered such questions as those by Duschak. Shifting from past to present, from descriptive to normative,

---

*Germany, 1871–1918*, Forschungen zum Alten Testament 1/122 (Tübingen, 2018), 109–16, 157–59, 268–70; cf. also idem, 'Of Lions, Arabs & Israelites: Some Lessons from the Samson Story for Writing the History of Biblical Scholarship', *Journal of the Bible and its Reception* 5/1 (2018), 31–48. Another argument – less sociological or anthropological than historical or chronological (as well as theological) – opposed, in the words of one interpreter, 'a general custom of elucidating the gaps of the biblical legal order through the rabbinic laws of centuries later' on the grounds that 'even Jewish law changed in the course of time thanks to various external and internal events and relations', meaning 'it is very questionable to explain the *beginnings*, the childhood stage, of a legal system through the *result* which emerged in the course of history': Max Mandl, *Das Sklavenrecht des alten Testaments. Eine rechtsgeschichtliche Studie* (Hamburg, 1886), 4–5; cp. also the review by M. v. O. in *Vierteljahrschrift für Volkswirtschaft, Politik und Kulturgeschichte* 25/1 (1888), 103–06. Michaelis had advanced this argument a century earlier, as plainly stated in a section title: 'The explanation of the laws of Moses is not to be taken from the Talmud and the rabbis' (Michaelis, *Mosaisches Recht*, 1:59, §18).

58    Moritz Duschak, *Das mosaisch-talmudische Eherecht, mit besonderer Rücksicht auf die bürgerlichen Gesetze* (Vienna, 1864), vi. Moritz was alone in neither topic nor title: cf. Samuel Holdheim, *Ueber die Autonomie der Rabbinen und das Princip der jüdischen Ehe. Ein Beitrag zur Verständigung über einige das Judenthum betreffende Zeitfragen* (Schwerin, 1843); Ignaz Graßl, *Das besondere Eherecht der Juden in Oesterreich nach den §§. 123–136 des allgemeinen bürgerlichen Gesetzbuches*, 2nd ed. (Vienna, 1849); Zacharias Frankel, 'Grundlinien des mosaisch-talmudische Eherechts', in *Jahresbericht des jüdisch-theologischen Seminars 'Fraenckelscher Stiftung'. Breslau, am Gedächtnisstage des Stifters, den 27. Januar 1860* (Breslau, 1860); P. Buchholz, *Die Familie in rechtlicher und moralischer*

the chapters concentrate on contemporary appropriations of Mosaic law. This part begins with '*Gesetz als Gegensatz*: The Modern Halachic Language Game' by Irene Zwiep. So as Santini, she scrutinizes Heymann Steinthal and *Völkerpsychologie*; along with Ilany, the author investigates themes of a Hebraic national legacy and a Judaic contribution to European culture; and with Wiedemann, Zwiep inquires into law, ethics, morality, and *Sittlichkeit*. Yet she turns away from the discourse on Mosaic law as dominated by Christians and populated by concerns with systems of belief, trust in historical claims, and faith in revealed doctrine. Instead, her analysis pivots on changes to the framing of Mosaic law – in its distinction, identification, and delineation – among Jewish thinkers.

Zwiep explores dynamic reassessments of Jewish communal law, or halacha, amid great change in legal corpora, practices, and systems and in arrangements of church, state, and society. As her essay argues, the foundational rules and principles of Jewish legal practice may have remained stable, but the perception and observance of Jewish law changed dramatically. Indeed, the traditional integration of law, devotion, and morality spelled trouble for Judaism in a new order that claimed a separation of church and state and a distinction between moral religion and national legislation. With this inherent combination of law and religion featuring as both essence and embarrassment for Judaism, a range of thinkers engaged in what she calls a 'halachic language game', deploying old vocabulary into new contexts. On the one hand, her chapter discerns a trend among Jewish thinkers to amplify the Hebrew Bible as ethical and to dampen the legal corpus as esoteric. On the other hand, Zwiep's essay detects internal differentiation, distinguishing three interpretative trajectories for reframing halacha. If Moses Mendelssohn and Hirschel Lewin (1721–1800) had cast it as a Jewish canon law outside competition with the state legal apparatus, Moritz Lazarus (1824–1903) followed Steinthal to depict it as collective morality that imitated a divine holiness, while Zacharias Frankel (1801–1875) portrayed it as Jewish jurisprudence, not only on par with Roman and German traditions but also a form of rational legislation. Examining a Judaism divided

---

*Beziehung nach mosaisch-talmudischer Lehre, allgemein faßlich dargestellt* (Breslau, 1867); Samuel Spitzer, *Die jüdische Ehe nach mosaisch talmudischen und den in Oesterreich bestehenden, besonders neuesten Ehegesetzen* (Essek, 1869); Ludwig Lichtschein, *Die Ehe nach mosaisch-talmudischer Auffassung und das mosaisch-talmudische Eherecht* (Leipzig, 1879); Joseph Bergel, *Die Eheverhältnisse der alten Juden im Vergleiche mit den Griechischen und Römischen* (Leipzig, 1881); Emil Fränkel, *Das jüdische Eherecht nach dem Reichscivilehegesetz vom 6. Februar 1875* (Munich, 1891).

by external, Western categories of religion, law, and morality, Zwiep contends that religious practice ultimately became internalized, moralized, and thereby anesthetized, with traditional Jewish law discarded as a viable normative system. The lawgiving Moses thus seemed to be less modern – more time-bound – than the moralizing one.

In 'The Truth Shall Abide: Samson Raphael Hirsch and Abraham Geiger on the Binding Nature of Torah', Judith Frishman focuses on two stars within this discursive firmament mapped by Zwiep. Her chapter, too, attends to notions of *Sittlichkeit* and Kantian morality, to historicising and mythological readings of the Bible, and to questions of law *vis-à-vis* the authentically, essentially, irreducibly, or unalterably Jewish. Frishman homes in on an internal debate at the heart of modern Judaism, between orthodoxy and reform. Through her close reading of a critical exchange between Hirsch (1808–1888) and Geiger (1810–1874), she demonstrates how both liberalising and traditionalising movements conceptualized obedience to Jewish law. That exchange grappled with how to read text and uphold tradition amidst an assimilating Jewish bourgeoisie and a reforming rabbinate, and its positions were often articulated in epistolary and dialogic form and mostly siloed in specifically Jewish channels. Her essay details the attempt to grapple with competing claims both between and among philosophy, theology, and history: to synthesize or stabilize, to suspend or suppress them. It also charts the effort to measure the interpretative significance of historical development and to adjudicate the analytical suitability of other methods for understanding the Torah: whether seemingly universal, 'scientific' modalities of reading were applicable to particular ancient texts. In doing so, Frishman traces the emergence of questions on election, whether this meant a people chosen or a people choosing obedience. Ultimately, her analysis targets debates over what it means to be true to the teachings of Moses.

'Law, generally speaking, to the average man is dull reading', wrote a man from Babylon (on Long Island) in 1906, 'and we need not be afraid to admit that this universal rule holds good with regard to the Law of Moses'.[59] But jurists, just like exegetes, are not normal people. For many, that reading was gripping. Lawyers from London to Leipzig thus offered their own studies of Mosaic law, including the barrister Harold M. Wiener (1875–1929), who sought 'to apply the ordinary methods of legal study to the solution of Biblical problems'.[60]

---

59    George Downing Sparks, 'The Law of Moses Historically Considered', *The Sewanee Review* 14/3 (1906), 281–87, at 287.

60    Harold M. Wiener, *Studies in Biblical Law* (London, 1904), vii; cf. also idem, *Essays in Pentateuchal Criticism* (Oberlin, 1909); idem, *The Origin of the Pentateuch* (London,

Well into the twentieth century, American law reviews – in Ann Arbor, New
Orleans, and elsewhere – devoted space to articles on biblical law, as did *The
Green Bag: An Entertaining Magazine for Lawyers*, based in Boston.[61] So too
journals for legal studies affiliated with both Cambridge universities, LSE, and
the French society for comparative law found room to publish book reviews on
Mosaic legislation.[62]

Legislators in Parliament, however, long betrayed a special interest in biblical law – especially on the matter of marrying sisters. With "'A Law for Jews
and Not for Christians"? Mosaic Law and the Deceased Wife's Sister Debate in
Victorian Britain', Michael Ledger-Lomas finds the fingerprints of Moses in the
making of modern law. Following Frishman, he highlights Christian views of
Judaism wed – in a spirit of supersessionism – to the notion of dead letters,
and with Wiedemann his chapter underscores how Christians could disavow
biblical law yet feel no small discomfort when discovering similar laws and

1910); Gerhard Förster, *Das mosaische Strafrecht in seiner geschichtlichen Entwickelung*,
Ausgewählte Doktordissertationen der Leipziger Juristenfakultät (Leipzig, 1900).

61   Clarence A. Lightner, 'The Mosaic Law', *Michigan Law Review* 10/2 (1911), 108–119; Louis
Binstock, 'Mosaic Legislation and Rabbinic Law', *Loyola Law Journal* 10 (1929), 13–19.
Further examples include the Swiss Harvard professor Walther Hug, 'The History of
Comparative Law', *Harvard Law Review* 45/6 (1932), 1027–1070, and the British Chief Rabbi
Joseph Herman Hertz, 'Ancient Semitic Codes and the Mosaic Legislation', *Journal of
Comparative Legislation and International Law* 10/4 (1928), 207–21. The *Green Bag* articles
came mostly in a series of series by David Werner Amram, a prominent UPenn law professor, promoter of Zionism, and student of Marcus Jastrow: 'Chapters from the Ancient
Jewish Law', 4/1 (1892), 36–38, 4/10 (1892), 493–95, 6/9 (1894), 407–08; 'Some Aspects of
the Growth of Jewish Law', 8/6 (1896), 253–56, 8/7 (1896), 298–302; 'Ancient Conveyance
of Land', 10/2 (1898), 77–78; 'Chapters from the Biblical Law', 12/2 (1900), 89–92, 12/4
(1900), 196–99, 12/8 (1900), 384–87, 12/9 (1900), 483–85, 12/10 (1900), 504–06, 12/11 (1900),
585–89, 12/12 (1900), 659–61, 13/1 (1901), 37–40, 13/2 (1901), 70–74, 13/4 (1901), 198–202, 13/
6 (1901), 313–16, 13/8 (1901), 406–08, 13/10 (1901), 493–96, 13/12 (1901), 592–94; 'A Lawyer's
Studies in Biblical Law', 14/2 (1902), 83–84, 14/5 (1902), 231–33, 14/7 (1902), 343–46, 14/
10 (1902), 490–93, 15/1 (1903), 41–44, 15/6 (1903), 291–94. Amram collected some of these
articles and added other material for his *Leading Cases in the Bible* (Philadelphia, 1905),
having previously published *The Jewish Law of Divorce According to Bible and Talmud, with
Some References to its Development in Post-Talmudic Times* (Philadelphia, 1896).

62   I.G., review of *Studies in Biblical Law*, by Harold M. Wiener, *Harvard Law Review* 18/5
(1905), 408–09; Nathan Isaacs, review of *The Origin and History of Hebrew Law* by J.M.
Powis Smith, *Harvard Law Review* 45/5 (1932), 949–52; T.W. Manson, review of *Studies in
Biblical Law* by David Daube, *Cambridge Law Journal* 10/1 (1947), 135–36; B. Grey Griffith,
review of *Studies in Biblical Law* by David Daube, *Modern Law Review* 11/2 (1948), 239–40;
Jean-Philippe Lévy, review of *Mosaïc Law in Practice and Study throughout the Ages* by
Pieter Jacobus Verdam, *Revue internationale de droit comparé* 12/4 (1960), 891–93.

stories in other ancient cultures. In line with Santini and Ilany, Ledger-Lomas shows Michaelis to have been enormously influential even beyond the German lands: here, in questions of the value and validity of Mosaic jurisprudence. His essay, like Zwiep's, underscores efforts by Jews to reconsider the place of Jewish law in a modern (Christian) state.

Yet departing from past work on the economic, social, and sexual spheres of marriage as well as the colonial dimensions to legal reform, Ledger-Lomas arrives at scriptural foundations. Spotlighting the interpretation and interpreters of biblical sources, he illuminates the textual basis of protracted debates on marriage and the family and elucidates the political cartography of those theological positions. Those debates, he argues, reveal a still more basic disagreement over the evaluation and application of Mosaic law. As his chapter expounds, moreover, the British discussion was tied to the German one, not only through the work of Michaelis or an *enquête* among professors of Hebrew but also via attention to marriage law abroad. Germans, like Jews, became beacons and bogeymen by turns. As Ledger-Lomas brings to light, marriage reformers could overcome their anxiety about German theology to find friends of convenience in biblical critics and Jewish commentators when certain arguments or expertise lent the right support. However, he also illustrates how figures on both sides of the issue in Britain found common ground: in different forms of anti-Judaism. Ultimately, his essay maps a feedback loop between ideas of marriage in the modern state and ideas of how to interpret and apply the law of Moses.

'Jews were not Socialists at heart', Theodor Herzl recalled telling the Prussian secretary of state, strategically. Herzl, remembering something he had read, elaborated: 'Through the Decalogue Moses created an individualistic form of society. And the Jews ... are and will remain individualists'.[63] As Carolin Kosuch shows, not everyone agreed. In 'Moses and the Left: Traces of the Torah in Modern Jewish Anarchist Thought', Kosuch traces the transformation of Mosaic law in radical political thought among German Jews. If together with Zwiep she considers Mendelssohn's view of Judaism as a rational religion of tolerance and humanity, she focuses like Frishman on responses to embourgeoisement and acculturation of German Jews and considers alongside Ledger-Lomas the politics involved in appropriating the Mosaic law.

Kosuch's chapter explores a post-Mendelssohn world, moving from a state project of Jewish emancipation and assimilation to a political project to turn

---

63     Patai, ed., *The Complete Diaries of Theodor Herzl*, trans. Zohn, vol. 2, entry for 18 September 1898, p. 667.

the world upside down. She disentangles the artistic, anarchist, and socialist strands all intertwined in the political theory of Gustav Landauer (1870–1919) and Erich Mühsam (1878–1934). Charting different responses to the problems of modernity and reactions to the place of Judaism in it, her chapter maps the interweaving of enlightened and romantic ideas in an age of formal legal equality for Jews. Kosuch considers why these two thinkers turned to anarchism and how, precisely as Jews and radicals, Landauer and Mühsam encountered tensions on both sides: raised in bourgeois Jewish families yet active in the mostly non-Jewish proletarian circles of their politics. However, their anarchistic ideas, she contends, entailed a reframing of central figures in the Jewish canon, including Moses. By fusing Mendelssohn and Romanticism, Herder and *Haskalah*, Landauer and Mühsam fashioned a new interpretative framework for Jewish law and particularity: as a duty to act – to ameliorate humanity and to consummate a new, true equality. This attempt at an anarchist reconfiguring of Jewish tradition, Kosuch further argues, constituted a dual form of resistance: to the Protestant bourgeoise as well as to their acculturated Jewish milieu.

The core of this volume comes from a workshop – international, interdisciplinary, intergenerational – entitled 'Mosaic Law among the Moderns: Constructions of Biblical Law in 19th-Century Germany' and hosted in Cambridge during the summer of 2019. Presenters and respondents from Germany and Belgium, Israel and Italy, as well as the US, UK, and Netherlands, engaged in thought-provoking papers and lively conversation for three days' time. That exchange was especially enriched by its diversity: in career stage, including predocs, postdocs, and professors, as in field, ranging from history and classics to studies religious, German, and Jewish. The delay between presentation and publication resulted from the usual holdups – major and minor, individual and collective – for such collaborative undertakings as well as the rather unusual one: namely a global pandemic.

Many of the chapters in *Moses among the Moderns* originated in that Cambridge conference, appearing here in revised form: those by Irene Zwiep, Felix Wiedemann, Carolin Kosuch, and Judith Frishman. Two others from that event have seen replacement. As Ofri Ilany's original presentation was committed elsewhere, he wrote a new piece specifically for this volume.[64] Suzanne L. Marchand delivered a striking keynote presentation under the title 'Greek Freedom and Mosaic Law in 19th-Century Germany', laying out what Greek

---

64      Ofri Ilany, 'Christian Images of the Jewish State: The Hebrew Republic as a Political Model in the German Protestant Enlightenment', in *Jews and Protestants: From the Reformation to the Present*, eds Irene Aue-Ben-David et al. (Berlin, 2021), 119–35; see further idem,

freedom meant to German thinkers in this period (a *geistige Freiheit* to imagine, learn, and create and to escape clerical and feudal legalism as well as aristocracy; a *bürgerliche Freiheit*, with citizens free to serve the state; a *völkische Freiheit*, championing individual tribes or peoples from domination by other, larger empires or states) – and in contrast to 'oriental' despotism, including the idea of Greek laws as universal and modern, with Mosaic law cast as 'oriental' and particular, a survival unfit for a new age. Instead of that paper, Marchand has contributed a retrospective on the volume as a whole, helping to weave the strands together.

Two other papers were delivered at the workshop but owing to other commitments could not, understandably, be revised for publication here. Both expanded the scope in transnational and colonial history. With 'Which Law for the Colonial Empire? Rule of Law and (Christian) Religion in German Colonialism', Nicola Camilleri (University of Padua) compared the tension between cultural and religious difference among inhabitants of the German empire and of colonial rule, on the one hand, and in the Italian colonial context, on the other. One tentative yet tantalizing result of the discussion was how legal history related to religion in the metropole transformed yet extended to the colonies: how procedures for marriage between Catholics and Protestants informed the same for citizens and subjects outside the borders in Europe. Annelies Lannoy (Ghent University) presented 'The Law and the Republic: Maurice Vernes and Aristide Astruc on the History of Mosaic Law and its Instruction in the Ecole Laïque'. After surveying the Protestant Vernes's (1845–1923) historical work on biblical law and political writings on the importance of the Old Testament for French secular education, Lannoy traced his strategic alliance with Rabbi Astruc (1831–1905) to integrate the history of religion – specifically Jewish history – into curricular programming. She thus illuminated biblical scholarship across national and confessional borders in the matrix of state and secularity. Furthermore, a third was set to expand the confessional, linguistic, and geographic perspective. Cristiana Facchini (University of Bologna) was meant to deliver 'Monitoring German Scholarship on the Bible: Jesuit and Catholic Counter-Narratives (1850s–1900s)'. Through the journal *La Civiltà Cattolica*, founded in 1850, the paper planned to map Jesuit interpretation of German Protestant interpretation from the foundation of the

---

'*Herr Zebaoth* and the German Nation: Bible and Nationalism in the anti-Napoleonic Wars', *Global Intellectual History* 5/1, Special Issue: 'Theology & Politics in the German Imagination, 1789–1848', ed. Ruth Jackson Ravenscroft (2019), 104–24. The title of the workshop paper ran 'The Israelites' *Nationalgeist*: Ethnography and Politics in Johann David Michaelis's Interpretation of Mosaic Law'.

journal to the modernist crisis (1907), focusing on the relationship between the nation state and the transnational Catholic community of faith. Unfortunately, Facchini had to withdraw from the program shortly before the workshop.

In addition to the papers revised or replaced for this volume, three were solicited to supplement the publication in breadth and depth. Carlotta Santini thus added her original contribution to include perspectives from German classics, while Michael Ledger-Lomas expanded the scope to interactions with German biblical studies in the British context. Yet another was planned to address Mosaic law in the historiography of Muhammad and formative Islam. David Moshfegh (IE University, Madrid) kindly agreed to write a chapter entitled 'Semitic Religion, Theocracy, and *Islamwissenschaft*', centered on Goldziher, Wellhausen, Nöldeke, and William Robertson Smith (1846–1894). This plan did not come to fruition, however.

This account of the prehistory to *Moses among the Moderns* aims not only to elucidate its becoming – the possibilities and actualities – or to illuminate latent ideas from unrecorded conversations or unrealized intentions. Rather, it also seeks to cast light on future pathways for work on the cultural history of Moses and reception of biblical law. In this way, it should signpost the roads travelled, those not taken, and those to be taken even further.

When Moses breathed his last on Mount Nebo – in the same region Hechler would hope to find the Ark of the Covenant and therein his very own writings – Joshua led the Israelites onward, to new vistas, new horizons. So too others should now lead the way to add further pieces to this Mosaic mosaic, to this reception of Moses and his many roles: in cultural, intellectual, and religious history, across the German lands, throughout the nineteenth century – and beyond (Cover Image; Figure 3).

### Acknowledgements

Generous support for work on this essay came from the Flemish Research Council (FWO). My gratitude extends to Laura Loporcaro, Suzanne Marchand, and Rebecca Van Hove, who kindly offered detailed feedback on the manuscript of this essay. As usual, translations are mine if not stated otherwise.

FIGURE 3    Cartoon, by Joseph Keppler and Frederick Opper, of 'The Modern Moses'
Uncle Sam parting the waters for Jewish immigrants amidst pogroms in the
Russian Empire. Published in the American magazine *Puck* in 1881. Further
information on this item, including controversy at the time of its antisemitic
caricature, available online through the Jewish Virtual Library, 'Judaic
Treasures: From the Lands of the Czars'.
IMAGE COURTESY OF CORNELL UNIVERSITY – PJ MODE COLLECTION OF
PERSUASIVE CARTOGRAPHY.

# PART 1

## Representations of the Past

∴

# CHAPTER 1

# 'The Early Speech of Nations'

*Biblical Poetry and the Emergence of Germanic Myth*

*Ofri Ilany*

In 1755, Johann Christoph Strodtmann (1717–1756) published an essay entitled 'Übereinstimmung der deutschen Alterthümer mit den biblischen, sonderlich hebräischen' ('Resemblances Between German and Biblical, or Hebrew, Antiquity'). In its 500 pages, the author, rector of the Gymnasium Carolinum Osnabrück, listed similarities between the people of Israel, as described in the Bible, and the Germanic peoples (*Deutschen*), as depicted by Tacitus and other classical authors. He found several parallels between these ancient peoples – among other things, their forms of dress, alimentation, law, sacrifices and liturgy as well as their warring practices and political structures. In his preface, Strodtmann explains the utility of his book:

> This tract may perhaps aid those who, though they may take no interest in German antiquities for their own sake, feel moved to do so because of the similarities with the biblical ones – lest they otherwise remain strangers in their own homeland, consigned to a bad understanding of their ancestors.[1]

This comment succinctly illustrates the relation between the scholarship of Hebrew and German history in the mid eighteenth century. In the mid-eighteenth century, the discipline of Germanic history was still considered a somewhat esoteric undertaking, garnering little interest in the German republic of letters. These scholars shied away from the history of their 'ancestors' (*Vorfahren*) and that of their 'homeland' (*Heimat*). Scholarship of Hebrew history, by contrast, enjoyed much greater prestige, and German scholars took the lead in the historical construction of the biblical past. Subsequently, the logic underlying Strodtmann's book is an attempt to legitimate the young discipline of Germanic history by folding it into scripture in general, and

---

1  Johann Christoph Strodtmann, *Übereinstimmung der deutschen Alterthümer mit den biblischen, sonderlich hebräischen* (Wolfenbüttel, 1755), ii. Unless noted, all translations are mine.

Hebrew history in particular. As I will argue in this chapter, scholarly and literary treatment of the antiquity of Germanic and other northern European peoples grew in this period alongside *Orientalistik* ('oriental studies') and biblical studies. Throughout the eighteenth century, the two disciples were intertwined, with research into the Germanic peoples developing, in many cases, as an outgrowth of scholarship on the Bible. Within this discourse, the question of Hebrew poetry played a central role: biblical poetry served as a model for epic poetry.

This chapter explores the role of epic in the emergence of nationalism by examining the intersection of Germanic myth and the Hebrew Bible. During the latter half of the eighteenth century, there was a growing fascination in and appreciation for the ancient customs of 'natural peoples'. Among this esteemed group, which included nations like the Celts and Germans, the Hebrews were prominently featured, showcasing a culture in which the epic held great significance. It was during this period that biblical interpreters started to interpret Hebrew poetry as a means of creative expression, providing insights into the worldview of ancient humanity.

In this period, German scholars sought to legitimize the study of Germanic history by drawing parallels between the ancient Germanic peoples and the people of Israel described in the Bible. This intertwining of Germanic history and biblical studies resulted in the formation of a complex relationship between the Bible and the emerging national myths in German literature. Figures like Johann Gottfried Herder (1744–1802) and Friedrich Gottlieb Klopstock (1724–1803) imagined the ancient Germanic forefathers as Nordic Hebrews, while biblical interpreters depicted the Hebrews as epic heroes akin to the Germanic leader Hermann, the Ossianic tradition, or even the Homeric heroes. These interwoven mythologies of the Hebrews and Germans provided mutual legitimacy, establishing a new ideal of naturalism and originality.

It was only in the nineteenth century that Germanic historical scholarship finally broke free from its dependence on biblical studies. While in the eighteenth century there was a fascination with the Hebrews and their connection to Germanic identity, the nineteenth witnessed a shift in attitudes towards Judaism within the German nationalist discourse. As nationalism gained dominance, German intellectuals began questioning the influence of Hebrew culture on German society and sought to establish an exclusive national mythology. Moreover, the emergence of racial theories and the consolidation of the Aryan narrative further marginalized the Semitic influence in the history of human civilization. Understanding this historical backdrop helps shed light on the changing dynamics between the German national movement and Judaism in the nineteenth century.

To make this argument, I begin the chapter by presenting the baroque sources of comparison between the Hebrew people and Germanic peoples as well as between the Hebrew and German languages. I survey the hypothesis of the German language's Hebrew origin as a language and the transformations this theory underwent in the early eighteenth century, followed by this conception's development in the work of Klopstock, who purported to identify a link between ancient Germanic religion and Hebrew monotheism –in service to the cultural-poetic project of establishing a German identity. Next, I turn to how this dynastic conception (which sought to tie customs and words to the biblical tradition) was supplanted by an ethnographic approach (which assimilated the Israelites into the natural history of ancient mankind), in effect equating them with other ancient peoples including the Celts and the Germans. It was in this context that the Hebrew language and Hebrew poetry were accorded the qualities of naturality and authenticity. Finally, I trace the influence of the 'discovery' of the Ossian epic – identified as an ancient Celtic poetic cycle – on the ethnographic interpretation of biblical poetry. I conclude by showing how the Bible was framed as myth in the writings of Herder and Wilhelm Martin Leberecht de Wette (1780–1849).

## 1    The Hebrew Source Theory

Linkage between the ancient Germanic peoples and the Hebrews appeared in German literature as far back as the sixteenth century. One popular baroque theory held that Hebrew was the parent language from which German derived.[2] The humanist Wolfgang Lazius (1514–1565) and Calvinist orientalist Theodore Bibliander (1509–1564) claimed that the Germans had descended from the biblical son of Gomer, Ashkenaz (*Ascenas*), grandson of Japhet, and that the ancient Germanic forefather Tuisco, mentioned by Tacitus, lived during the time of Abraham.[3] Lazius went so far as to try to link Ascenas (the Saxonian Askanier dynasty) and the city of Aschersleben, in Sachsen-Anhalt.[4] Later,

---

2    See Andreas Gardt, ed., *Nation und Sprache. Die Diskussion ihres Verhältnisses in Geschichte und Gegenwart* (Berlin, 2000), 178–179.

3    William J. Jones, 'Early Dialectology, Etymology and Language History in German-Speaking Countries', in *History of the Language Sciences / Geschichte der Sprachwissenschaften / Histoire des sciences du langage*, eds Sylvain Auroux et al., 3 vols, Handbücher zur Sprach-und Kommunikationswissenschaft 18 (Berlin, 2000–06), 2:1105–1115.

4    See, for example, Casper Abel, *Teutsche und Sächsische Alterthümer* ... (Braunschweig, 1729), 6.

Johannes Micraelius (1597–1658), even proposed that the name 'Germans' (*Germanen*) originated in the Hebrew word *ger* ('foreigner').[5]

A more cautious theory, which held 'Celto-Germanic' and Hebrew were sister-languages of a common ancestry, gained prominence throughout the seventeenth century. Its most popular variation was formulated by the Huguenot orientalist Samuel Bochart (1599–1667) in his *Geographia Sacra seu Phaleg et Canaan*, published in 1646.[6] There, he claimed that many of the peoples of the Old and New Worlds were descendants of the Canaanites, forced to flee their land and resettle elsewhere around the world. Bochart and his followers did not stop at identifying the indigenous peoples of the Americas with the Canaanites, however. To them they added the Celts, who – they claimed – were dispersed from England as far afield as Germany. Bochart even identified Tuisco himself with Canaan, son of biblical Ham.[7]

Around the end of the seventeenth century, leading figures such as Gottfried Wilhelm Leibiniz began casting doubt on Hebrew's status as the mother of all tongues.[8] The Dutch orientalist Albrecht Schultens (1686–1750), for instance, claimed in 1724 that Hebrew was not a divine language given to Adam but rather a branch on the tree of the Arabic language.[9] Schultens laid the groundwork for a systematic orientalist interpretation of the Bible based on the use of Arabic. Simultaneously, he and others undermined the hypothesis of the Hebrew-Celtic origin. Voltaire (1694–1778), for his part, pulled no punches when describing this idea as 'one of the greatest follies of the human spirit'.[10]

Consequently, over the first third of the eighteenth century, the hypothesis of German's Hebrew origin gradually disappeared from those antiquarian essays concerned with Germanic antiquity. But scholars continued to compare these two peoples.[11] Thus, the Saxon Caspar Abel (1676–1763), a pioneer of

---

5    Johannes Micraelius, *Antiquitates Pomeraniae, Oder Sechs Bücher vom Alten Pommerlande* ..., 6 vols (Leipzig, 1723), 3:205.

6    Samuel Bochart, *Geographia Sacra, seu Phaleg et Canaan* (Caen, 1646). See further Graham Parry, *The Trophies of Time: English Antiquarians of the Seventeenth Century* (Oxford, 1995), 308–13. It is perhaps unsurprising that as a Huguenot, Bochart had images of fleeing and expulsion on his mind exactly at this time. I thank Paul Michael Kurtz for this observation.

7    Bochart, Geographia Sacra, 1:505.

8    Sigrid von der Schulenburg, *Leibniz als Sprachforscher* (Frankfurt, 1973), 26.

9    Edward Breuer, *The Limits of Enlightenment: Jews, Germans, and the Eighteenth-Century Study of Scripture* (Cambridge, MA, 1996), 92.

10   Voltaire, *Essai sur les mœurs et l'esprit des nations* (Geneva, 1756), 169.

11   See, for example, the aforementioned Abel, *Teutsche und Sächsische Alterthümer*; Johann Ehrenfried Zschackwitz, *Erläuterte Teutsche Alterthümer, Worinnen* ... (Frankfurt, 1743); Christian Ulrich Grupen, *Teutsche Alterthümer, zur Erleuterung des Sächsischen auch Schwäbischen Land-und Lehn-Rechts* ... (Hannover, 1746). Regarding writing on the

Germanic scholarship published *Teutsche und Sächsische Alterthümer* (*German and Saxonian Antiquities*) in 1729, followed, several years later, by *Hebräische Alterthümer* (*Hebrew Antiquities*). Abel demonstrates the significant link between *Germanistik* and *Hebraistik* at the time. In his book on Hebrew antiquities, Abel suggested the possibility that Arioch, king of Ellasar (Gen 14:1), was in fact a Scythian and the Germans were descendants of Canaanites who had fled the Israelites.[12] Although he ended up renouncing the theory, his engagement with it at all indicates the kind of currency such ideas enjoyed well into the mid-eighteenth century.[13]

While Strodtmann himself compared, alongside many other laws and customs, stoning and decapitation among the Germans and the Hebrews, he rejected the hypothesis of Hebrew origins for the German language, as espoused by early modern German humanists. Against what he called the 'prevailing belief', Strodtmann doubted that Germans could have inherited their customs from the Hebrews, seeing as they were geographically so far removed and, in general, did not come into contact with other peoples. He suggested an alternative theory, instead: the similarity in customs had come from the Germanic people migrating from the East just like others yet maintaining the ancient customs of humanity's 'ancestors' (*Patriarchen*) well before the Jews appeared on the scene.[14] On the one hand, Strodtmann's enumeration of similarities in customs was meant to legitimize the scholarship of German antiquarian scholarship. On the other hand, it relied on the conceptual framework of contemporaneous ethnography, centered around the idea of 'ancient' or 'natural' peoples. The underlying assumption was that the customs of ancient peoples had been both similar and simple, before they veered off in different directions over the course of their historical development.

A similar stance was formulated by the orientalist and biblical scholar Johann David Michaelis (1717–1791), a key figure in the rise of historical criticism.[15] Michaelis addressed the claim of Tacitus that Germanic tribes had no

---

Germanic peoples during the early modern period, see Ludwig Krapf, *Germanenmythos und Reichsideologie. Frühhumanistische Rezeptionsweisen der taciteischen 'Germania'* (Tübingen, 1979); Dieter Mertens, 'Die Instrumentalisierung der "Germania" des Tacitus durch die deutschen Humanisten', in *Zur Geschichte der Gleichung 'germanisch-deutsch'. Sprache und Namen, Geschichte und Institutionen*, eds Heinrich Beck and Dieter Geuenich (Berlin, 2004), 37–102.

12    Caspar Abel, *Hebräische Alterthümer, Worinnen ...* (Leipzig, 1736), 477.

13    Ibid., 219; idem, *Teutsche und Sächsische Alterthümer*, 6.

14    Strodtmann, *Übereinstimmung der deutschen Alterthümer mit den biblischen*, 4–5.

15    For more on Michaelis, see the contributions by Carlotta Santini, Michael Ledger-Lomas, and Irene Zwiep in this volume.

written records of their past – only songs by which they passed their fathers' heroic deeds down through the generations. Comparing this practice to those of the Hebrews, Michaelis thought it hardly surprising that no monuments of poetry from 'our forefathers' had survived, given the Germanic peoples' dispersal throughout history, in contrast to the poetry of Scandinavia and Iceland, which were much more preserved. For him, the evidence of a poetry similar to that of the Hebrews in 'peoples so far removed' pointed to 'the overwhelming identity of mortal ingenuity (*ingenium mortalium*) in every place on earth'.[16] In this anthropological understanding, the Hebrews and the Germans represented two 'natural peoples' who shared similar characteristics, namely a reliance on poetry and oral traditions – a condition Michaelis identifies, in his *Mosaisches Recht* (*Mosaic Jurisprudence*), as the natural state that is the infancy of all peoples.[17]

## 2      Klopstock: Between David and Hermann

The poetry of Klopstock, considered the father of modern German verse, includes especially pertinent comparisons between Hebrews and Germans. Klopstock arrived at a poetico-theological synthesis of German identity, Protestant faith, and fierce adherence to the Old Testament and to the 'Hebrew model'. While some rationalist theology strained to suppress the Old Testament so as to bring Christianity closer to 'natural religion', Klopstock championed a German Christianity in which Hebrew myth played a pivotal role.

In 1769, Klopstock published his drama *Hermanns Schlacht* (*Herman's Battle*), a decisive contribution to the formulation and circulation of the national myth of 'Hermann's battle' and of the Teutons more generally.[18] Hermann, a Cherusci military leader, is presented as a patriotic hero who defends German soil and safeguards its independence. *Hermanns Schlacht*, and the two national plays that followed, ultimately had a much more pronounced influence on German culture than the biblical tragedies Klopstock wrote in the same period as well. But poetically, the Hermann dramas often constitute developments in or variations on his biblical dramas, their worldviews and dramatic notions.[19] Both in

---

16      Robert Lowth, *De sacra poesi Hebræorum, prælectiones academicæ Oxonii habitæ*, ed. Johann David Michaelis (Göttingen, 1758), 72.

17      Johann David Michaelis, *Mosaisches Recht*, 6 vols (Frankfurt am Main, 1770–75), 1:308–309.

18      See, for instance: Hans Kohn, *Die Idee des Nationalismus. Ursprung und Geschichte bis zur französischen Revolution* (Frankfurt, 1950), 563.

19      Gerhard Kaiser, *Klopstock. Religion und Dichtung* (Mainz, 1975), 280.

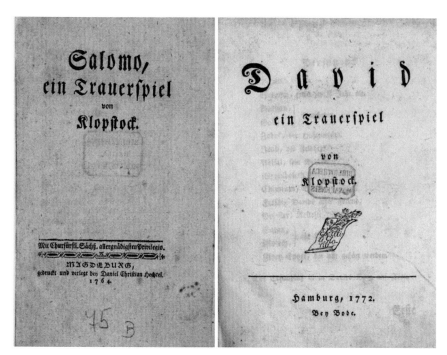

FIGURE 4     Title pages to Klopstock's Salomo (1764) and David (1772)
             IMAGE COURTESY OF THE MUNICH DIGITIZATION CENTRE OF THE BAVARIAN
             STATE LIBRARY

style and in plot, the national-patriotic play is similar to the two biblical plays by Klopstock: *Salomo* (*Solomon*, 1764) and *David* (1772) (Figure 4). Furthermore, as his letters reveal, Klopstock wrote *Hermanns Schlacht* primarily to win support from those German tastemakers who despised biblical plays.[20]

As already suggested, Klopstock's generation was under the full sway of Enlightenment worldviews, with their affinity for neoclassicalism, especially around the Berlin circle of publisher and editor Christoph Friedrich Nikolai (1733–1811). Against this background, Klopstock stood out for his religious identity, and his biblical plays were greeted with open derision by critics. However, enlightened audiences showed greater tolerance for the Germanic plots, and critics accepted his Hermann dramas with relative excitement.

In the preface to an anthology of his national and biblical plays, Klopstock expresses wonderment at their icy reception of his pieces dealing with Hebrew

---

20    Friedrich Gottlob Klopstock, *Klopstocks sämmtliche Werke*, vol. 10, *David, Hermanns Tod* (Leipzig, 1823), 237.

material. Some critics' blind loyalty to the classical style, he writes, constrains their taste so severely that they refuse to accept any work that fails to conform to the ideal of Greek or Latin poetry. Other German critics, Klopstock ruefully adds, are willing to tolerate plays about their idolatrous forefathers even as they derisively reject the Bible. Nathan the prophet, who features as the protagonist of *Salomo*, Klopstock admonishes, merits no less interest or respect than Brenno, the wise Druid of *Hermanns Schlacht*:

> They who believe revealed religion just as little as the polytheism of our forefathers are therefore in the wrong if they consider Nathan and Brenno (I will name those only) as unequally worthy objects for a poet. If they do otherwise, they are indicting Nathan over Brenno. Their judgment of poetry is now enchained to their opinions, and perhaps also to their passions; this kind of clanging can only be heard with so much displeasure.[21]

It was for this reason, Klopstock explains, that he chose to publish the biblical and national dramas together, so 'one has constant cause for comparison' (*daß man fortwährenden Anlaß zur Vergleichung hat*).[22]

As Gerhard Kaiser observes, rather than a purely tactical decision, the binding of the Hebrew and Teutonic plots in fact gave expression to Klopstock's theology.[23] Like the biblical dramas, the Hermann ones invoke the reader's 'forefathers' gods' (*Götter seiner Väter*). Hermann's campaign is seen not only as safeguarding the homeland but even saving the ancient Teutonic religion. In the second play of the trilogy, *Hermann und die Fürsten* (*Hermann and the Princes*), the Germanic princes deride the words of the gods and are punished in battle. By contrast, Druid Brenno, like Nathan the Prophet, has faith in the people's eternity and in Germany's future salvation. The Germanic religiosity underpinning these plays seems, at first sight, to be at odds with Klopstock's devout Protestant faith, a tension he addresses in a note added to the play *Hermanns Tod* (*Herman's Death*). As for the Teutonic worship of the god Wodan, he clarifies:

> [o]ur ancient forefather the Scythians had neither ancillary gods nor demigods. They worshiped one god. Their colonies in Europe changed the concept of the Supreme Being by additions, though not so profoundly

---

21 Friedrich Gottlob Klopstock, *Klopstocks sämmtliche Werke*, vol. 8, *Der Tod Adams, Hermanns Schlacht* (Leipzig, 1823), v.

22 Ibid.

23 Kaiser, *Klopstock. Religion und Dichtung*, 276–77.

as those of the worshipers of Zeus or Jupiter [...] Because they loved war above all else, the supreme God was above all engaged in war.[24]

In its pure form – as brought from the East by the Celts – the ancestral religion was therefore based on the worship of one warlike god, not many idols. The Celts, and then the Germans, were only corrupted after European settlement, and even then, their religion remained purer than the idol worship of the Greeks and Romans.

Celto-Germanic religion thus represents a relic of oriental monotheism inside pagan Europe. According to Strodtmann, the name among ancient German tribes for the god of their fathers was clearly 'reserved from the old true religion'.[25] This image influenced other writers, among them members of the 'Göttingen Grove' (*Göttinger Hain*), who claimed the Germans were a god-fearing people devoted to a single invisible god.[26] In this way, Klopstock brought Hebrew monotheism closer to Germanic religion. From the viewpoint of the Christian German, the two ancient peoples praying to the god of their forefathers stand on equal footing: both on the threshold of redemption, 'in the vestibule of the sanctuary' (*Vorhof zu dem Heiligthume*).[27]

To a large extent, Klopstock here revives the baroque humanist hypothesis on the Scythians' and Hebrews' shared origins – the same hypothesis rejected by the biblical and orientalist scholars of his day. He was not alone. Drawing on the epic of Ossian, for example, other writers in this period argued that the ancient Celts still bore certain traits they had brought with them when migrating westward after the confusion of languages after the Tower of Babel.[28] At any rate, the intellectual context in which this theological-historical move by Klopstock appears is notably distinct from that of sixteenth-and seventeenth-century antiquarians and chroniclers. In the 1770s, this theory grew within a new discourse of patriotism and originary nationalism, which was then only beginning to emerge in the German republic of letters.

The historian Hans-Martin Blitz has countered a prevailing narrative in scholarship on German nationalism. Where others have tended to trace the origins of this movement back only as far as to the French Revolution and

---

24    Ibid., 246.
25    Strodtmann, *Übereinstimmung der deutschen Alterthümer mit den biblischen*, 250.
26    For instance, Johann Heinrich Voß, "Deutschland. An Friedrich Leopold, Graf zu Stolberg [1772]', in idem, *vermischte Gedichte und prosaische Aufsätze* (Leipzig, 1784), 20–21.
27    Klopstock, *Klopstocks sämmtliche Werke*, vol. 8, 9.
28    See: Howard D. Weinbrot, *Britannia's Issue: The Rise of British Literature from Dryden to Ossian* (Cambridge, 2007), 537–39.

the anti-Napoleonic wars, he contends that German writers and poets were in fact expressing aggressive, exclusive, and even bellicose views about the German homeland as far back as the mid-eighteenth century.[29] He views Klopstock as the outstanding figure of this generation of national poets, his poems and plays exhibiting real national zeal and passion for the fatherland (*Vaterlandschwärmerei*). This particular form of nationalism, however, was not formulated through identification with any extant state or political power: Klopstock himself was the protégé of the king of Denmark. Rather, this was a nationalism whose main ingredient comprised a rejection of French-style enlightened absolutism, the Enlightenment, and modernity altogether. In this patriotic discourse, worship of archaic Germanic culture functioned as a tool of wider civilizational critique. Thus, Hermann's campaign itself is portrayed as a war against Rome, 'the world tyrant' (*Welttyrannin*). Blitz identifies in Klopstock's work a shift from religious thematics in the early poems and plays to a national thematics in the later works. Exemplifying the tension between these two elements is the poem 'Mein Vaterland' ('My Fatherland', 1768), where the poet moves from fervent love of the German homeland to his search for a more abstract religious homeland, 'the fatherland of the human race' (*Vaterland des Menschengeschlechts*).

Yet analysis of Klopstock's work only through its Christian Hebraistic element obfuscates the distinction between its two aforementioned phases. In terms of content, Klopstock often structures his writings as a series of dichotomies between two elements. At one pole lie the local, the religious, the biblical, and the Germanic, which Klopstock identifies with himself; at the other lie the foreign, the classicist, the cosmopolitan, and the Greco-Roman, often associated with France. Klopstock's overtly sentimental and pietistic poetry, which highlights explicitly biblical elements and lauds David at the expense of Pindar, is structured as a parallel to the paeans to Hermann, striker of the Romans. As a result, the biblical prophet is extracted from his former religious status, becoming prophet of the nation, while the Teutonic hero is cast in

---

29    Hans-Martin Blitz, *Aus Liebe zum Vaterland. Die deutsche Nation im 18. Jahrhundert* (Hamburg, 2000); idem, "'Gieb, Vater, mir ein Schwert!" Identitätskonzepte und Feindbilder in der "patriotischen" Lyrik Klopstocks und des Göttinger "Hain", in *Machtphantasie Deutschland. Nationalismus, Männlichkeit und Fremdenhaß im Vaterlandsdiskurs deutscher Schriftsteller des 18. Jahrhunderts*, eds Hans Peter Herrmann et al. (Frankfurt, 1996), 80–122.

biblical garb.[30] Klopstock uses the Hebraic to develop a model of separatist religious zeal linked intimately to the Germanic fatherland.

### 3    The Hebrews, a Naturvolk

In the second half of the eighteenth century, the Hebrews featured among the select group of 'peoples of a national epic'. Biblical interpreters began reading Hebrew poetry as a form of creative expression for ancient humanity's *Weltanschauung*. They saw in the Hebrew language and its poetry a new quality of naturalism and emotion.[31] The ancient customs of 'natural peoples' were now raising interest and even garnering admiration.

The prevailing sentiment of the Enlightenment was unsympathetic towards 'natural peoples'.[32] This derisive characterization was applied to the Hebrews by the deists and Voltaire, for instance. Yet the Germanic tribes were depicted with no less derision. Many English and Scottish writers compared the Germans, as described by Tacitus, with the indigenous peoples of the Americas – an unflattering comparison on both sides. David Hume (1711–1776) considered the Germanic peoples 'the most rude and barbarous of the whites', even though he found them slightly better than the native peoples of America and Africa.[33] Edward Gibbon (1737–1794), William Robertson (1721–1793), and others articulated similar opinions, using the Germanic peoples as ancient literature's main example of a barbarian lifestyle.[34] However, the middle of the eighteenth century, the neoclassical ideal of 'rational nature' began to transform in meaning, taking on new significations centered on originary uniqueness, creativity, spontaneity, sensuality, and simplicity.[35]

---

30    On later transformations of the prophets in German intellectual history, see Paul Michael Kurtz, 'Is Kant Among the Prophets? Hebrew Prophecy and German Historical Thought, 1880–1920', *Central European History* 54 (2021), 34–60.

31    Joachim Dyck relates the description of Hebrew as a natural language to French theologian Antoine Augustin Calmet (1672–1757), whose main tome on the Bible was published by Mosheim in 1745: see Dyck, *Athen und Jerusalem. Die Tradition der argumentativen Verknüpfung von Bibel und Poesie im 17. und 18. Jahrhundert* (Munich, 1979).

32    Ronald Meek, *Social Science and the Ignoble Savage* (Cambridge, 1976), 37–68.

33    David Hume, 'On National Characters', in idem, *Essays and Treatises on Several Subjects*, 4 vols (London, 1753), 1:277–300, at, 291.

34    Ronald L. Meek, *Social Science and the Ignoble Savage* (Cambridge, 1976), 34, 138–41; Christopher J. Berry, *Social Theory of the Scottish Enlightenment* (Edinburgh, 1997), 62–64, 81–86.

35    Arthur O. Lovejoy, '"Nature" as Aesthetic Norm', repr. in idem, *Essays in History of Ideas* (Baltimore, 1948 [1927]), 76.

In his essay on the ideal of the noble savage during the German Enlightenment, Johann Reusch describes a 'German obsession' for this topos, which rose to prominence in tandem with the expansion of bourgeois culture.[36] According to Reusch, despite the centrality of Montesquieu (1689–1755) and Jean-Jacques Rousseau (1712–1778) for formulating the idea of 'the noble savage', this intellectual fashion was linked, in Germany, with resistance to the political models of French absolutism, applied in Prussia at the time by Friedrich the Great. The philosophical and poetic discussion on this topic was imported into Germany mainly from English and Scottish essays. Indeed, British and German intellectuals shared in an unprecedented zeal for their local peoples' 'pre-colonial' (that is, pre-Roman) histories.[37] Reusch identifies this trend with the rising questions among middle classes across the German lands regarding a 'German' identity, dissent of assimilation and 'Romanization', and subsequent anxieties surrounding nascent industrialization and the weakening of traditional social structures.

The celebration of Hebrew ideals by German poets was linked, then, to an entire array of aesthetic values that developed in the mid-eighteenth century and revolved around a new relation to the concept of the 'natural'. The most potent fuel for this movement was supplied by lectures from Robert Lowth (1710–1787) on Hebrew poetry. Thanks to the influence of this essay, Hebrew poetry became a central topic of German aesthetic-poetic discourse.[38] For Lowth, the biblical text was a kind of ideal expression of the natural powers themselves: an expressive, unmediated overflowing of sublime forces given form by the ancient soul's own tumultuous means of expression. Therefore, the uniqueness of Hebrew poetry was not necessarily in its hidden truths but in its very simplicity and rawness.

At the time of their publication, Lowth's ideas defied the dominant opinions in the European republic of letters. Enlightenment philosophers tended to describe Hebrew poetry as coarse, unintelligible, and nebulous texts written under the influence of the hot 'oriental' climate. Even those scholars who did not undermine the Bible's religious authority agreed, in most cases, that

---

36  Johann J. K. Reusch, 'Germans as Noble Savages and Castaways: Alter Egos and Alterity in German Collective Consciousness During the Long Eighteenth Century', *Eighteenth-Century Studies* 42 (2008), 91–129.

37  Ibid., 96.

38  On Lowth's influence on German Bible scholarship, see Anna Cullhed, 'Original Poetry: Robert Lowth and Eighteenth-Century Poetics', in *Sacred Conjectures: The Context and Legacy of Robert Lowth and Jean Astruc*, ed. John Jarick (New York, 2007), 25–47; Rudolf Smend, 'Lowth in Deutschland', in idem, *Epochen der Bibelkritik*, Gesammelte Studien 3, Beiträge zur evangelischen Theologie 109 (Munich, 1991), 43–62.

it never approximated classical poetry's aesthetic quality.[39] Lowth, on the contrary, rehabilitated that selfsame 'Oriental poetry': the Hebrew poets now became 'natural poets' – their emotions and expressions unbridled and free.[40] Among the *Sturm und Drang* writers, a literary and philosophical movement that celebrated naturalism, sentimentality, and particularism, Lowth's ideas contributed to the Bible being read as a sublime poetic creation conserving primal and natural forms of expression. In a 1778 essay written by Herder on the influence of poetry on different peoples' 'way of life', he contends this kind of poetry can express ideas no form of philosophical *Abstaktion* could ever convey precisely because it is natural, pastoral, earthly, and simple.[41]

Herder developed this idea at length in his *Vom Geist der Ebräischen Poesie* (*On the Spirit of Hebrew Poetry*), which further develops Lowth's ideas. The book is set up as a dialogue between two speakers – one skeptical of Hebrew poetry, the other sympathetic. Alciphron, the skeptical of the two fictional interlocutors, opens the essay by describing Hebrew as 'a poor barbarous tongue' (*arme barbarische Sprache*) – meager, coarse and unpleasant.[42] He goes on to compare Hebrew to the language of the indigenous peoples of North America, calling the former 'those Hurons of the Orient' (*morgenländischer Huronen*).[43] Herder's own stance is presented by the second speaker, Euthyphron. He does not deny that Hebrew is simple or lacking in sophistication but does claim, on the contrary, that those very characteristics give the language its best qualities. Hebrew, he claims, should be compared to 'a poor country girl, but beautiful and pure' (*schönes und reines Landmädchen*).[44] That Hebrew is noun-poor yet verb-rich is exactly what makes it, for Herder, a living, young, and poetic language.

---

39    See Jonathan Sheehan, *The Enlightenment Bible: Translation, Scholarship, Culture* (Princeton, 2005), 153.
40    Dyck, *Athen und Jerusalem*, 95–98.
41    Johann Gottfried Herder, 'Ueber die Wirkung der Dichtkunst auf die Sitten der Völker in alten und neuen Zeiten. Eine Preißschrift (1778)', repr. in *Johann Gottfried von Herder's sämmtliche Werke*, Section 2, *Zur schönen Literatur und Kunst*, vol. 9, ed. Johann von Müller (Tübingen, 1807), 378–79.
42    Johann Gottfried Herder, *Vom Geist der Ebräischen Poesie. Eine Anleitung für die Liebhaber derselben, und der ältesten Geschichte des menschlichen Geistes*, 2 vols (Dessau, 1782–83), 1:1–6.
43    Ibid., 1:4.
44    Ibid., 1:11–12.

4    Ossian and the Bible

In the field of literature, the circulation of a savage ideal had much to do with the publication, in 1760, of the young Scottish poet James Macpherson's (1736–1796) *Fragments of Ancient Poetry, Collected in the Highlands of Scotland, and Translated from the Gaelic or Erse Language*, colloquially referred to as the Epic of Ossian. Old and blind, the Ossian in question – a Celtic warrior-poet and kind of 'Homer of the North' – relates the exploits of his father, Fingal, his dead son, Oscar, and Oscar's lover, Malvina, who looks after him in his old age. As we will see below, figures like Herder were fascinated with the Ossian epic, which, as its title attests, was presented as a translation of pre-Christian and pre-bardic Scottish poetry, inasmuch as Macpherson had collected his material from folk tradition. Today, however, it is known as one of the greatest fabrications in literary history. Notwithstanding that history, recent decades has seen scholarly focus shift away from the piece's authenticity and towards the immense influence it has had on European culture and the growth of nationalism in the second half of the eighteenth century. As Maike Oergel has suggested, Macpherson's 'invention' of the ancient Scottish epic can be seen as an extreme, though indicative, example of national and other traditions – a constituent element of modern nationalism.[45]

Several years following its publication, Ossian and his style quickly became cult objects for poets and other writers. The epic, and the reactions to it, played a significant role in circulating of ideals like sentimentality, naturalism and expressiveness, as well as shifting the focus from classical literature to the local histories of Nordic European peoples.[46] Much of the excitement that the epic

---

45    Maike Oergel, *The Return of King Arthur and the Nibelungen: National Myth in Nineteenth-Century English and German Literature* (Berlin, 1998), 9.

46    On the influence of the epic of Ossian, see Lesa Ní Mhunghaile, 'James Macpherson und Ossian. Eine literarische Kontroverse zwischen National- und Universalkultur', in *Aufklärung zwischen Nationalkultur und Universalismus*, ed. B. Wehinger (Hannover-Laatzen, 2008), 155–66; Joep Leerssen, 'Ossian and the Rise of Literary Historicism', in *The Reception of Ossian in Europe*, ed. Howard Gaskill (London, 2004), 109–25; Kristine Louise Haugen, 'Ossian and the Invention of Textual History', *Journal of the History of Ideas* 59 (1998), 309–27; Burton Feldman and Robert D. Richardson, *The Rise of Modern Mythology, 1680–1860* (Bloomington, IN, 1972), 201–03. On the reception of the epic of Ossian in German literature, see Wolf Gerhard Schmidt, *'Homer des Nordens' und 'Mutter der Romantik'. James Macphersons Ossian und seine Rezeption in der deutschsprachigen Literatur* (Berlin, 2003); Howard Gaskill, 'German Ossianism: A Reappraisal?', *German Life and Letters* 42 (1989), 329–41; Rudolph Tombo, *Ossian in Germany: Bibliography, General Survey, Ossian's Influence upon Klopstock and the Bards* (New York, 1901).

of Ossian raised had to do with its positive portrayal of humanity's ancient state. This was the first time European poets and other writers were exposed to northern European writing identified with the pre-Christian epoch. Critics identified its protagonists' customs as belonging to the earliest epochs of human existence – exactly because Macpherson himself fashioned his heroes in keeping with the received images of his period regarding the lives of 'natural peoples'.[47]

In a similar vein, Ossian's customs and forms of expression were compared, from the moment the epic was published, to those of biblical heroes. Edinburgh critic and rhetoric professor Hugh Blair (1718–1800), who was instrumental in the epic's dissemination, marvels at the fact that the piece's style 'is remarkably similar to that of the Old Testament'. A friend of Macpherson's, Blair goes on to present the stylistic resemblance as proof of the epic's authenticity – it was written in 'the early speech of nations'.[48] The similarity is, in fact less surprising when one takes into account the fact that Blair apparently participated in the epic's writing, drawing stylistically mostly on the King James Version of the Bible.[49] Blair in fact quotes Lowth, and seems to have assimilated the same 'primitive' qualities the Anglican bishop had perceived in the poetry of the Hebrews – notably, the use of parallelisms – into the epic.[50]

The epic of Ossian, then, is constructed on the model of biblical poetry; simultaneously, its publication had a decisive impact on the way ancient texts, headed by the Bible itself, were read and interpreted at the time. The new ideal, embodied by the Celtic heroes, resulted in a reorganization of ethnographic categories. Ossian was simultaneously compared with Homer, the Bible, and the poetry of the Iroquois people. The epic's translation into German in 1768 was a decisive moment in the growth of the discourse on the German people's ancient roots. The fact that Ossian was a Celt, rather than a German, did not stop Herder, Klopstock, and others from seeing him as representing their ancient forefathers. Klopstock rewrote his early odes, replacing their classic motifs with ones having to do with Ossianic figures. He republished a poem

---

47 Haugen, 'Ossian and the Invention of Textual History', 309–310.
48 Hugh Blair, *The Works of Ossian, the Son of Fingal* (London, 1765), 345.
49 See, for instance, Howard D. Weinbrot, Britannia's *Issue: The Rise of British Literature from Dryden to Ossian* (Cambridge, 2007), 537; Lopa Senyal, *English Literature in Eighteenth Century* (New Delhi, 2006), 276.
50 Hugh Blair, *A Critical Dissertation on the Poems of Ossian, the Son of Fingal* (London, 1763), 63.

FIGURE 5    'Ossian Receiving the Ghosts of French Heroes', or 'Apotheosis of the French
Heroes Who Fell in the War for Liberty', by Anne-Louis Girodet de Roussy-Trioson,
commissioned by Napoleon, ca. 1800. HELD IN THE MUSÉE NATIONAL DES
CHÂTEAUX DE MALMAISON ET BOIS-PRÉAU; WORK IN PUBLIC DOMAIN;
IMAGE BY SOROURKE25, COURTESY OF WIKIMEDIA COMMONS

originally printed in 1748 under the title *Daphnis und Daphne* – in 1771 it featured *Selmar und Selma*.[51] Johann Wolfgang von Goethe's (1749–1832) Werther

---

51    Klaus Düwel and Harro Zimmermann, 'Germanenbild und Patriotismus in der deutschen
Literatur des 18. Jahrhunderts', in *Germanenprobleme in heutiger Sicht*, ed. Heinrich Beck
(Berlin, 1986), 358–95.

expresses the intellectual fashion that had taken root in German literary circles when he exclaims: 'Ossian pushed Homer out of my heart!'[52] (Figure 5)


## 5    The Infancy of Humankind

As the most eloquent ideologue of the *Sturm und Drang* movement, Herder unsurprisingly also voiced unbridled admiration for Ossian. By his own account, it was the reading of the epic that liberated him from the yoke of classical tradition. The value of Ossian is no less, he claimed, than that of Homer and Pindar, to whom the Germans were 'enslaved' as if living in a Greek colony.[53] As against classical uniformity, Herder extolls the 'uniqueness' (*Eigenheit*) of those peoples heretofore considered 'barbaric', and he celebrates the 'unrefined' (*unverkünstelte*) qualities of the Nordic 'nature peoples', as expressed in Ossian's epic.

The ancient customs and political structures expressed in bardic poetry represents, according to Herder, humankind in its infancy (*Menschengeschlecht in seiner Kindheit*), understood as being similar to the world of the biblical patriarchs (*Patriarchenwelt*).[54] Herder especially cites the great proximity between the spirit of Ossian and that of the Song of Songs: 'Even Ossian, whether genuinely happy or simply mimicking the emotion, sings – as do all the love songs of those naïve, unrefined peoples – in the tones of the Song of Songs'.[55] Influenced by Herder and Klopstock, several interpretive texts seeking to explain the Bible in general – and its poetic parts in particular – in comparison with Ossian, began appearing in the following decades. Theologian and poet Wilhelm Nicolaus Freudentheil (1771–1853) published an essay on 'The Hebrews' Victory Hymns' in the German art encyclopedia *Allgemeine Theorie der schönen Künste (General Theory of the Fine Arts)*. It leaned heavily on an Ossianesque interpretation of the biblical Song of the Sea (Exod 15:1–18), the Song of Deborah (Judg 5:2–31), and of the Psalms.[56] Several years later, Karl

---

52    Johann Wolfgang von Goethe, *Leiden des jungen Werthers* (Leipzig, 1774), 11.
53    See Karl Menges, 'Particular Universals: Herder on National Literature, Popular Literature, and World Literature', in *A Companion to the Works of Johann Gottfried Herder*, eds Hans Adler and Wolf Koepke (Suffolk, 2009), 189–214.
54    Johann Gottfried Herder, *Aelteste Urkunde des Menschengeschlechts* (Riga, 1774), 83.
55    Johann Gottfried Herder, *Lieder der Liebe* (Leipzig, 1778).
56    Wilhelm Nicolaus Freudentheil, 'Ueber die Siegslieder der Hebräer', in *Nachträge zu Sulzers allgemeiner Theorie der schönen Künste. Charaktere der vornehmsten Dichter aller Nationen; nebst kritischen und historischen Abhandlungen über Gegenstände der schönen Künste und Wissenschaften von einer Gesellschaft von Gelehrten*, vol. 4/2 (Leipzig, 1795), 253–70.

Wilhelm Justi (1767–1846) offered a similar interpretation of David's Lament (2 Sam 1:17–27). David's mourning of his slain foe Saul, he claims, is a phenomenon also seen in the cultures of Nordic people:

> Nevertheless this song should be seen as proof of David's generous character [...] the ancient oriental – and several Nordic – peoples would honor their enemies who had fallen in battle. Thus exactly did the splendid Ossian lament in tender tones at his enemy's grave; and so did Larthmor weep for his son Uchal.[57]

Taking the Bible's considerable influence on the writers of Ossian's epic, its protagonist's lament on his enemy's tomb was itself fashioned according to David's Lament. But contemporary interpreters sought alternative explanations for the similarity. Göttingen theologian Johann Caspar Velthusen (1740–1814) also marveled at the great similarity between the biblical lament and the one in Ossian. In a strange essay on the subject, published in 1807, he concludes that there can be no other explanation other than that Hebrew poetry must have been disseminated among the 'savage nations' (*rohe Nationen*) in general, and among Scottish bards in particular.[58]

By juxtaposing the Hebrews with the Celtic bards, Herder imbued the Bible with a new meaning: no more Christian scripture but instead ancient national epic. In doing so, he moved away from scholastic and classical poetic conventions towards signifying natural popular poetry as an independent frame of reference with its own set of values: 'more ancient, more popular, more alive' (*je älter, je volkmässiger, je lebendiger*).[59] Herder was the first serious writer of the eighteenth century to place myth (*Mythos*) – an idea that became the key term in the discourse on early poetry – at the center of his social and cultural theory. The term's graduation to widespread use in the 1760s – as a partial substitute for the derogatory term *Fabel* – signifies a new array of relations between religion, history, and fiction.

Anthropologist Talal Asad has pointed to the link between the rise of the term 'myth' and the emergence of new cultural consumption customs, headed

---

57    Karl Wilhelm Justi, *National-Gesänge der Hebräer* (Marburg, 1803), 78.

58    Johann Caspar Velthusen. *Einfluss frommer Juden und ihrer Harfe auf den Geist roher Nationen, insonderheit auf Ossians Bardenlieder* ... (Leipzig, 1807).

59    Johann Gottfried Herder, 'Ueber Ossian und die Lieder der alten Völker; Auszug einiger Briefe 1773. Aus der Sammlung von deutscher Art und Kunst', repr. in *Johann Gottfried von Herder's sämmtliche Werke*, Section 2, *Zur schönen Literatur und Kunst*, vol. 8, ed. Johann von Müller (Tübingen, 1807), 31.

by the reading of imaginative literature. According to Asad, the 'disenchant-ment' that Max Weber (1864–1920) identified as a feature of the modern world gave rise, in the nineteenth century, to a new imaginative literature industry, which established the modern distinction between the real and the fictional.[60] This literature dealt mostly with the pre-modern past, now described as won-drous and magical, distinguishing it from the instrumental and scientific pres-ent. The pre-modern text was banished from its status as factual truth but took on a kind of modern meaning and enchantment, now understood as a kind of ancient, belletristic literature.[61]

## 6      The Bible as National Myth

Herder's texts on myth were the first to explicitly formulate ideas already implicit in the writings of Johann Georg Hamann (1730–1788), Rousseau, Montesquieu and others.[62] He repurposed, for his own ends, the rationalist historical theories of David Hume and Göttingen scholar Christian Gottlob Heyne (1729–1812) regarding the birth of myth, which described it as savage peoples' sensual and coarse mode of thinking.[63] However as opposed to these writers, Herder's affirmative approach to myth tightly links myth and people. Myth is described as a people's priceless trove of memories – a vital element in its coming-into-being and existence.[64]

Herder, who never garnered the professorship at Göttingen he desired and openly derided Michalis' dry and uninspired biblical scholarship, used the Hebrew language to formulate his antagonistic, *Sturm-und-Drang*-inspired, Hebraist-poet stance. In this understanding, while Hebrew was in fact unfit for abstract thinking, that was exactly the quality that made it the language of poets. More than any other writer of his period, Herder's innovative stance

---

60    Talal Asad, *Formations of the Secular: Christianity, Islam, Modernity* (Stanford, 2003), 13–14.
61    Ibid., 14.
62    Feldman and Richardson, *The Rise of Modern Mythology 1680–1860*.
63    On Heyne's theory of myth, see Marianne Heidenreich, *Christian Gottlob Heyne und die Alte Geschichte*, Beiträge zur Altertumskunde 229 (Leipzig, 2006), 166–169; Christian Hartlich and Walter Sachs, *Der Ursprung des Mythosbegriffes in der modernen Bibelwissenschaft* (Tübingen, 1952), 11–19. On Enlightenment conceptions of myth, see Joseph Mali, *The Rehabilitation of Myth: Vico's 'New Science'* (Cambridge, 2002), 124–26.
64    See, for example, Sonia Sikka, *Herder on Humanity and Cultural Difference: Enlightened Relativism* (Cambridge, 2011), 234–236; George S. Williamson, *The Longing for Myth in Germany: Religion and Aesthetic Culture from Romanticism to Nietzsche* (Chicago, 2004), 33.

focused on the uses of myth. He expressed this idea as early as his 1767 essay, 'Vom neuern Gebrauch der Mythologie' ('Of New Uses of Mythology'), in which he called for myth to be used to breathe new life into literature. As Karl Menges has noted, this idea was quite extraordinary upon its publication, as against the period's more dominant rationalist writers' stance on archaic beliefs.[65] Herder is adamant that correct study of the stories of heroes and the gods would allow writers to create their own myths. He calls on writers to leave be the question of believability and to seek rather the poetic creation's political, no less than sensual-aesthetic, value. The poet 'creates the people around him' and has the power 'to lead souls' into the world he sees with his mind's eye.[66]

It was exactly for that reason, Herder claims, that Moses' poetry had an effect no less powerful on his people's souls and customs than that of his law; David's hymns brought him glory no less than his military battles.[67] However, it does bear noting that when referencing the Bible, Herder on the whole preferred to use the terms 'poetry' (*Poesie, Dichtung*) or 'epic' (*Epos*) over 'myth' (*Mythos*). A preacher and theologian, he must have feared classification of the Old Testament as 'myth' in the same vein as the narratives of pagan gods. His aspirations for a position at the University of Göttingen, where he was suspected of unorthodox views, may have also contributed to this reticence.[68] Subsequently, in his 1778 essay he distinguishes between Hebrew poetry and the mythologies of other ancient peoples.[69] The Hebrews are distinct among other peoples in that their poems are 'divine', rather than mythological and simply fictional. The term 'mythology' carries, in this case, negative valuation, as synonymous with fairytales or old wives' tales.

However, as opposed to Heyne and other writers of the Enlightenment, Herder did not view ancient 'oriental' poetry as a lower stage in the development of world consciousness, but rather sought in it the source (*Ursprung*) containing all of culture's future developments. He describes ancient text as containing a fresh creative quality that cannot be attained by modern, cerebral means. Ancient poetry contains elements that only later times were

65    Menges, 'Particular Universals,' 191–92.

66    'Ein Dichter ist Schöpfer eines Volkes um sich; er gibt ihnen eine Welt zu sehen, und
      hat ihre Seelen in seiner Hand, sie dahin zu führen': Herder, 'Über die Wirkung der
      Dichtkunst', in *Sämmtliche Werke*, 488.

67    Ibid., 372.

68    Heyne, who sought to help Herder a position in the Theology Faculty, reported in a letter
      to him on the obstacles to finalizing the (move): 'wenn Sie nur mehr orthodox wären!'
      Herder in fact never attained the position. Quoted in: Rudolph Haym, *Herder nach seinem
      Leben und seinen Werken*, 2 vols (Berlin, 1880–85), 1:713.

69    Herder, 'Über die Wirkung der Dichtkunst', 376.

able to develop and imbue with meaning.[70] Herder is also loath to eschew the more traditionally religious aspect of Hebrew poetry, 'suffused with the divine spirit'.[71] Hebrew poetry enjoys a clear advantage over 'national poetries' (*Nationalpoesien*), he claims, in that it was a divine poetry sung in the Temple (*daß sie Gottes-daß sie reine Tempelpoesie ward*).[72]

Herder subsequently describes 'oriental mythology' as a certain substrate of the Bible. The narratives of the Garden of Eden,[73] and of Job,[74] for instance, are according to Herder, mythical stories. But even as he employs all the cautiousness he can muster when using the explicit term 'myth' in the biblical context (a cautiousness that was destined to disappear from the texts of biblical interpreters in the generation after him), it is easily evident that Herder's understanding of Hebrew poetry in fact employs a mythical interpretation. This can be seen in the ways he invokes the figure of the Hebrews' national god. This figure appears at first in fanciful 'oriental' garb, but when investigated in the context of Hebrew national culture, its political value – and its ability to rally the people and link it to its homeland – is fully revealed.

To fully understand the ways Herder and other writers of his period refashioned understanding of the Old Testament as myth, we should bear in mind that this story was created only after the image of the Hebrews as a savage 'oriental' people had already been set. Only having been fully transformed into story, into ancient fiction, could the biblical text be transformed into national mythology. The mythical category was therefore the interpretive medium through which Herder bridged the problematically removed world of the 'oriental' Hebrews and that of his contemporary German readers'. Using it, Herder transforms the Bible's exotic character from a liability to an advantage.

Biblical poetry is the ancient Hebrews' national myth, but the German nation could only derive advantage from it by re-interpreting it – as was the case with the poetry of Klopstock and Herder's own texts. More than any other text, the poetry of the Hebrews sets a model for the correct relation between a people and its national literature – a model in which the people is brought into being through poetry, as led by poets.

---

70  Oergel, *The Return of King Arthur and the Nibelungen*, 24–25.
71  See, for example, Herder, 'Von die Wirkung der Dichtkunst', 365.
72  Herder, *Vom Geist der Ebräischen Poesie*, 1:360.
73  Ibid., 1:153.
74  Ibid., 1:136.

7      De Wette and the End of Ethnographic Interpretation

Reading the Hebrew Bible as Hebrew myth would, however, lead to a real revolution in German biblical scholarship in the next generation. It was Wilhelm
Martin Leberecht de Wette who signified the dawn of a distinct new period
in this field.[75] De Wette had studied under Herder in Weimar in his youth and
went on to study at the University of Jena, where he was deeply influenced
by the work of Immanuel Kant (1724–1804) and Friedrich Wilhelm Joseph
von Schelling (1775–1854). These philosophical influences brought him to a
theological-interpretive stance well outside the binary of rationalism and
orthodoxy that so characterized eighteenth-century biblical interpretation.
Based on a new, critical methodology, de Wette profoundly undermined the
historical interpretation of the Bible – reinscribing Christian religion into new
meanings of emotion and nationalism.

De Wette's main critical innovation is considered one of the foundations
of modern biblical scholarship to this day: the identification of Deuteronomy
with 'the book of the law' discovered by Hilkiah, the high priest, 'in the house
of the Lord' during the reign of Josiah (2 Kings 22). In his 1806 *Beiträge zur
Einleitung in das Alte Testament* (*Contributions to an Introduction to the Old
Testament*), de Wette sought to prove that Deuteronomy was, in fact, written
shortly prior to its 'discovery' and may have been authored by Hilkiah himself.[76] Given there is no biblical mention of the Torah's existence prior to the
period of Josiah, he concluded that all of the books of the Old Testament were
assembled and edited by scribes during this period.

Discussions on the identity of the author of the Pentateuch had taken place
since the seventeenth century, if not before. But de Wette bases his source theory on a completely novel way of reading the Bible. Written hundreds of years
following the events they narrate, the books of the Old Testament, he argues,
were worthless as historical sources. The writer of the Old Testament – transformed by de Wette into a 'narrator' (*Erzähler*) – was not even interested in
describing the ancient past itself but rather fashioned the text in keeping with

---

75    On de Wette's innovations see, for example, Thomas Albert Howard, *Religion and the
      Rise of Historicism: W. M. L. de Wette, Jacob Burckhardt, and the Theological Origins of
      Nineteenth-Century Historical Consciousness* (Cambridge, 2006), 43–56; Daniel Weidner,
      'Politik und Ästhetik: Lektüre der Bibel bei Michaelis, Herder und de Wette', in *Hebräsiche
      Poesie und jüdischer Volksgeist. Die Wirkungsgeschichte Johann Gottfried Herders im
      Judentum Mittel- und Osteuropas*, eds C. Schulte et al. (Hildesheim, 2003), 57–63.
76    Wilhelm Martin Lebrecht de Wette, *Beiträge zur Einleitung in das Alte Testament*, 2 vols
      (Halle, 1806–07), 1:179.

the political needs of the Late Kingdom period. Therefore, the Bible cannot be used for the investigation of Hebrew history. He claims as early as his 1805 book that

> the historical treatment of the Old Testament, which some have tried to carry out recently, seems to me to be inadmissible and unfruitful in for the expansion of Bible scholarship. The Bible is a sacred book, and as such given to us at hand as it were, and we as theologians should seeks in it religion, not secular history, politics, and jurisprudence.[77]

With these two sentences, half a century of German biblical interpretation characterized by historical criticism – seeking to reconstruct the biblical events' historical context – drew to a close. Within the new interpretive school inaugurated by de Wette, no scientific conclusion regarding Hebrew history, politics, and law can be made based on reading the Bible. If the focus of Michaelis's and Herder's project was the reconstruction of the Hebrew people and its culture while comparing biblical narrative to ethnographic descriptions, de Wette presents biblical narrative as an actual roadblock to historical knowledge. For him and his ilk, from now on the Bible can only be investigated as text. Therefore, there is no point in using ethnographic methods to reconstruct the figures of biblical heroes.

De Wette compares the Bible to classical epics such as Virgil's *Aeneid*. Like the Latin epic written under Augustus, the Bible's role was to supply the residents of the Kingdom of Judah with answers to questions pertaining to their collective existence. How did we appear? How was our state founded under Jehovah's reign? And how did our laws come into being?[78] Having been written explicitly to supply these questions with a mythological answer, the Bible could not possibly supply information to contemporary scholars of Hebrew history. In fact, de Wette's interpretation trod a path lined out by Herder, Johann Gottfried Eichhorn (1752–1827), and Johann Philipp Gabler (1753–1826) in their essays on Hebrew poetry and Hebrew myth.[79] But de Wette does not even attempt to distinguish between the Bible's poetic and historical elements. Eichhorn used the mythical interpretation method apologetically, to bridge the gap between the Hebrews' ancient terms and those of his own time. De Wette, by contrast, imbues myth with a distinctly positive sense.

---

77    Wilhelm Martin Leberecht de Wette, *Aufforderung zum Studium der hebräischen Sprache und Literatur* (Jena, 1805), 27.

78    De Wette, *Beiträge zur Einleitung in das Alte Testament*, 1:32.

79    See also Carlotta Santini's essay, infra.

Although works on biblical history continued to be written throughout the nineteenth century and beyond, De Wette signifies a significant shift: the Bible's mythic role supplants its historical one. This element of de Wette's method signifies its break with rationalist interpretation, tying it to Romanticism and nationalism. De Wette does not deny scripture's historicity to detract from its importance but rather to imbue it with new meaning. In his conception, understanding this work demands grasping it in religious and artistic terms. While Herder vacillated between treating the Bible as a universal work and stressing its national elements, de Wette primarily describes it as the Hebrew national epic (*hebräisches Nationalepos*). Written and edited according to the Hebrew theocracy's 'national interest' (*Nationalinteresse*) at a given point in time, the Bible,

> [c]onsidered as poetry and myth, now appears as the most important and richest object [...] It is the product of the patriotic religious poetry of the Israelite people, in which it reflects its spirit, its way of thinking, its patriotism, its philosophy and religion, and it is for this reason that it is one of the foremost sources of the history of culture and religion.[80]

In the history of biblical scholarship, de Wette's interpretive approach is usually identified with Kant's philosophy.[81] And it is true that, like Kant's Copernican turn in his *Critique of Pure Reason* – reorienting metaphysical inquiry away from the object and to the structure of consciousness – de Wette reoriented Bible scholarship away from historical event and to the structure of the text itself. However, another context forming the backdrop to this revolution should not be ignored: the expanding work, by poets and philologists of de Wette's period, to establish national literature and mythology by collecting and editing medieval epics and popular stories. The creation of the Hebrew national epic as narrated by de Wette is eerily similar to the Romanticists' project of creating a 'German mythology', as embodied for instance in a rising interest in the Nibelungen cycle and German folktales.

De Wette himself was one of the main voices calling for the creation of a Protestant-German mythology. After finding his place at Berlin University in 1810, he developed ardently nationalistic views. In his *Über Religion und Theologie* (*On Religion and Theology*) of 1815, de Wette lays out a vision of a new popular liturgy, a religious epic that would be based on a set of mythological

---

80    De Wette, *Beiträge zur Einleitung in das Alte Testament*, 2:98.
81    See for example, John William Rogerson, *W.M.L. de Wette, Founder of Modern Biblical Criticism: An Intellectual Biography* (Sheffield, 1992), 27–30.

symbols drawn from nature and history. He mentions Herder as the one who, in his openness to and love of the popular poetries of all peoples – Hebrew and Greek, southern and northern – laid the groundwork for such a mythology. Since then, he continues, this project has expanded and 'many of the most beautiful flowers of antiquity' (*viele der schönsten Blüthen des Alterthums*) have grown on German soil.[82] In this way, according to de Wette, the Germans are revealed as the most religious among the nations. This vision shows the clear stamp of Romantic influence: this religion will reach full fruition when natural philosophy (*Naturphilosophie*) is fully embedded into epic poetry, ushering in a new 'epic of cosmogony' (*Epos der Kosmogonie*).

German Romanticism's turn to nature and folktales becomes a new Christian religion. But as George S. Williamson has noted, while this religion parallels the rituals of the ancient Hebrews or the Catholic Church, it seeks to supersede them and replace them with a German-Protestant mythology.[83] De Wette and other fans of myth based themselves on Herder, but Herder himself had shown universal interest in the poetry of all ancient peoples, while they increasingly confined themselves to the German people's ethnic roots alone.

## 8    Celebrating the Vorzeit

In the eighteenth century, Bible reading underwent a reorientation into the newly founded sciences of the time, such as the natural history (*Naturgeschichte*), ancient history (*Altertumskunde*), and ethnography. The Israelites' assimilation into early humanity's natural history led to their identification as a people with characteristics similar to other ancient peoples'. In this fashion, a new image emerged, of the Patriarchs in particular and the ancient Hebrews in general. The Israelites' theological role as 'God's people' (*Volk Gottes*) was transformed into its characterization as a 'nature people'.

Within the dominant strand of the Enlightenment, the 'savage' was perceived more as 'ignoble' than noble.[84] But in the mid-eighteenth century, many European writers began to characterize the state of nature, and ancient peoples' forms of expression, differently. The *Vorzeit*, a nebulous term signifying the ancient time of the patriarchs, became an aesthetic and political ideal. The

---

82    Wilhelm Martin Leberecht de Wette, *Ueber Religion und Theologie. Erläuterungen zu seinem Lehrbuche der Dogmatik* (Berlin, 1815), 64. See further, Williamson, *The Longing for Myth in Germany*, 92–98.

83    Williamson, *The Longing for Myth in Germany*, 96.

84    Meek, *Social Science and the Ignoble Savage* (Cambridge, 1976).

Germans' ancient lifestyle, heretofore described by Enlightenment writers as coarse and barbarian, was now lauded as a naïve, heroic, and poetic epoch.[85] The enduring popularity of Ossian's epic exemplified the admiration of natural peoples.

However, the reception of Ossian, as well as other northern myths, in the mid-eighteenth century cannot be understood without this phenomenon being contextualized into the novel emergent forms of Bible reading. The works presented above exemplify the many intersections between the two influential texts of this period. One could say that MacPherson, Blair, Klopstock, Herder, and their interpreters established an aesthetic discourse celebrating the natural and the authentic by adopting and mimicking ancient texts. The works of these writers, who mainly belong to the same generation, correspond with and cite each other. But one of the underlying fundaments of their entire discourse is a poetic reading of the Bible.

The idealization of national poetry in Herder, as well as in Klopstock and others, is tied to the values and sensitivities of *Sturm und Drang*. This aesthetic transvaluation was mainly fueled by an attempt to move away from Neoclassical taste dictates. In the context of the establishment of a national German culture, Hebrew poetry and Celto-Germanic myth together made up an alternative to the classical corpus, exemplifying qualities such as uniqueness and originality. The positioning of the Bible as 'ancient poetry' played a crucial role in dismantling the boundaries between sacred and secular text. The transformation of revelation into creative-natural inspiration allowed Romantic poets to view their own work as a form of new revelation. As scripture was being transformed into literature, so too were secular poetry and literature transformed into a kind of new sacred text.

Eighteenth-century philosophers and biblical scholars sought to find in the Hebrew regime the fundaments of a kind of Egyptian-derived enlightened absolutism. Herder, by contrast, downplayed the importance of the Egyptian influence and regarded the fundamental elements of the Hebrew constitution as an authentic creation.[86] He still held, however, that Moses was intimately aware of the Egyptian priests' secrets as well as of the Egyptian state constitution, which 'would become the cradle of several peoples' political

---

85    See further Michael M. Carhart, *The Science of Culture in Enlightenment Germany*, Harvard Historical Studies 159 (Cambridge, MA, 2007), 139–145; Williamson, *The Longing for Myth in Germany*, 72–74.

86    Cf. also William Warburton, *The Divine Legation of Moses Demonstrated, On the Principles of a Religious Deist, From the Omission of the Doctrine of a Future State of Reward and Punishment in the Jewish Dispensation*, 2 parts (London, 1738, 1742).

organization'.[87] Many of the regulations having to do with Priests and Levites, in particular, betrayed Egyptian influence, he claimed.

As opposed to Michaelis, Eichhorn, Schiller, and many other German Enlightenment writers, though, who depicted Moses' constitution as an Egyptian set of rules adapted to the customs and needs of a nomadic clan, Herder focused on poetry instead. What formed the essential core of the Hebrew nation was its organic communitarianism, rather than those diktats of universal reason, and it was this core that was worthy of imitation. Herder even marvels at the way Moses was able to fashion an entire people (*seines ganzen Volks gewirkt*) through his poetry.[88] When poetry is presented as the main political medium, poets become folk leaders. Moses and David are compared with a poet such as Klopstock, rather than with monarchs such as Friedrich II (1712–1786) or Joseph II (1741–1790).

Helmut Walser Smith sees in Herder's writing an expression of the epistemic shift from the visual to the aural.[89] He claims that while classical Enlightenment thinking organized the world according to a visual paradigm (in which the object of representation is external to the subject), Herder signifies a shift to an affective, linguistic-musical discourse which collapses the inside/outside binary. The Nation is carried by the reverberation of linguistic sound as cast in poetic meter. Subsequently, while Michaelis catalogues the Hebrews' customs and rituals at a lexical remove, Herder strains to *listen* to their poetry. Thus, for instance, he writes in 1778 that throughout the Hebrews' history, the voices of their 'singers and patriots' (*ihre Sänger und Patrioten*) accompanied them:

> The Psalms of David are actually national psalms: even if only sung by the people itself, they rang of music the nature and effect of which we really have no idea [...] The people sang him [David?], and the prophets awakened the spirit of his chants as the spirit of Moses had awakened him. He is still alive. [...] The spirit that hovered around his harp has done great work on earth and can do so again, even if the poetry of other nations is but a dream.[90]

---

87  Herder, *Vom Geist der Ebräischen Poesie*, 1:350.
88  Ibid., 1:358.
89  Helmut Walser Smith, *The Continuities of German History: Nation, Religion, and Race across the Long Nineteenth Century* (Cambridge, 2008), 55–56.
90  Herder, 'Von der Wirkung der Dichtkunst', 222.

As Benjamin Redekop notes, for Herder the Hebrews were an example of a people living its public life through the medium of religious poetry.[91] And Anthony La Vopa stresses that in Herder's understanding, it was only an oral culture that could give birth to an authentic national public.[92] Speech, rather than writing, is what established the Hebrews' 'national public' (*Nationalpublikum*) and safeguarded the vitality of their body politic – from the revelation at Mount Sinai, through the prophets and up to the Psalms. This public comes into being only through unmediated speaking, an immediate connection between tongue and ear (*das Band der Zunge und des Ohrs knüpft ein Publicum*).[93] By the same token, the Germans could approximate the 'Hebrew public' (*Publicum des Ebräer*) and become a unified being only if they remain true to their own tongue.[94]

Hermeneutically, Herder's method posits that the main obstacle facing the biblical scholar is overcoming the gulfs imposed upon him by the distance of time, space, and custom existing between himself and the text's writer. The Hebrews' style and images, he claims, are derived from a world of meanings familiar to the writer and the readers he was addressing, but alien to the modern reader. Like Michaelis, Herder thus follows Lowth, who claimed that understanding Hebrew poetry depends on 'seeing everything through their eyes [...] ascertain everything according to their own opinions and make all efforts to become Hebrews ourselves while reading the creation of the Hebrews'.[95] Unlike Michaelis, Herder does not call for the scientific reconstruction of the oriental world of objects in the service of text interpretation but for an 'empathetic merging' (*Einfühlung*) with the biblical writer and his work. As he had written in one of his earlier essays, the translator and interpreter must undergo a transformation (*Umwandlung*) to become 'a Hebrew among Hebrews, an

---

91    See Benjamin W. Redekop, *Enlightenment and Community: Lessing, Abbt, Herder, and the Quest for a German Public*, McGill-Queen's Studies in the History of Ideas 28 (Montreal, 2000), 202–03.

92    Anthony J. La Vopa, 'Herder's Publikum: Language, Print, and Sociability in Eighteenth-Century Germany', *Eighteenth-Century Studies* 29 (1995), 5–24.

93    Johann Gottfried Herder, ed., *Briefe zur Beförderung der Humanität*, 10 vols (Riga, 1793–97), 5:59.

94    Ibid.

95    On Herder's interpretive method, see Hans W. Frei, *The Eclipse of the Biblical Narrative: A Study of Eighteenth and Nineteenth Century Hermeneutics* (New Haven, 1974), 183–202; Bernd Auerochs, 'Poesie als Urkunde. Zu Herders Poesiebegriff', in *Johann Gottfried Herder. Aspekte seines Lebenswerks*, eds Marin Keßler and Volker Leppin (Berlin 2005), 93–114.

Arab among Arabs'. Only thus can he identify with the times of 'Moses, Job and Ossian'.[96]

The rise of nationalism as the dominant ideology among the liberal elite of early nineteenth-century Germany opened the door to criticism of 'Hebrew influence' on German culture. As opposed to the situation in the eighteenth century, the question on the agenda was not that of the Bible's rationalism but of its relevance to German culture – or lack thereof. German intellectuals were celebrating myth but sought an exclusive national mythology. G.W.F. Hegel (1770–1831), for instance, complained of the Germans' 'national fantasy' (*die Phantasie des Volkes*) having been superseded by Christianity – and its replacement by the foreign fantasy of the Hebrews.[97]

At this point, German philology began liberating itself from Herder's legacy and the categories it had established. From the second third of the nineteenth century onward, the European conception of the history of civilization would lean towards the Aryan track, appropriating the Greek sources of European culture as well and finally distancing itself from the theory of the Hebraic-Semitic source. The Great 'Aryan Family' now lumped India, Greece, and Germany into one group, from which the 'Semites' were banished altogether.[98] The more the conception of progress and cultural development became set as a narrative of gradual improvement and refinement, the more Semites were removed from their central and privileged site in the drama of early history – and ultimately excluded from the history of human civilization. The idea of the Aryan nation was fused to an understanding of all of humanity as differing along racial lines of physiognomy and character. Thus, while racist histories of mankind had not gained total dominance yet during the 1850s, the following decade saw the rigidification of the demarcation lines between Aryans and Semites, now

---

96   Johann Gottfried Herder, *Kritische Wälder, oder Betrachtungen, die Wissenschaft und Kunst des Schönen betreffend, nach Maasgabe neuerer Schriften*, 3 vols (Riga, 1769), 2:18.

97   Georg Wilhelm Friedrich Hegel, 'The Positivity of Christian Religion', in idem, *Early Theological Writings*, trans. T. M. Knox (Chicago, 1948 [1795/96]), 146–47.

98   German philological research explicitly contrasted Hebrew origin hypotheses with Indian-Sanskrit ones. As I have shown above, the origin of these arguments was not philological but rather ethnic-biblical. For more, see Ofri Ilany '"Alle unsere Wanderungen im Orient". Die deutsche Sehnsucht nach dem Orient – Theologie, Wissenschaft und Rasse', in *Tel Aviver Jahrbuch für deutsche Geschichte* (2017), 41–68; Bruce Lincoln, *Theorizing Myth: Narrative, Ideology, Scholarship* (Chicago, 1999), 76–100; Maurice Olender, *The Languages of Paradise: Race, Religion, and Philology in the Nineteenth Century*, trans. Arthur Goldhammer (Cambridge, MA, 1992); Pascale Rabault-Feuerhahn, *Archives of Origins: Sanskrit, Philology, Anthropology in 19th Century Germany*, trans. Dominique Bach and Richard Willet (Wiesbaden, 2013); Tuska Benes, *In Babel's Shadow: Language, Philology, and the Nation in Nineteenth-Century Germany* (Detroit, 2008).

hardening along 'biological' aspects.[99] The spirit of Hebrew poetry, like that
of Moses' law, was increasingly exorcised from the drive of European history.

---

99    Suzanne L. Marchand, *German Orientalism in the Age of Empire: Religion, Race,
      and Scholarship*, Publications of the German Historical Institute, Washington D.C.
      (Cambridge, 2009), 129.

# CHAPTER 2

# The Rise of Jewish Mythology

## Biblical Exegesis and the Scientific Study of Myth

*Carlotta Santini*

Inevitably, any comparative study aimed at investigating the contacts, exchanges, and crossovers between two of the most important and richest interpretative traditions of modernity – the philological of classical erudition and exegetical of biblical learning – cannot help but issue a long list of caveats: theoretical, empirical, circumstantial. Biblical exegesis in Germany (Protestant, Jewish, Catholic) is an old and deeply rooted tradition. Its disciplinary autonomy is perhaps one of the most important factors that allow us to speak of a certain exchange between the mythological study of ancient Greece and the interpretation of the Bible, but never a true assimilation. There are also other causes – simpler ones but, for this very reason, more difficult to evade.

Let us start with something immediate and perhaps naïf. In their handbook *Hebrew Myths* (1963), one of the most widely diffused and popularized texts on Jewish mythology, Robert Graves and Raphael Patai introduced their attempt to define a mythological corpus within the book of Genesis as follows:

> The word 'myth' is Greek, mythology is a Greek concept, and the study of mythology is based on Greek examples. Literalists who deny that the Bible contains any myths at all are, in a sense justified. Most other myths deal with gods and goddesses who takes sides in human affairs, each favouring rival heroes; whereas the Bible acknowledges only a single universal God.[1]

Far from ignoring the achievements of the history of religions, which has recongized the origins of Hebraic religion in the melting pot of the ancient Middle East, this distinction between monotheism and polytheism in the opening pages of a book on Jewish mythology hints at a fundamental interpretative problem. For if Jewish *Urmonotheismus* is not a historical fact, it is certainly a cultural one, reiterated in different historical stages from antiquity

---

1  Robert Graves and Raphael Patai, *Hebrew Myths: The Book of Genesis* (New York, 1964), 11.

to modernity, which served the most diverse purposes: ideological, political, cultural. And this bias (in many senses still alive today), this cultural construction, has been considered a fact – a fact of faith – which has influenced the most enlightened scholars.

While the study of ancient texts has mostly been guided by historical or antiquarian interests, the study of biblical texts is not unfamiliar with doctrinal and dogmatic concerns. This is particularly true in the case of mythological studies. Recall that even in field of classics, which due to the character of its materials was mostly free from this kind of concern, a certain aversion to myth arose and strongly influenced the beginnings of classical mythological studies. The ancient prejudice against the *Graecia Mendax*, mother of poets, which represented myth as nothing more than a fable and childish invention, could not fail to manifest itself even more strongly in the case of the sacred texts of the Judeo-Christian tradition. If myth itself already does not enjoy good publicity, how daringly and with what strategies can it be applied not only to the study of a religion of the past, but also to the analysis of the living sources of the dominant religion of the present?

In this chapter, I will address the crucial question of the relationship between the tradition of classical studies and that of Jewish studies, in particular biblical studies, focusing on the first attempts at a mythological approach to the Bible by the authors of the second half of the eighteenth century and the beginning of the nineteenth century. By no means is this contribution intended to be an exhaustive report on the modern history of Jewish mythology studies, nor even on the Science of Mythology as such. What I would like to show, through some chosen readings, are the conscious or unconscious frictions – conceptual, doctrinal, and interpretive frictions – that permeate the work of the scholars who first applied themselves to the definition of this hinge field.

The starting point of our analysis is given by the introduction in the scholarly panorama at the end of the eighteenth century of a new theoretical and hermeneutical instrument, which determined a turn in the methodological approach to ancient texts: namely, the concept of myth. The term was first adopted by Christian Gottlob Heyne (1729–1812) to designate the tradition of tales about the gods and heroes of antiquity handed down in poetic texts. It was thus meant to replace the more generic and epistemologically less pregnant Germanic term *Fabel* (from the Latin *fabula*: fable or tale). This innovation sparked a heated debate in classical studies, dividing scholars between those who accepted to use the term and those who rejected it.

This *querelle* lasted more than a century: a telling instance comes from one of the last actors in that debate, Ulrich von Wilamowitz-Moellendorff (1848–1931), who refused to adopt the term *Mythos* and used the even more specific – and in this case imprecise – German term *Sage* (legend, but used mostly for German and Nordic epic sagas) or simply the generic term *Geschichte* (history). In the panorama of Jewish studies, Micha Josef Bin-Gorion (1865–1921) chose the term 'Saga', like Wilamowitz, and thereby underlined the epic and historical narrative character of biblical tales. Louis Ginzberg (1873–1953), for his part, preferred to adopt the term 'legend', which reinforced the historical value of the biblical tradition.

Yet hidden within this terminological debate was a much more substantial, and fundamental, which impacted philosophical and theological circles alike: the question of the very essence of the religious phenomenon expressed by mythical accounts and forms. With Heyne, we witness a re-evaluation of mythological materials that now convey a deeper, true content. Translating the language encoded in images of myth, he argued, could uncover ancient *Weltanschauungen* and expose the original dimension of ancient religious feeling. The University of Göttingen became the centre of diffusion for this approach to myth, where Heyne was active and scholars like Karl Otfried Müller (1797–1840) and Friedrich Gottlieb Welcker (1784–1868) inaugurated a new discipline within the field of classical philology. Scientific mythology – focused above all on ancient Greece – aimed to investigate the laws that governed myth and to fashion, as a result, a hermeneutical key able to explicate its contents.

For this analysis, I will discuss a series of case studies with special attention to scholars at work in both areas – biblical and classical studies – who were involved in different ways (for or against) in what can be called the 'mythological shift'. This particular scope will allow me to highlight the changing fortunes of myth as an applied concept together with its epistemological implications for biblical exegesis. First, I discuss one of the most influential exegetes in the history of scholarship, Robert Lowth (1710–1787). Though active before what we might call Heyne's 'mythological shift', Lowth was among the first to impose methodological demands and open up hermeneutical spaces in the context of biblical scholarship, which the mythological approach soon challenged.

Second, I turn to an early manual on classical mythology, written by Martin Gottfried Hermann (1755–1822) and prefaced by Heyne himself. Here the terminological innovation as well as the scale of Heyne's proposal for an axiological re-evaluation of ancient myth can also be measured by examining the strategies of adoption of this term in the work of his contemporaries and closest collaborators. In fact, 'myth' seems to have morphed into an inescapable point of reference – a concept nearly everyone was compelled either to accept or to reject. A shining example of the kind of diatribe that the decision for or against the use of 'myth' – and, by extension, for or against the Göttingen

School of Mythology – appears in disagreements between Johann Gottfried Eichhorn (1752–1827), one of the most celebrated biblical scholars – also trained in Göttingen – and his editor/commentator Johann Philipp Gabler (1753–1826). Their complicated, and controversial, relationship serves as a sort of mirror for all successive (mis)understandings in Jewish studies and further testifies to the unavoidable ambiguity, which still persists, on the meanings necessarily implied in the case of 'Jewish myths'.

Finally, I discuss in brief two parallel and somewhat alternative approaches to myth in Jewish antiquity: by Georg Lorenz Bauer (1755–1806) and Friedrich August Carus (1770–1807). Bauer implemented the rigorous methods of the new scientific mythology and presented a coherent system of ancient Jewish myth, thus doing for the Hebrews what Karl Otfried Müller did for the Greeks. Carus, thought less known among scholars in classics, is in reality a core figure in the intellectual history of the century. His method, also tied to the positions of Eichhorn, integrated a psychological and anthropological approach to ancient Jewish mythology and inaugurated a would-be discipline that enjoyed great success during the nineteenth century, with Heymann Steinthal (1823–1899) and Moritz Lazarus (1824–1803), and gained full acknowledgment into the beginning of the twentieth century, with Wilhelm Wundt (1832–1920): *Völkerpsychologie*.[2]


## 1    Historical and Theoretical Context

This selection of some authors (and omission of others) cuts a path, necessarily circumscribed, to foreground critical issues that can account for the complex process of introducing the concept of myth into the biblical tradition. Before advancing through my reconstruction, it is perhaps useful to sketch a map, however rough, of the social and intellectual context in which this itinerary will move and through which the case studies will acquire depth and coherence. As recently reiterated by Heinz Wissmann, the development of 'criticism' and, in particular, philological exactitude (*Akribie*) in textual study traces back to that fertile season of biblical interpretation which the Protestant

---

2  Commonly translated into English as 'cultural psychology' or 'folk psychology,' *Völkerpsychologie* studied the psychology of those collectives usually called 'peoples', regardless of whether they belong to a nation or not.

Reformation made possible from the sixteenth century onwards, especially in the Netherlands and the Protestant German lands – a development to which the progressive refinement of classical philology is also indebted.[3] In contrast to the Italian humanists, who were more interested in Greek culture (literature, rhetoric, and politics), in Germany the study of ancient languages – especially Greek and Hebrew – aimed to grant an access to the sacred texts of the Judeo-Christian tradition that was free from the prejudice and mediation (and translation) of Catholic tradition. This philological mode of reading the biblical texts was clearly, often self-consciously, opposed to medieval exegesis, which followed a variety of hermeneutical paths: from figural and allegorical to moral and dogmatic.

Throughout the history of its transmission, the Bible has been the object of exegetical approaches just as diverse as the historical, social, and religious contexts of their application over the centuries, even millennia. The stabilization of biblical texts into a canon, moreover, did not constitute single, univocal, isolated moment but rather a long process of arranging divergent components, infusing them with coherence, converging them into holy writ, and, not least, implementing them for moral instruction. Interpretative traditions within Judaism in particular greatly expanded the boundaries of 'sacred scripture', through rich phases of transmission, study, and commentary: from the Midrash to the Talmud to the later Kabbalah. Taking these different traditions and their varied viewpoints into account remains essential for understanding 'the biblical text(s)'. Indeed, long before transforming into a narrow target for the scientific instrument of 'myth', materials in 'the Bible' that showed a distinctly narrative, imaginative, or poetic character had been the object of study by such exegetical traditions. In some cases, those same modes of Jewish interpretation had helped expand the frontiers of marvelous or fantastic element in the sacred texts themselves, reinterpreting, expanding, and enriching the heritage of biblical legends.[4]

It is therefore towards all these exegetical traditions that the science of myth had to show its innovativeness. Yet the study of ancient myth also took its first steps on unstable ground. Despite its claims to being neutral and scientific – following Enlightenment ambitions and ideals, which culminated in the positivism of the later nineteenth century – the mythological approach has still not been fully secularized. The reason is a persistence of confessional premises as its very foundation. Those premises were built into the discipline of humanist

---

3   Heinz Wismann, *Penser entre les langues* (Paris, 2012).
4   Cf. Anthony Grafton and Joanna Weinberg, *'I Have Always Loved the Holy Tongue': Isaac Casaubon, the Jews, and a Forgotten Chapter in Renaissance Scholarship* (Cambridge, MA, 2011).

philology itself, which, into the eighteenth century, was conceived as *ancilla fidei*, the 'handmaid of faith'. The first challenge presented to the new mythological method was related to basic theological claims, according to which science cannot impose itself on faith. This means that, paradoxically, the scientific study of mythology should not criticize faith, but neither can it claim to confirm it. Rather, the mythological approach sought to strike a balance – difficult, often unequal – that held faith and science together. The perception of any balance struck, however, proved quite subjective indeed, based on the specific intellectual and moral needs of the individual scholars involved.

The participants of this debate over mythology were diverse on any number of levels: Protestant, Catholic, and Jewish thinkers (on a very large spectrum of religious observance and disciplinary affiliation), philosophers, historians, and philologists, as well as scholars interested in anthropology and psychology. Bringing high stakes to the *Wissenschaft des Judentums*, which began to consolidate at the start of the nineteenth century, were the comparison of cultures, the delimitation of Germanness, with its philhellenism, and the recognition of Jewish tradition as a field worthy of study.[5] It was a complex confrontation, both cultural and social, one that involved the elites of German and Jewish communities.

By increasingly orienting themselves toward classicism, opting to convert to Christianity, and embracing German nationalist positions, especially at the end of the century, scholars of Jewish heritage gradually gained a place alongside German colleagues in the institutions of cultural and civil life. Movements of Jewish assimilation or accommodation – associated with terms like *Deutsches Judentum* (German Judaism) or *Jüdische Aufklärung* (Jewish Enlightenment) – often bore a secularizing connotation, including the rejection of ancestral faith and proselytism, and the integration into the dominant non-Jewish society. In other cases, Jews made fewer concessions. Representing this second stance, which affirmed a distinction as well as fruitful exchange between Jewish and German cultures, was the eclectic philologist Jacob Bernays (1824–1881). Representing one of the final acts in social and cultural recognition of the Jewish intelligentsia in modern European culture was the opening of the Berlin *Hochschule für die Wissenschaft des Judentums* (Higher Institute for Jewish Studies), in 1872, where Heymann Steinthal (1823–1899) was also engaged.

These contours remain essential for understanding the conditions and perspectives in which the study of Jewish myth developed from the late eighteenth

---

5   For more on Steinthal and Lazarus as well as the *Wissenschaft des Judentums*, see the chapter by Irene Zwiep in this volume.

to early nineteenth century. On one hand, scholars felt a distinct need to press the Bible into the domain of science and free it from the exclusive prerogative of faith, manifested in both Protestant and Jewish orthodoxy. They aimed to recognize in biblical writings not only, and above all, 'sacred scripture' but also the textual artifacts of a culture as much Jewish as it was Christian – which deserved serious intellectual attention no less than classical antiquity, if not an even higher rank. In this context, Athens and Jerusalem became an extremely significant pairing, as did Homer and Moses – the 'authors' of the stories of Greeks and Jews, respectively.[6] For Jewish thinkers, this juxtaposition became all the more explicit. As Arnaldo Momigliano showed in *Pagine ebraiche*, they used such pairings to reassert and legitimate their double belonging, or rather their double cultural confession: German and Jewish alike and, for this reason, all the better scholars of antiquity.[7] It was precisely the entry afforded by ancient Judaism to Greek and Roman antiquity – the temple of classicizing German culture – that inspired Bernays, equally skilled in Greek and Hebrew, to utter the famous words: 'How sad that Goethe didn't know Hebrew like I do!'[8]

---

6    Bernd Witte, *Moses und Homer. Griechen, Juden Deutsche: Eine andere Geschichte der deutschen Kultur* (Berlin, 2018). On the polemical use of this structuring in Protestant historiography of ancient Judaism, see Simon Goldhill, 'What Has Alexandria to Do with Jerusalem? Writing the History of the Jews in the Nineteenth Century', *Historical Journal* 59 (2016), 125–51.

7    Arnaldo Momigliano, *Pagine Ebraiche* (Rome, 1987); see also Theodor Dunkelgrün, 'The Philology of Judaism: Zacharias Frankel, the Septuagint, and the Jewish Study of Ancient Greek in the Nineteenth Century', in *Classical Philology and Theology: Entanglement, Disavowal, and the Godlike Scholar*, eds Catherine Conybeare and Simon Goldhill (Cambridge, 2020), 63–85.

8    Jacob Bernays, *Jugenderinnerungen und Bekenntnisse* (Berlin, 1900), 104. For more on Bernays, see Paul Michael Kurtz, 'Defining Hellenistic Jews in Nineteenth-Century Germany: The Case of Jacob Bernays and Jacob Freudenthal', *Erudition & the Republic of Letters* 5 (2020), 308–42. The aversion of the father of *Deutsche Klassik*, Goethe, toward Jewish culture is the subject of a recent study by Karin Schutjer, *Goethe and Judaism: The Troubled Inheritance of Modern Literature* (Evanston, 2015). The topic bears direct relevance here since it is precisely Goethe's approach – his use of certain biblical motifs – that would have an important legacy over the *longue durée* in mythological studies. One need only recall the *Flight into Egypt*, the famous incipit of his *Wanderjahre*. In the economy of the work, this scene plays the role of not only a mystery or sacred representation but also a mythical archetype, in the sense the mythologist Karoly Kerényi gave to this term: an eternal model, pre-constituted, to which the characters, the carpenter Joseph and his Mary of the novel, conform themselves. Furthermore, one could consider Goethe's intention to rewrite the history of Joseph (Gen 37–50) expressed in his *Aus meinem Leben. Dichtung und Wahrheit*, 4 vols (Leipzig, 1811–33), 1:333. Such a project was finally realized by Thomas Mann, in the 'most mythological' of his books, the tetralogy *Joseph und seine Brüder* (Berlin, 1933–43). The mediation between Goethe and Mann on the story of Joseph was made possible by Micha Josef Bin-Gorion (Berdyczewski), one of the most important scholars of the

A final node for mapping this context is the status of the Bible. The authors under review never truly questioned the privileged position of biblical texts, which impacted how they understood Jewish myths. Although in 1861 Benjamin Jawett could scandalously state that scripture had to be interpreted like any other book,[9] this statement was still problematic enough more than one century later, at the time Arnaldo Momigliano spoke in front of a specialized public of biblical exegetes in Dallas:

> Let me admit from the start that I am rather impervious to any claim that sacred history poses problems, which are not those of profane history. As a man trained from early days to read the Bible in Hebrew, Livy in Latin and Herodotus in Greek, I have never found the task of interpreting the Bible any more or any less complex than that of interpreting Livy or Herodotus. Livy is of course less self-assured about the truth of what he tells us about Romulus than the Pentateuch is about Abraham. But the basic elements of a sacred history are in Livy as much as in the Pentateuch.[10]

With this equation of profane and sacred narratives, Momigliano provocatively contradicts the theoretical premise that operated so powerfully for so many scholars in the late eighteenth and early nineteenth centuries.[11] Indeed, their inquiries made them feel all the more obliged to reflect on the legitimacy of applying the same critical instruments to these two corpora from antiquity – the biblical and the classical – which are as different *de jure* as *de facto*. The question of 'legitimacy' for the method arises precisely from the difficulty in establishing a homogeneity between the objects of analysis themselves.

---

push to rediscover Jewish mythology at the beginning of the twentieth century, with his most interesting *Joseph und seine Brüder. Ein altjüdischer Roman*, repr. (Berlin, 1933 [1917]).

9       Benjamin Jowett, 'On the Interpretation of Scripture', in *Essays and Reviews* (London, 1860), 338. I am grateful to one of the anonymous reviewers of this volume, who reminded me of this reference, which serves to reinforce Momigliano's argument.

10     Arnaldo Momigliano, 'Biblical Studies and Classical Studies: Simple Reflections upon Historical Method', *Annali della Scuola Normale Superiore di Pisa*, 3rd Series, 11 (1981), n. 1, 25. Address at the centennial conference of the Society of Biblical Literature in Dallas, 6 November 1980.

11     See also Arnaldo Momigliano, 'Religious History Without Frontiers: J. Wellhausen, U. Wilamowitz, and E. Schwartz', *History and Theory*, 21/4, Beiheft 21: 'New Paths of Classicism in the Nineteenth Century' (1982), 49–64.

2      The Status of Biblical Poetry: Robert Lowth

Turning to key moments in the development of a mythological approach to biblical texts, we begin in Oxford with Robert Lowth and his *Lectures on the Sacred Poetry of the Hebrews*, first published in Latin in 1753. Johann David Michaelis (1717–1791), the great orientalist and interpreter of Moses, annotated and republished these lectures in Göttingen five years later, thus introducing Lowth's formative ideas to German intellectual circles.[12] Two centuries before Momigliano, Lowth expressed the same demand for an analytical consistency for the biblical and classical traditions, in almost the same words:

> That Poetry which proceeds from divine inspiration is not beyond the province of criticism. Criticism will enable us to account for the origin of the art, as well as to form a just estimation of its dignity. ... It would not be easy, indeed, to assign a reason, why the writings of Homer, of Pindar, and of Horace, should engross our attention and monopolize our praise, while those of Moss, of David, and Isaiah, pass totally unregarded. Shall we suppose that the subject is not adapted to a seminary in which sacred literature has ever maintained a precedence? Shall we say, that it is foreign to this assembly of promising youth, of whom the greater part have consecrated the best portion of their time and labour to the same department of learning? Or must we conclude, that the writings of those men who have accomplished only as much as human genius and ability could accomplish, should be reduced to method and theory; but that those which boast a much higher origin, and are justly attributed to the inspiration of the Holy Spirit, may be considered as indeed illustrious by their native force and beauty, but not as conformable to the principles of science, not to be circumscribed by any rules of art?[13]

Preceding Heyne's claim for work on myth, Lowth shows the same methodological insistence for scientific study – in this case, on a poetic analysis of the

---

12   The Latin edition of his *De sacra poesi Hebræorum, prælectiones academicæ Oxonii habitæ* was first published in Oxford, in 1753, and republished in two volumes, by Michaelis, in Göttingen, in 1758. This latter 'German edition' was the more important for the legacy of this work, since it underwent many editions and wide circulation, especially in Germany. It also became the basis for the first English translation, of 1787, by George Gregory. On Michaelis, see further the chapters by Ofry Ilany, Michael Ledger-Lomas, and Irene Zwiep in this volume.

13   Robert Lowth, *Lectures on the Sacred Poetry of the Hebrews*, trans. George Gregory, with notes by Michaelis and others, 3rd ed. (London, 1835), 21–22.

FIGURE 6   Relief of Moses (left) and Homer (right) on the Louvre, Paris.
On the east façade of the Lemercier Wing and west façade of the Square Court,
respectively. Sculpture by Jean-Guillaume Moitte and Antoine-Denis Chaudet,
1806.

IMAGES BY MARIE-LAN NGUYEN; COURTESY OF WIKIMEDIA COMMONS, CC
BY 4.0. HTTPS://COMMONS.WIKIMEDIA.ORG/WIKI/FILE:MOSES_MOITTE_
COUR_CARRÉE_LOUVRE.JPG; HTTPS://COMMONS.WIKIMEDIA.ORG/WIKI/
FILE:HOMER_CHAUDET_COUR_CARRÉE_LOUVRE.JPG

Bible. A distinctive British tradition – still observable today, in some of the most
prestigious chairs on both sides of the Anglophone Atlantic – has long framed
classical studies within the broader *cursus studiorum* of rhetoric and poetry.[14]
In fact, Lowth himself was bishop and professor of poetry at Oxford. Within
this tradition of erudition, the interpretation of texts (ancient and modern
alike) proved thoroughly formal, with a focus on issues of poetic genre, style,
composition, and vocabulary. It was precisely this mode of investigation –
addressing poetry in an ostensibly unhistorical way – that allowed Lowth to
apply the same tools of analysis to Moses and Homer equally (Figure 6).

---

14   It is in this proto-comparitivism of literature that we first encounter famous Homerists
Alexander Pope (1688–1744) and Thomas Blackwell (1701–1757). But one could continue
into the present day, with figures like Robert Fagles (1933–2008), translator of Homer and
Sophocles and expert of William Shakespeare (1564–1616) and John Milton (1608–1674).

Lowth interests us because he establishes parallels (however specious at times) between classical images and biblical descriptions. Reading the former as metaphor and the latter as allegory, he proposes a comparative study to identify elements of a common epic genre in both the *Iliad* and the Bible. Leaving aside his formal textual analysis and his pseudo-historical thesis of a shared genesis for Hellenic and Hebraic poetic themes (both, he argued, derived from Egyptian fables), there is one intuition that can be considered substantial. Lowth speaks of the scriptures, unsurprisingly, as the texts of revelation, the result of divine inspiration. But even the Greeks, he claims, believed that poetry had derived from the gods, particularly the Muses. This parallel, which was only meant to reinforce the rights of the literary study of holy texts, is likely to suggest – and this is how it will be read by later scholars – that the same status of 'revelation' can be assigned to the Bible and the *Iliad*. Fifty years later, for example, Bauer would take this point much further, stating that the concept of a scripture inspired by God – like the idea of inspiration from the Muses – was itself a myth. The works of Homer and Moses should therefore enjoy the same legitimacy since both were inspired by God. Any difference is quantitative, not qualitative. The ancient works differ in the degree to which they provide a witness to God: an obscure, confused revelation given to the Greeks; a clear, distinct one granted to the Hebrews through Moses: textbook deism, so to speak. For Lowth, this relationship between classical and biblical poetry cannot be separated from the stylistic simplicity of the Bible, which contrasts the complex constructions of Homer.

Though of clear biblical and, more broadly, theological derivation, the question of 'revelation' – translated into an epistemological theory – would prove to be a foundational idea in the new mythological approach to the classical tradition. The idea of a progressive revelation of the divine, passing through the centuries as well as the different perceptive and cognitive capacities of humanity, became central in Friedrich Wilhelm Joseph von Schelling's (1775–1854) 1842 *Philosophie der Mythologie*. The philosopher also dedicated an entire work to elucidating the different degrees of apprehending revelation in his *Philosophie der Offenbarung* (1854).[15] Friedrich Creuzer (1771–1858), one of the key figures in Heidelberg romanticism – an intellectual current that called for a rediscovery of myth at the start of nineteenth century – also made revelation

---

15    Friedrich Schelling, 'Philosophie der Mythologie', in *Friedrich Wilhelm Joseph von Schellings sämmtliche Werke*, section 2, vol. 2, ed. Carl Friedrich August Schelling (Stuttgart, 1857); idem, 'Philosophie der Offenbarung', in *Friedrich Wilhelm Joseph von Schellings sämmtliche Werke*, section 2, vols 3–4, ed. Carl Friedrich August Schelling (Stuttgart, 1858).

(*Offenbarung*) the crucial concept in his epistemology of religious perception, which he articulated in the degrees of symbol, myth, and allegory.


3      A New Science of Myth: Christian Gottlob Heyne and Martin
       Gottfried Hermann

If Lowth attempted to place Homer and Moses on the same level in order to justify his literary and critical approach to the Bible, Christian Gottlob Heyne reversed the perspective. This expert on Homer, friend of Johann Gottfried Herder (1744–1802), and founder of the Göttingen School of Mythology justified the study of ancient myth by resorting to the *auctoritas* of the Bible.

In 1787, Heyne composed a foreword to one of the first modern manuals for the study of mythology: the *Handbuch der Mythologie aus Homer und Hesiod als Grundlage zu einer richtigen Fabellehre des Altertums* (*Handbook of the Mythology from Homer and Hesiod as Basis for a proper Teaching of Fables from Antiquity*) by Martin Gottfried Hermann.[16] Presenting this book that consciously proposed a new theory of myth in ancient Greece, Heyne invokes nothing short of the Bible itself to set out the science of mythology:

> If mythology is nothing more than the epitome [*Inbegriff*] of fables, fictions and unrhymed fairy tales, or even pieces of pagan superstition, its usefulness is very limited, and it is to be deplored that the reading of the poets and the study of the ancients makes it necessary not to be completely foreign to these fairy tales. But matters run somewhat differently. Mythology is in itself the oldest history and oldest philosophy; the essence [*Inbegriff*] of the old folk and tribal sagas expressed in the old raw language; and viewed from this side it receives a new value, as a remnant of the oldest imaginations and expressions. [...] This is what the experience of the legends collected by Moses teaches to us.[17]

The position here clearly reflects Euhemerism, common in the eighteenth century and especially popular in France, with figures like Antoine Banier (1673–1741) and Noël-Antoine Pluche (1688–1761).[18] But as this last sentence suggests,

---

16    Martin Gottfried Hermann, *Handbuch der Mythologie aus Homer und Hesiod als Grundlage zu einer richtigen Fabellehre des Alterthums* (Berlin-Stettin, 1787).
17    Christian Gottlob Heyne, 'Vorrede', in ibid., a3.
18    Cf. Antoine Banier, *Explication historique des fables, où l'on découvre leur origine & leur conformité avec l'histoire ancienne, & où l'on rapporte les époques des héros & des*

Heyne uses biblical *auctoritas* as a pretext, as a form of *captatio benevolentiae* (the rhetorical technique of 'winning goodwill'). He submits an uncontroversial argument only to reverse it, employing that argument as a picklock to burglarize the very authority which seemed to hold it secure.

A few pages later, he goes so far as to argue that the dangerous Greek mythology (i.e. Greek poetry, as with Lowth) even provides the key to interpreting scripture. In a passage worth quoting at length, Heyne claims:

> A well-articulated presentation of mythology, in which the young scholar gets to know it after its origin, its first form, then after its formation and transformation; and thus he receives just by these means reasonable concepts about the early state of the peoples, about the first steps of culture, about the modes of imagining of the ancient world, which are the seeds of their concepts of religion and philosophy, which once again emerged all the more gloriously and brilliantly among the Greeks, and which at the same time provided the right basic concepts for the interpretation of the ancient writers, and consequently of the holy writers, in whom there are still pieces and remnants from that early age, whose misunderstanding had so many sad consequences. Such a lecture deserves a recommendation, and so far I hope to see myself justified if I have considered the first attempt of this kind not unworthy of public approval.[19]

In his foreword, Heyne proves himself a master at three-card monte. He uses the Bible to legitimate the study of myth and, vice versa, fashions mythology into a hermeneutical lens for the Bible. Hermann, from his side, could not avoid openly dealing with the difficulties of definition and systematic organization of the mythical materials, which are inescapable in the conception of a handbook. When defining the object of his treatise, he cannot hide behind rhetorical ambiguity, as did Heyne. Rather, he must reaffirm, all the more firmly, that Homer was not Moses, that the gods of Greek myth are not the one and only God of the Bible. His manual thus begins with an 'Abhandlung über die Götter Homers' ('Treatment on the Gods of Homer'), which constitutes nothing less than an essay on the concept of divinity.

---

*principaux événemens dont il est fait mention*, 2 vols (Paris, 1711); idem, *La mythologie et les fables expliquées par l'histoire*, 8 vols (Paris, 1738); Noël-Antoine Pluche, *Histoire du ciel, considéré selon les idées des poëtes, des philosophes, et de Moïse*, 2 vols (Paris, 1739).

19    Christian Gottlob Heyne, 'Vorrede', a8.

In this *Abhandlung*, Hermann maintains that the concept of divinity (*Gottheit*) is not the same among all peoples, just as it is not the same in childhood and adulthood of a single person. The idea of divinity among the ancient Greeks, so Hermann, was no more than a glorified man: stronger, faster, more beautiful.[20] But Greek gods, he continues, could be wounded. They are neither eternal nor immortal, according to the sense Christian theology gave to these terms. At most, they can be said to be everlasting, in that they persist: lasting longer and enduring more than mortals but still subject to fate (*Thyche*) and destiny (*Moira*). Hermann does not spare the gods of Greece the traditional accusation of ancient philosophers either, as by Epicurus and others: that they are slaves to the passions and devoid of morals. But according to Hermann, the more humanity progresses and civilization advances, the more the idea of divinity develops. In the end, Hermann believes he and his contemporaries – the sons of an advanced age in human history – cannot rightly define the Greek gods as *Gottheiten* (divinities) but should, instead, call them *göttliche Wesen* (divine beings).[21]

4      Dialogue of the Deaf: Johann Gottfried Eichhorn and Johann
        Philipp Gabler

A noticeable gap stood between Heyne's foreword and Hermann's manual. Whereas Heyne supplied a conceptual innovation – however much he hid

---

20    Friedrich Nietzsche, in a famous fragment, referred to this same idea that conceives of the gods as 'Supermen': 'Living on mountains, travelling a lot, getting around quickly – in all this we can already equate ourselves with the Greek gods. We also know the past and almost the future. What would a Greek have said if he could see us?': Friedrich Nietzsche, *Nietzsche Werke. Kritische Gesamtausgabe*, section 4, vol. 1, *Richard Wagner in Bayreuth* (*Unzeitgemäße Betrachtungen IV*), *Nachgelassene Fragmente, Anfang 1875 bis Frühling 1876*, eds Giorgio Colli and Mazzino Montinari (Berlin, 1967), 5[116] (pp. 146–47).

21    The validity of these divine beings was not – so Hermann – to be considered panhellenic but dependent on local circumstance, linked as they were to particular regions or cities. Religiosity thus took the form of polytheism in Greece, since no other system could have developed in such a fragmented society. This position will be developed later, and much more consciously, by Wilamowitz, for whom a Greek deity, such as Artemis, could not rightly be considered the same god in all places in Greece (mainland, islands, Ionian coast; or Thrace) simply because it bore the same name everywhere. Wilamowitz himself investigated the local specificities for each epiphany of great mythological figures, through the various epithets they received in any given place and argued for their possible derivation from local deities that belonged to the different traditions of individual lineages that ultimately constituted the great jumble that we call 'Greece'.

himself behind the ambiguities afforded by rhetoric – Hermann always feared the opposition his work might stir, not only from philologists but, worse, from theologians. Providing another, still more striking example of the kind of divergence that can arise between an introduction and a treatise, between an author and an editor (truth be told, a phenomenon as common now as in the eighteenth century) is Johann Gottfried Eichhorn and his famous *Urgeschichte. Ein Versuch* (*Primeal History: An Essay*), a book that has been read as a manifesto for the science of myth as applied to the study of the Bible.

Eichhorn's *Urgeschichte* has a remarkable editorial history. The work, not a tome at first but a treatise in a specialist journal, provided a new interpretation to the first three chapters of Genesis. When first published in 1779, moreover, it appeared anonymously: this already offers an important clue to the kind of concerns the author had about the potential reception of his work.[22] Yet so successful was *Urgeschichte* that its author was soon revealed. In 1790, the essay underwent augmentation for a new edition, becoming a sizable work three volumes in length.[23] The first half of volume one contains an introduction and commentary by Johann Philipp Gabler, a theologian trained by Eichhorn and Johann Jakob Griesbach (1745–1812). The second half then transitions to part one of Eichhorn's original *Urgeschichte*, supplemented with extensive notes by Gabler. While volume two comes entirely from Gabler, volume three includes, again, an extensive introduction by Gabler followed by the second and last part of Eichhorn's *Urgeschichte*, once more with copious notes from Gabler.

This work requires, then, the greatest of caution for analysis. When a statement is attributed to Eichhorn (whether by his contemporaries or today), it is not always, in fact, by Eichhorn himself. Rather, often it is Eichhorn through Gabler or, perhaps, even only Gabler. Indeed, the editor may well be speaking through the mouth of the author at times. Gabler's own theses frequently come in more assertive formulation than those of Eichhorn, and in many cases, he forces and distorts their meaning.

A very telling example of this kind of distortion is that Eichhorn – clearly – does not speak of 'myth' in his text. It is rather Gabler himself who subscribes to Heyne's terminological reform and deliberately places Eichhorn in this same current as well. According to Eichhorn, scripture poses many hermeneutical problems – with a style far from any historiographical narrative and an abundance of tropes – but it cannot, and should not, be considered myth: 'So

---

22    Anonymous [Johann Gottfried Eichhorn], 'Urgeschichte. Ein Versuch', in 2 Parts, *Repertorium für Biblische und Morgenländische Litteratur* 4 (1779), 129–256.

23    Johann Gottfried Eichhorn, *Urgeschichte*, ed. Johann Philipp Gabler, 3 vols (Altdorf-Nürnberg, 1790–95).

not mythology, not allegory, but true history'.[24] Strikingly, the explanatory note by Gabler to this explicit assertion by Eichhorn misses the point entirely and proposes a completely different interpretation, one in line with Heyne:

> N. 37: I too do not accept any allegory in this text, since all, even the most subtle, allegorical interpretations have so much against them; the mythical explanation alone is certainly the only true one.[25]

As this brief quotation clearly demonstrates, Gabler reads Eichhorn and comments on his arguments based on his own understanding and his own theoretical positions. Gabler's notes can be considered as a book within the book, which carefully and consciously diverges from the thesis expressed by Eichhorn. The result is a book with two heads, a dialogue of the deaf.

Although he refuses to use the term myth, making his own the concern to preserve the credibility of sacred history, Eichhorn does not turn a deaf ear to the new instances raised by the science of myth. How does he explain the inconsistencies and the many poetic, even fantastic, images of the sacred text if he refuses to regard them as myths? He refers to what 'must have been' the typical way of thinking at the time they were conceived (*Denkungsart seines Zeitalters*). The biblical author, so Eichhorn, lets God act effectively and unmediated, as a personified natural force. This line of thinking – an argument for accommodation, where simpler people needed explanation in their own terms – does, indeed, trace back even earlier, into the seventeenth century, but it nonetheless continued to contrast much of Christian theology even in Eichhorn's time, which held that God did act in the world but primarily in a mediated way, through the forces of nature he himself created.

Are we to conclude, then, that mythological images and narrative fragments were inserted for etiological purposes, as imaginative explanations without any actual link to fact? Eichhorn proposes a long series of questions ultimately directed at one basic concern: 'is there a place in Revelation for error and falsehood?'[26] For there is no doubt in Eichhorn's mind that such fanciful ideas and anecdotes cannot be true in the strict sense of the word. The classic example commented on by Eichhorn occurs in Genesis, the story of the snake and the tree of the knowledge of good and evil. If one were to concede that such episodes contain at least some kind of religious content – as Eichhorn did – and

---

24    Ibid., 3:79.
25    Ibid., 3:79–80, n. 37 by Gabler.
26    Ibid., 3:81.

reveal some kind of truth (allegorically, symbolically, whatever), how could the adoption of images (the tree, the fruit, the snake), which have nothing to do with the divine, be justified?

Eichhorn explicitly denies any arbitrariness in the biblical text and eliminates any place for this new 'mythological' fashion in biblical interpretation, which seeks mysterious meanings behind sacred stories and images. But precisely in this apologetic effort, he becomes more royalist than the king. Paradoxically, his response turns more radical than that of any ancient exegete, or even modern mythologist. In the opinion of Eichhorn, the snake must be taken literally. Should interpreters therefore believe in a talking snake? Not necessarily, he argues, but certainly at the time the book of Genesis was conceived, such a marvelous event would not have seemed unbelievable. Consequently, if the story of the snake is not true in itself, it was true from viewpoint of the biblical author. Since the marvelous corresponds to the original mentality that made the biblical text, the very presence of such fantastic elements only further confirms, for Eichhorn, the great antiquity and credibility of sacred scripture itself. Ultimately, Eichhorn's explanation centered on the mentality of biblical authors, whereas Heyne focused on a form of historical understanding.

## 5    Biblical Mythology: Georg Lorenz Bauer

The strange work of two souls – Eichhorn and Gabler – contradictory in their approaches, assumptions, and purposes, *Urgeschichte* nonetheless exerted a tremendous influence on subsequent studies of myth. That influence went in two directions. Eichhorn's approach, which declined to use the term 'myth' and reflected on epistemological legitimacy and internal coherence in the alternative mentality of ancient *Weltanschauungen*, was taken over by representatives of *Völkerpsychologie*, as we will see in the following section. Gabler's approach, which ultimately called into question the hermeneutical tools of the Göttingen School of Mythology, can rather be recognized in the work of Georg Lorenz Bauer, the heir of Heyne for biblical studies and an advocate for the concept of myth.

Let us start small. In 1802, Bauer published *Hebräische Mythologie des alten und neuen Testaments, mit Parallelen aus der Mythologie anderer Völker, vornemlich der Griechen und Römer* (*Hebrew Mythology of the Old and New Testaments, with Parallels from the Mythology of Other Peoples, Especially the Greeks and Romans*). Therein he described biblical mythology as follows:

The name mythology ... would have been rejected as profane in the past if it had been applied to the biblical writers, and if a mythology would have been discovered in their books. And even now there are certainly still many who find this name offensive, and who receive no little irritation if the same value has to be recognized to Hebrew legends of the past, as to the legends of all other peoples.[27]

Earlier, in 1794, Heinrich Corrodi (1752–1793) had posed an urgent question with the title of his essay in *Beiträge zur Beförderung des vernünftigen Denkens der Religion*: 'Whether myths are to be found in the Bible' – a question he answered in the negative.[28] Against Corrodi, Bauer stated, 'I felt it necessary to propose a theory of the biblical myths'.[29] He then proposed a division of biblical myths into three kinds: (1) philosophical myths, (2) historical and historical-philosophical myths, and (3) poetic or mixed myths. For each biblical narrative, he argued, one should ask, first, whether or not it is a myth and, if so, what kind of myth it is. Thereafter, and only thereafter, one can consider whether a fact (*Faktum*) – or some other kind of content – might lie behind that myth.

With respect to the Göttingen School, Bauer's position tends toward greater theoretical abstraction, an approach to myth rather typical of German idealism, especially Schelling, but one also nourished by a genuine historicist intuition. In fact, to Bauer belongs the famous description of myth formation as a great snowball, which advances through oral tradition and accumulates new material all the while.[30] Bauer proved himself adept at refined, complex analysis. Working at the intersection of theoretical currents that would come to define the history of the study of myth in the nineteenth century, he represents the first author to formulate a univocal method for interpreting mythical images and their spiritual contents. As Bauer asserted, 'These philosophemes have their own manners, in which they are disguised'.[31] To unravel the

---

27    Georg Lorenz Bauer, *Hebräische Mythologie des alten und neuen Testaments, mit Parallelen aus der Mythologie anderer Völker, vornemlich der Griechen und Römer*, 2 vols (Leipzig, 1802), 1:21.

28    Heinrich Corrodi, 'Ob in der Bibel Mythe zu finden sind?', *Beiträge zur Beförderung des vernünftigen Denkens der Religion* 18 (1794), 1–73.

29    Bauer, *Hebräische Mythologie des alten und neuen* Testaments, 1:iv.

30    This vision of historical development as a qualitatively homogeneous accumulation and stratification, in which nothing is lost and which can, in principle, be retraced and explored, reappears in the theories of Friedrich Carl von Savigny's (1779–1861) Historical School and becomes crucial for the Brothers Grimm and their approach to German myth.

31    Bauer, *Hebräische Mythologie des alten und neuen Testaments*, 1:8.

intertwining of mythical formulations, to understand how myth 'thinks', so to speak, Bauer adopted an analogical method to establish correspondences, that is, stable laws for translating mythical images into logical propositions. With this insight, he anticipates the more famous Karl Otfried Müller, who translated the relationships of filiation and, more broadly, the systems of mythical causality into mechanical laws of physical causality (essentially relationships of cause and effect).[32]

The identification of analogies and correspondences between mythical and other language (poetic, sacred, historical) was thus by no means new in biblical exegesis. What distinguished Bauer as an interpreter is his eminently formalistic, almost proto-structuralist method. He opposes any easy comparativism – a position also adopted by Lowth, to a certain extent – which might claim merely external or arbitrary identities between various elements of disparate cultures. Typical examples of this approach appear in the famous *Demonstratio Evangelica* by Bishop Huet (1630–1721), who recognized in every mythical divine couple a recasting of the ancient couple Moses and Sarah.[33] Another example is the attempt by Gerhard Johannes Vossius (1577–1649) to find a correspondence between the figures of Moses, Asclepius, and Mercury because of the common snake staff (*caduceus*).

Instead, Bauer advocates, with great conviction, for the autonomy and relative independence of each individual culture, as would Müller. This position, however, does not prohibit any and all comparison *per se* but recognizes such endeavors as a method for studying the epistemology of humanity in general, not for judging particular cultures as such. As he himself declared, 'Nothing seems to be more useful than comparing the myths of other peoples, where one sometimes encounters a striking similarity that is not based on an identity of facts, but on the same sentiments and ways of thinking of people at certain levels of culture'.[34] Just as Eichhorn acknowledged difference in the sensibilities and the thought structures (*Denkungsart*) of ancient authors who spoke of God and sacred history in the most 'natural' way in this youthful stage of human development, so too, *mutatis mutandis*, Bauer insists on asking about the gnoseological stage of primitive humanity. He distinguishes different degrees of the sensibility (*Empfinden*), intuition (*Anschauen*), and judgment (*Urteilen*), by which the religious consciousness of each individual people progressed. In his own words, he argued,

---

32    Karl Otfried Müller, *Prolegomena zu einer wissenschaftlichen Mythologie* (Göttingen, 1825).
33    Pierre-Daniel Huet, *Demonstratio Evangelica* (Paris, 1679).
34    Bauer, *Hebräische Mythologie des alten und neuen Testaments,* 1:iv.

The distinction of these epochs teaches not only what poets have added arbitrarily, as a fruit of their creative imagination, through which they proved their talent as poets, but, above all, what is far more important, how ideas have expanded and how one can therefore tie in with the old ideas, newer and better ones.[35]

With its anthropological perspective, Bauer's position pioneered an approach to ancient myth that foreshadows the theories of cultural polygenists (with their ideas of 'convergences') as well as the most refined investigations into the migration of cultures and the survival of ideas, as developed in the late nineteenth century. Bauer's interpretation thus allowed for myth to be conceived not only as a universal form, which necessarily emerges from every people in every place at every time, but also as a historical form, linked to the development of culture and the laws of tradition. Far from producing contradictions, his framework comes very close to the history of ideas today. Bauer made it possible to maintain a dual analytical method that can hold together both historical and philosophical demands.

## 6      Hebrew Myth and Völkerpsychologie: Friedrich August Carus

One epistemological constant is common to all scholars who adopted 'myth' as a critical category for understanding the biblical tradition. That constant appeared yet again in the final stage of our itinerary through key developments in the study of ancient myth: namely *Völkerpsychologie*. An intellectual stream that flowed alongside the currents of anthropology, philology, and linguistics, *Völkerpsychologie* (translated, only with difficulty, as folk psychology, peoples' psychology, and cultural psychology) proved to be hugely important for the development of Jewish studies, in addition to the earlier movements of *Haskalah* (Jewish Enlightenment) and *Wissenschaft des Judentums*. Its chief representatives, too, belonged to the Jewish intelligentsia who lived and worked at cultural crossroads in modern Europe and beyond (Figure 7).

   *Völkerpsychologie* upheld a central assumption, both anthropological and epistemological: the 'human type', in its physical aspects and spiritual faculties, was unique. According to this premise, established by Adolf Bastian (1826–1905), founder of the Berlin School of Anthropology, variation in humankind

---

35    Ibid., 1:41.

was not physical, physiological, racial, or the like. Rather, it arose through historical developments, in processes of differentiation that separated peoples and nations from one another. Crucially, this principle affirmed an equality of human nature yet an inequality of peoples – an argument already advanced with force by Montesquieu (1689–1755) in his discussion of different legal traditions.

From this theoretical perspective, the phenomenon of differentiation in humankind has to be seen as an epiphenomenon in culture. It came as a consequence of accumulated experience, confrontations between people from disparate *milieux*, progressive changes in various *Weltanschauungen*, and diverse customs and conceptions: in a word, of culture. Studying the psychology of peoples promised access – through the testimonies of different populations – to the larger history of cultural development, ancient and modern alike. Heymann Steinthal described the enterprise as follows:

> Through the world of the mind, a chain of causal connections runs as rigorously as through nature. [...] Even in the spiritual world, one might say, no atom is lost; whatever was, remains indestructible; in our spirits live the spirits of all the deceased of all time. This is what is called tradition, transmission, namely the fact that each generation takes up the spiritual inheritance of its fathers. The elements of thought, which are thus transmitted, may, after all, experience various destinies; but they are never destroyed.[36]

Reformulating in cultural terms the Law of the Conservation of Mass, by Antoine Lavoisier (1743–1794), Steinthal brings the world of history and culture into the realm of 'science'.

The choice of Jews as a privileged object of study in *Völkerpsychologie* was, in a sense, unavoidable: if not obligatory. Like Greeks, Chinese, Iranians, and many others, Jews constitute a people with a rich corpus of writings. Yet more than the Greeks or any other people, Jews had maintained an extraordinary cultural coherence and continuity, even without any territorial or political support system. As with early-modern notions of the Hebrew Republic, and as enabled by certain texts in the Hebrew scriptures themselves, the Jewish people once again afforded scholars with an opportunity to study the development

---

36    Heymann Steinthal, *Mythos und Religion* (Berlin, 1870), 3, translation and comments by the author.

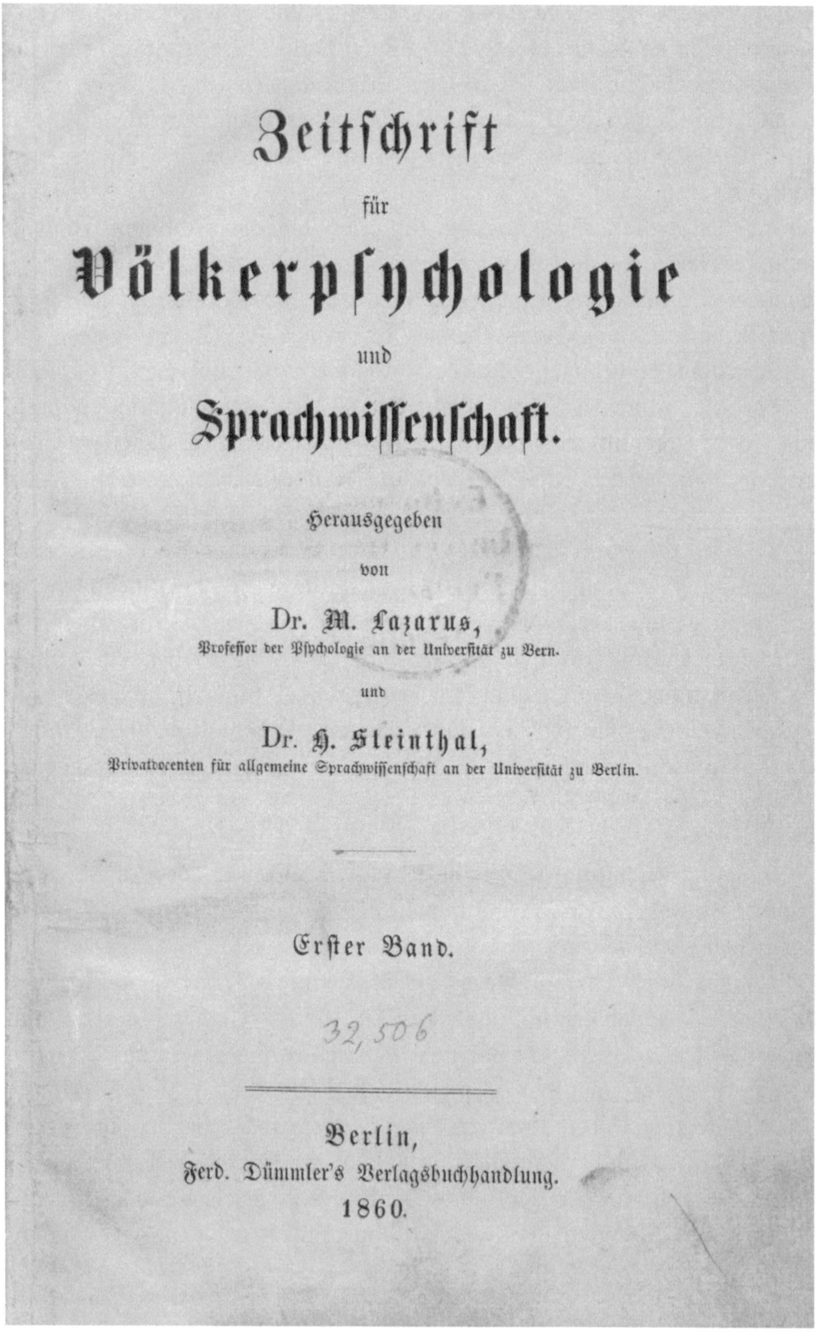

Zeitschrift

für

Völkerpsychologie

und

Sprachwissenschaft.

Herausgegeben

von

Dr. M. Lazarus,
Professor der Psychologie an der Universität zu Bern.

und

Dr. H. Steinthal,
Privatdocenten für allgemeine Sprachwissenschaft an der Universität zu Berlin.

Erster Band.

*32,506*

Berlin,
Ferd. Dümmler's Verlagsbuchhandlung.
1860.

FIGURE 7    Title page of the first volume of *Zeitschrift für Völkerpsychologie und Sprachwissenschaft*, 1860

of the very concept of 'a people' and to follow that idea backwards without ever losing the thread, so to speak.

The great success of *Völkerpsychologie* from the end of the nineteenth into the early twentieth century is, of course, well known, and the final act of this intellectual current – Wilhelm Wundt's *opus magnum* – influenced major psychologists, anthropologists, and philosophers, like Ernst Cassirer (1874–1945).[37] Less familiar, however, is the origin of this science, which owes its birth to a contemporary of Heyne and Bauer: Friedrich August Carus (1770–1807). With his 1809 *Psychologie der Hebräer* (*Psychology of the Hebrews*), Carus sought neither to 'moralize' the biblical material nor to offer a poetic or aesthetic analysis.[38] Rather, he aimed to understand 'how deeply the writers [of the Bible] themselves only (not their explicators) looked into human nature, how far they observed it strictly. The Bible deserves more than any book to be treated with this historical fidelity'.[39] For Carus, there is a 'historicity' to the Bible, but not one of facts or content. Instead, as Eichhorn already pronounced, it is an undeniable historicity of the ancient gaze, that is, the viewpoint of its authors. For the first time, an approach to mythology enhances the psychology of the writer. Accordingly, those who conceived and later wrote down the biblical texts can serve as a kind of psychologist, insofar as these ancient researched and explicated the soul (*Seelenforscher* and *Seelenerklärer*), the *Weltanschauung* of their time.

By analyzing language, belief, and the texts themselves as the work of ancient writers, Carus ventured to reconstruct the history of the Jews as a group who traversed history for thousands of years, coming to define themselves as a people. How did Jews become self-aware as a people, as a nation, from Abraham to the Diaspora and beyond? In Carus's terms, Jewish history constituted the 'History of the Self-Perceiving Psychological Culture of the Nation' *par excellence*. Like Eichhorn and Bauer, he too conducts his inquiry by identifying the 'epistemological' stages of human consciousness: from primary reflexes (*Triebe*) to consummate reason (*Vorstand*), passing through dream images, early etiology, and the development of emotions. Just as much as the Göttingen School of Mythology, exponents of *Völkerpsychologie*,

---

37    Wilhelm Wundt, *Völkerpsychologie*, 10 vols (Leipzig, 1900–20). For an overview on the rise and fall of this discipline, see Egbert Klautke, '*Völkerpsychologie* in 19th-Century Germany: Lazarus, Steinthal, Wundt', in *Doing Humanities in Nineteenth-Century Germany*, ed. Efraim Podoksik, Scientific and Learned Cultures and Their Institutions 28 (Leiden, 2020), 243–63.

38    Friedrich August Carus, *Psychologie der Hebräer* (Leipzig, 1809), published posthumously.

39    Ibid., 23.

especially Steinthal, fully believed in the progress of knowledge, a progress that manifested itself – like the transition from infant unconsciousness to adult knowledge – in the development of peoples from the very beginning of humanity up to the present day. The study of human consciousness as well as the psychic past of individuals thus became a means for understanding the history of psychology for all humanity: a statement later adopted by Jungian psychoanalysis. In this conceptual framework, moderns who undertake the hermeneutical task of interpreting biblical texts are the adults of humankind: adults who read the testimonies from the childhood of humanity.

## 7      Conclusion

The positions of *Völkerpsychologie* do not stand all too distant from those already taken by Hermann in his essay on the concept of divinity and the meaning of ancient religion. For Steinthal, myth is a form of consciousness: it certainly can contain kernels of truth but cannot become fully aware of them. Religion is something more, however, as Hermann stated. It belongs to a higher, more advanced stage in an awareness of the divine. From this point of view, applying the concept of myth as a tool to interpret the biblical texts can ultimately aim only at enhancing the value of the Bible itself, to better understand its stratifications and the different levels of signification contained within the scriptures.

Yet throughout the nineteenth century, there was little willingness to value myth as myth in and of itself, at least in relation to the Bible. Many efforts were undertaken to save the Bible from so-called rationalists or Pyrrhonists, as before. But during this period of hermeneutical innovation, a real risk to sacred texts lay precisely in the value assigned to the new and controversial study of myth.[40] If scholars did not accept that the Bible could be read through the lens of myth, it ran the risk of failing to stand the test of modern science, i.e. history, philology, mythology. Were it not scientifically explained, those unstable, marvelous, and fantastic materials would deprive the sacred book of all credibility and historical legitimacy. The principal objective of these diverse interpretative practices was to save the biblical tradition from sinking into the

---

40      Michael Legaspi, *The Death of Scripture and the Rise of Biblical Studies*, Oxford Studies in Historical Theology (Oxford, 2010).

prejudicial darkness of its 'mythical' materials, precisely by accepting myth as an instrument of analysis. Only in this way could the legitimacy and credibility of the Bible be preserved: that is, by making it a legitimate object of scientific inquiry.

On the other hand, if mythological analysis has since gained recognition, thanks in no small part to its use in studying canonical texts of positive religions, its transfer from the domain of dead to living religions – with all their emotional, confessional, and ideological baggage – has given rise to the first substantial criticism against this would-be science. As Steinthal observed already,

> There are philologists [and the thought goes here to Creuzer and Welcker] who have made religion and myth so identical that they measure the power of a people's religiosity on the mass of mythical figures or recognize the power of religion in the creation of myths. No, once again: religion is eternal, it is the supreme sanctuary of man; myth, on the other hand, is a finite form, and destroying the form so that the content may shine all the more pure and brightly is a commanded deed, is the task of our time. By eliminating myth, however, and then mainly through the all-round care of spiritual health, we are also working against those aberrations which are not the cause but the consequence and outbreak of spiritual illness. The unnatural, unhappy marriage of religion and myth would have long since been torn apart if everything connected with it had not had a particularly conservative force. [...] Getting rid of it is what makes us feel most like we have detached ourselves from our parents.[41]

From the very beginning of the new mythological science, and precisely in the context of its first scientification and systematization, we can see *in nuce* the same resistance that would be reiterated and reformulated over a century later, in the process of 'demythologization' of religious science by Rudolf Bultmann (1884–1976).

Despite subsequent efforts of nineteenth-century and even present-day mythologists, myth has never fully emancipated itself from the hold of theological commitments: its fixation on the truth-content of myth, how others could believe them, or why we should. Although authors like Creuzer and,

---

41    Steinthal, *Mythos und Religion*, 148–49.

later, Karoly Kerenyi (1897–1973) recognized its intrinsic value to religion, every time myth has been transferred from ancient to modern religions – still alive and active – the hermeneutical force of mythological analysis has been neutralized and relativized by scholars. Whether then or now, adherence to a particular confession entails undying commitments that inevitably affect the type of scientific inquiry practice on the objects esteemed by faith.

The consequences of such an unresolved ambiguity within academic approaches to Jewish myth manifest themselves still today. The manual *Hebrew Myths* (1963) by Patai and Graves, which opened this chapter, offers another excellent example of this perspective. The theoretical positions of its authors express far too much mutual independence for it to be a two-person work. Where we recognize the hand of Graves, a trained classical scholar and ethnologist educated within the Protestant faith, we find an interpretation of biblical narratives according to different criteria, ranging from anthropological structuralism to morphological study of mythical archetypes to comparison of myths. Graves' aim is clearly to treat the biblical material in the same way as classical materials, highlighting the constant forms and dynamics of all myths and their superhistorical and potentially universally valid meaning. When, instead, the pen moves on to Patai, a prominent ethnologist who grew up in an international and at the same time conservative Jewish milieu, those same 'myths' receive a well-circumscribed historical interpretation – almost euhemeristic – inscribed with a conscious process of past elaboration by the Jewish nation. The term 'myth' is barely preserved in the sections of Patai, but it loses all structural, archetypal, and philosophical connotations.

If taken literally, the two interpretations – each perfectly defensible – can only contradict one other. They leave the reader with the task of answering the underlying enigma: whether biblical tales are a necessary product of the human imagination, which manifest some kind of epistemological truths, or whether they constitute more or less faithful accounts of the development – as much sacred as historical – of Jews ancient and modern into a people. The work of Moses, as ancestor, as deliverer, as lawgiver, may fall on either side of the equation.

### Acknowledgements

This research was completed through funding from the European Commission: EURIAS 2018 (CRASSH Cambridge) and Marie Curie Horizon 2020 (University of Copenhagen). Unless cited otherwise, all translations are mine.

# CHAPTER 3

# Moses or Hammurabi?

## Law, Morality & Modernity in Ancient Near Eastern Studies

*Felix Wiedemann*

In December 1901, French archaeologists excavating the ancient site of Susa, in southern Iran, came across a remarkable object: the fragment of a black diorite stone covered with text in small cuneiform letters in the Akkadian (Babylonian) language. Two months later, in January 1902, two more fragments were found, completing one of the most important archaeological artefacts from the ancient Near East: the stele of Hammurabi, the Old Babylonian king from the eighteenth century BCE, now on display in the Louvre in Paris. A list of over 100 laws and regulations is inscribed on the stele, which is why the text has come to be called the Code (or Codex) of Hammurabi. Originally erected in the south Babylonian city of Sippar, the stele was probably brought by Elamites to their capital, Susa, when these archenemies of the Babylonians looted the region in the twelfth century BCE.[1]

Vincent Scheil (1858–1940), a French assyriologist and member of the Susa expedition, immediately recognised the importance of the find and began translating the cuneiform text at a breath-taking pace. His French translation was published in summer 1902, only a few months after the object was excavated.[2] Translations into other modern languages followed apace: the Berlin assyriologist Hugo Winckler (1863–1913) published a German version in his journal *Der Alte Orient* that very same year, and the first English translation, by Robert Francis Harper (1864–1914), followed in 1904.[3] Two years later, the

---

1   See on the historical context among others: Marc van Mieroop, *King Hammurabi of Babylon: A Biography* (Malden, 2005); Dominique Charpin, *Writing, Law, and Kingship in Old Babylonian Mesopotamia* (Chicago, 2010).

2   V. Scheil, 'Code des lois (droit-privé) de Hammurabi, roi de Babylone, vers l'an 2000 avant Jésus-Christ', in idem, *Mémoires publiés sous la direction de M. J. de Morgan*, 2nd series, vol. 4, *Textes élamites-sémitiques*, Ministère de l'instruction publique et des beaux-arts. Délégation en Perse (Paris, 1902).

3   Hugo Winckler, *Die Gesetze Hammurabis, Königs von Babylon um 2250 c. Chr. Das älteste Gesetzbuch der Welt*, Der Alte Orient 4/4 (Leipzig, 1902); Robert Francis Harper, *The Code of Hammurabi, King of Babylon About 2250 B.C.: Autographed Text, Transliteration, Translation, Glossary, Index of Subjects, Lists of Proper Names, Signs, Numerals, Corrections, and Erasures with Map Frontispiece and Photograph of Text* (Chicago, 1904).

orientalist Felix Peiser (1862–1921) and legal historian Josef Kohler (1849–1919) issued the first of a series of volumes on the law of Hammurabi: including a new critical edition, transcription, and translation of the famous code and other Old Babylonian legal documents.[4] While this became the standard scholarly reference edition, Scheil and Winckler also published the first popular versions of the Hammurabi Code for what was commonly called 'the educated public'.[5] In doing so, they were obviously reacting to a remarkable demand. But why did the publication of an ancient Near Eastern text from the early second millennium BCE provoke such a great public interest? (Figure 8)

To understand the fascination that the Hammurabi Code held for people in the early twentieth century – who believed it to be the oldest written law in history – one must look to the historiographical writings of the time and examine the prevailing narratives on the origins of civilization and its great achievements. The issue at the centre of debate related to the origins of law, and it arose early on: specifically, the relationship between the Old Babylonian law and the biblical or Mosaic law. This question seemed inevitable since the parallels between the two sources were too obvious to be overlooked. The most famous of these – the law of retaliation (*lex talionis*) articulated in the biblical phrase 'an eye for an eye and a tooth for a tooth' – is also the basic principle of the penal regulations in the Old Babylonian law.[6] Then, too, both codes claimed to be of divine or revealed character. The relief at the top of the stele represents King Hammurabi himself standing before the seated Babylonian

---

4  With the second volume, Peiser was replaced as editor by Arthur Ungnad; after Kohler's death, in 1919, the last volume was co-edited by the historian of law Paul Koschaker. See Josef Kohler and Felix Peiser, *Hammurabi's Gesetz*, vol. 1, *Übersetzung, juristische Wiedergabe, Erläuterung* (Leipzig, 1904); Josef Kohler and Arthur Ungnad, *Hammurabi's Gesetz*, vol. 2, *Syllabische und zusammenhangende Umschrift nebst vollständigem Glossar* (Leipzig, 1909); Josef Kohler and Arthur Ungnad, *Hammurabi's Gesetz*, vol. 3, *Übersetzte Urkunden, Erläuterungen* (Leipzig, 1909); Josef Kohler and Arthur Ungnad, *Hammurabi's Gesetz*, vol. 4, *Übersetzte Urkunden, Erläuterungen (Fortsetzung)* (Leipzig, 1910); Josef Kohler and Arthur Ungnad, *Hammurabi's Gesetz*, vol. 5, *Übersetzte Urkunden, Verwaltungsregister, Inventare, Erläuterungen* (Leipzig, 1911); Paul Koschaker and Arthur Ungnad, *Hammurabi's Gesetz*, vol. 6, *Übersetzte Urkunden. Mit Rechtserläuterungen* (Leipzig, 1923).

5  Jean-Vincent Scheil, *La loi de Hammourabi (vers 2000 av. J.-C.)* (Paris, 1904); Hugo Winckler, *Die Gesetze Hammurabis in Umschrift und Übersetzung* (Leipzig, 1904), 988.

6  See for instance §§ 196, 197, 200 of the Hammurabi Code. On this aspect, see Jan Dirk Harke, *Das Sanktionssystem des Codex Hammurapi*, Würzburger rechtswissenschaftliche Studien 70 (Würzburg, 2007), 24–25; on the *lex talionis* in the Hebrew Bible, see Sandra Jacobs, *The Body as Property: Physical Disfigurement in Biblical Law*, Library of Hebrew Bible/Old Testament Studies 582 (London, 2014), 68–189.

FIGURE 8      Photograph at the excavation of the stele of the Code of Hammurabi, 1901; held
              by the Louvre, Department of Oriental Antiquities, Paris, inventoried under SB8;
              AS 6064
              IMAGE COURTESY OF THE LOUVRE.

god Samas, who is giving him the law. The parallel to the revelation of biblical law on Mount Sinai, as narrated in the Bible, seemed unmistakable.

For this reason, the relationship between Babylonian and biblical law became a subject of intense debate immediately after the publication of the Code of Hammurabi. Far from being restricted to specialists in the fields of ancient Near Eastern studies, biblical studies, or the general history of law, the matter was widely discussed. With the authenticity and character of the Mosaic law at stake, Christian and Jewish scholars were very active in the discussion. Another reason for the prominence of the Hammurabi Question was the so-called Babel–Bible Controversy (*Babel-Bibel-Streit*), which was keeping German society in suspense in this same period. In a series of lectures between 1902 and 1904, Friedrich Delitzsch (1850–1922), the leading German assyriologist of the day, not only laid out the parallels between the Hebrew Bible and certain cuneiform texts (already well-known to scholars of the time) but also argued for the dependence of the former on the Babylonian sources.

Over time, Delitzsch's questioning of the authenticity and revelatory character of the Bible became increasingly charged with anti-Jewish prejudice, starting the scholar down the path towards the radical antisemitic ideology expressed in his later writings.[7] He gave his first *Babel-Bibel* lecture in January 1902, the very month in which the final fragments of the Hammurabi stele were discovered. The Hammurabi Code is already explicitly mentioned in his second, and in some respects more radical, lecture a year later.[8] Modern scholars, like Klaus Johanning or Yaakov Shavit and Mordechai Eran, have put the debate on Moses and Hammurabi in this context.[9] However, 'Moses and Hammurabi' was more than just a chapter of 'Babel versus Bible', as I hope to demonstrate in the following. Delitzsch was far from the first to raise the question, and most of the pamphlets and articles devoted to this subject did not even mention him. Thus, notwithstanding the obvious parallels, 'Moses and Hammurabi' was a discourse in its own right – and one deserving of a separate examination.

Looking at German publications on the subject released in the years immediately after the code's discovery, I will concentrate on certain aspects of the debate particularly worthy of discussion. Firstly, this chapter investigates the question of dependence of the Mosaic law on Babylonian sources or a direct relationship between them. Secondly, it examines the claims of the supposed modernity of the Hammurabi Code as compared to biblical law. And thirdly, it explores the case put forward in Moses' defence, so to speak, by those who continued to insist on the exceptional nature of biblical law and its special role in history. Finally, I will show how the debate fit in within the more general discourse on the history of law, morality, and *Sittlichkeit* among German scholars in the nineteenth and early twentieth centuries.

---

7   See Friedrich Delitzsch, *Babel und Bibel. Ein Vortrag* (Leipzig, 1902); idem, *Zweiter Vortrag über Babel und Bibel* (Stuttgart, 1903); idem, *Babel und Bibel. Ein Rückblick und Ausblick* (Stuttgart, 1904); idem, *Babel und Bibel. Dritter (Schluss-)Vortrag* (Leipzig, 1905). For more on the debate, see Klaus Johanning, *Der Bibel-Babel-Streit. Eine forschungsgeschichtliche Studie* (Frankfurt, 1988); Reinhard G. Lehmann, *Friedrich Delitzsch und der Babel-Bibel-Streit*, Orbis Biblicus et Orientalis 133 (Freiburg, Switzerland, 1994); Yaacov Shavit and Mordechai Eran, *The Hebrew Bible Reborn: From Holy Scripture to the Book of Books. A History of Biblical Culture and the Battles over the Bible in Modern Judaism,* trans. Chaya Naor, Studia Judaica 38 (Berlin, 2007), 193–352; Bill T. Arnold and David B. Weisberg, 'A Centennial Review of Friedrich Delitzsch's "Babel und Bibel" Lectures', *Journal of Biblical Literature* 121 (2002), 441–57; Eva Cancik-Kirschbaum and Thomas L. Gertzen, eds, *Der Babel-Bibel-Streit und die Wissenschaft des Judentums*, Investigatio Orientis 6 (Münster, 2021).
8   Delitzsch, *Zweiter Vortrag über Babel und Bibel*, 21–25.
9   Johanning, *Der Bibel-Babel-Streit*, 291–316; Shavit and Eran, *The Hebrew Bible Reborn*, 342–48.

1       The Historical Relationship

From the beginning on, there was no question but that the Hammurabi Code
was much older than the Mosaic law. The commonly accepted chronology (no
longer considered accurate today) placed the era of Hammurabi in the early
third millennium BC. For this reason, the obvious similarities between the
two codes were potentially grist to the mill for those of the 'furor orientalis'
(to use a felicitous phrase of Suzanne Marchand)[10] who, rejecting the heavily
Christianocentrism of German scholarship, were eager to bring the ancient
Near East into sharper focus. In this respect, there was certainly a kind of tri-
umphalism in the subtitle of Winckler's edition of the Hammurabi Code –
the 'world's oldest statute book' (*Gesetzbuch*) – since before 1902 this epithet
would normally have been reserved for the Mosaic law. Delitzsch's rhetorical
question as to whether the 'Israelite laws' had been influenced or even shaped
by the much older Babylonian law went in a similar direction.[11]

       Since all debates about the role of the Old Testament and the contribution
of the ancient Israelites to the history of civilization affected their supposed
heirs (i.e. modern Jews), the debate on Moses and Hammurabi was inevitably connected to that on the so-called Jewish question. Hence, Winckler's
and Delitzsch's triumphalist tone was certainly motivated to a large degree by
their antisemitism (expressed by both more openly in other publications). But
it would be a mistake to overemphasise this point and reduce the scholarly
enthusiasm for Hammurabi solely to antisemitic biases. Some of those scholars took up Hammurabi's part, so to speak, were themselves Jews or of Jewish
background, like Peiser or Carl Friedrich Lehmann-Haupt (1861–1938), while
most of the Christian defenders of Moses were not, by any means, defenders
of the modern Jews.

       Nonetheless, one might expect those Christian and Jewish scholars who
insisted on the uniqueness and authenticity of the Hebrew Bible and the historical truth of its narratives to have taken a hostile view of assertions concerning parallels between Babylonian law and biblical law, given the potential of
such claims to detract from the glory of the latter. Quite the opposite was true,
however: initially at least, conservative theologians were among those who
particularly welcomed the discovery of the Code of Hammurabi.

---

10     Suzanne L. Marchand, *German Orientalism in the Age of Empire. Religion, Race,
       and Scholarship*, Publications of the German Historical Institute, Washington, D.C.
       (Cambridge, 2009), 212–51.
11     Delitzsch, *Zweiter Vortrag über Babel und Bibel*, 25.

To understand these positive reactions, it is important to take into account the relationship that existed between assyriology and the Bible *before* the eruption of the Babel-Bible Controversy (*Babel-Bibel-Streit*): theologians usually welcomed the sensational discovery of ancient Near Eastern monuments and texts in the second half of the nineteenth century because they saw these as corroborating the authenticity and historical truth of the biblical narratives.[12] The excavations of ancient Assyrian sites in what is now Iraq provided general confirmation for the existence of biblical places like Nineveh – the city to which God sent his prophet Jonah to warn the residents there of impending divine wrath. What is more, discoveries like the famous Lachish relief, which represents the story of the Assyrian siege of the Judean city in 701 BC (now on display in the British Museum) seemed to shed new light on incidents that the Bible only briefly mentions (2 Kings 18, 13–15) and thus were viewed as extra-biblical sources for the interpretation of the text.[13]

At the turn of the twentieth century, the new discipline of biblical archaeology was established, whose purpose was clearly to uncover material evidence corroborating the text.[14] Moreover, the new knowledge about the ancient past was of particular importance to conservative Christians of the time because it could be used as a weapon against the major enemy in the field of biblical studies: philological or 'higher' criticism, as represented by liberal scholars like Julius Wellhausen (1844–1918). Most prominent in this respect was the Anglican cleric and assyriologist Archibald Sayce (1845–1933), who wrote several monographs (some of them translated into German) on the 'recent discoveries' in the Middle East that were aimed at refuting biblical criticism. According to Sayce, the 'verdict of monuments' came down entirely on the side of the biblical narratives.[15] The most influential German scholar with a similar

---

12    With a focus on Britain, see Steven W. Holloway, 'Biblical Assyria and Other Anxieties in the British Empire', *Journal of Religion & Society* 3 (2001), 1–19; Shawn Malley, *From Archaeology to Spectacle in Victorian Britain: The Case of Assyria 1845–1854* (Farnham, 2012). For continuing religious motives in German ancient Near Eastern studies, see Marchand, *German Orientalism in the Age of Empire*.

13    See David Ussishkin, 'The "Lachish Reliefs" and the City of Lachish', *Israel Exploration Journal* 30 (1980), 174–95.

14    See Barbara Zink Machaffie, '"Monument Facts and Higher Critical Fantasies": Archaeology and the Popularization of Old Testament Criticism in Nineteenth-Century Britain', *Church History* 50 (1981), 316–28. On the history of biblical archaeology, see Thomas L. Thompson, *The Mythic Past: Biblical Archaeology and the Myth of Israel* (New York, 1999); Thomas W. Davis, *Shifting Sands: The Rise and Fall of Biblical Archaeology* (Oxford, 2004); Eric H. Cline, *Biblical Archaeology: A Very Short Introduction* (Oxford, 2009).

15    See Archibald Henry Sayce, *Fresh Light from the Ancient Monuments: A Sketch from the Most Striking Confirmations of the Bible from Recent Discoveries in Egypt, Palestine, Assyria,*

agenda was the Munich assyriologist Fritz Hommel (1854–1936), a former student and close friend of Delitzsch. Hommel was a proponent of so-called pan-Babylonism[16], and as such his exegesis of the Old Testament was much less literal than those of other conservative scholars. Nevertheless, he rejected the decontextualized methods of philological criticism completely.[17] Asserting that assyriology provided external evidence 'testifying' to the authenticity and veracity of the biblical narratives, he broke off all contact with his former teacher when the Babel-Bible Controversy arose, accusing him of 'raping' the holy scriptures.[18]

Against this backdrop, the initially positive reactions of conservative Christians to the discovery of the Code of Hammurabi are not surprising. To Eduard König (1846–1936), for instance, an Old Testament scholar and also fierce opponent of Wellhausen and Delitzsch, the Hammurabi stele proved not only that a complex law system had existed in the very early periods of Near Eastern history but also that the ancient Hebrews had not been primitive

---

*Babylonia*, 2nd ed., By-Paths of Bible Knowledge 11 (London, 1884); idem, *The Witness of Ancient Monuments to the Old Testament Scriptures*, Proceedings of the Conference of the German Association of University Teachers of English 32 (London, 1884); idem, *Alte Denkmäler im Lichte neuer Forschungen. Ein Überblick über die durch die jüngsten Entdeckungen in Egypten, Assyrien, Babylonien, Palästina und Kleinasien erhältlichen Bestätigungen biblischer Tatsachen* (Leipzig, 1886); idem, *The 'Higher Criticism' and the Verdict of the Monuments* (London, 1894); on Sayce, see Roshunda Lashae Belton, 'A Non-Traditional Traditionalist: Rev. A. H. Sayce and His Intellectual Approach to Biblical Authenticity and Biblical History in Late-Victorian Britain' (PhD thesis, Louisiana State University, Baton Rouge, 2007); for more on these writings, see Machaffie, '"Monument Facts and Higher Critical Fantasies"'.

16   On pan-Babylonism see Michael Weichenhan, *Der Panbabylonismus. Die Faszination des himmlischen Buches im Zeitalter der Zivilisation* (Berlin, 2016); further, Marchand, *German Orientalism in the Age of Empire*, 236–43.

17   See his attack on Wellhausen, which was published in German and English at the same time: Fritz Hommel, *Die Altisraelitische Überlieferung in inschriftlicher Beleuchtung. Ein Einspruch gegen die Aufstellungen der modernen Pentateuchkritik* (Munich, 1897), trans. by Edmund McClure and Leonard Crosslé as *The Ancient Hebrew Tradition as Illustrated by the Monuments: A Protest Against the Modern School of Old Testament Criticism* (London, 1897). On Hommel, see Felix Wiedemann, '"Apologie der Semiten". Der Münchner Semitist und Assyriologe Fritz Hommel zwischen Philo- und Antisemitismus,' *Zeitschrift für Religions-und Geistesgeschichte* 75 (2023). For more on Wellhausen's biblical criticism, see Paul Michael Kurtz, *Kaiser, Christ, and Canaan: The Religion of Israel in Protestant Germany, 1871–1918*, Forschungen zum Alten Testament 1/122 (Tübingen, 2018).

18   Fritz Hommel, *Die altorientalischen Denkmäler und das alte Testament. Eine Erwiderung auf Prof. Fr. Delitzsch's "Babel und Bibel"*, 2nd ed. (Berlin, 1903), 9.

nomads before settling in Canaan.[19] In this respect, there was no difference between his views and those of Jewish Orthodox scholars like Seligmann Meyer (1853–1925), another active participant in the Babel-Bible dispute, who expressed the hope that the Babylonian codex would contribute to a better understanding of 'Jewish antiquity' and confirm the historical truth of the Hebrew Bible.[20] Hommel went even further: whereas Wellhausen and other liberal biblical scholars put the Law later than the Prophets and argued for its very late origin in the mid-first millennium BC, Hommel took the Hammurabi Code as evidence for a very old tradition of written law in the ancient Near East.[21] Early on, in his polemical attack on Wellhausen, he had identified Hammurabi with the biblical king Amraphel who, according to the book of Genesis, was involved in a war against the city of Sodom during the time of Abraham (Gen 14).[22] Once the code was found, Hommel became convinced that this biblical patriarch with Mesopotamian origins was responsible for bringing elements of Babylonian law to the Holy Land. This argument fit in with his pan-Babylonist convictions very well, since adherents of this school of thought contended that almost all cultural achievements stemmed from ancient Mesopotamia.

However, the highly ambivalent character of the ancient Near Eastern material had already become evident before the *Babel-Bibel-Streit*, specifically in the context of the British assyriologist George Smith's (1840–1876) discovery of the so-called Flood Tablet (the eleventh tablet of the Epic of Gilgamesh) in 1872. While it was true that the similarities between the flood stories in the Mesopotamian Epic and the Bible could be used as textual evidence that the event did occur, and thus of the veracity of the biblical narrative, the similarities also opened the possibility of regarding the Bible as merely reproducing an older Babylonian (or Sumerian) myth.[23] The antiquity of the

19    Eduard König, 'Hammurabis Gesetzgebung und ihre religionsgeschichtliche Tragweite', *Der Beweis des Glaubens. Monatsschrift zur Begründung und Verteidigung der christlichen Wahrheit für Gebildete* 39 (1903), 169–80.

20    Seligmann Meyer, *Contra Delitzsch! Die Babel-Hypothesen widerlegt* (Frankfurt, 1903), 8.

21    Fritz Hommel, *Grundriss der Geographie und Geschichte des Alten Orients*, part 1, *Ethnologie des Alten Orients. Babylonien und Chaldäa*, Handbuch der klassischen Altertums-Wissenschaft in systematischer Darstellung 3/1.1 (Munich, 1904), 238.

22    See the references to 'Khammurabi' in the index of Hommel, *The Ancient Hebrew Tradition as Illustrated by the Monuments*.

23    George Smith, *The Chaldean Account of Genesis: Containing the Description of the Creation, the Fall of Man, the Deluge, the Tower of Babel, the Times of the Patriarchs, and Nimrod. Babylonian Fables, and Legends of the Gods. From the Cuneiform Inscriptions* (London, 1876). On the contemporary debate, see Vybarr Cregan-Reid, 'Discovering Gilgamesh: George Smith and the Victorian Horizon of History', in *The Victorians and the Ancient*

Hammurabi Code was similarly problematic, since it could be seen as calling into question the originality of biblical law: as König put it, at stake were the *Ursprungsverhältnisse der Pentateuchgesetzgebung*, or the circumstances of the origin of the laws laid down in the Pentateuch.[24] Thus, thwarting the argument that Moses was mere copyist, adorned with laurels that rightly belonged to Babylonia, was of great importance to the (Christian and Jewish) defenders of the Bible: they had to demonstrate that the biblical code did not depend on the Babylonian code.

The most influential contribution to this line of argument came in the form of the thorough investigation published by the Austrian-Jewish orientalist David Heinrich von Müller (1846–1912).[25] To facilitate comparison, he created tables juxtaposing the provisions contained in the codes of Hammurabi, Moses, and the Roman Twelve Tables. At first glance, his findings were contradictory: he emphasized the close connection and the strong parallels between the two codes while at the same time arguing the Code of Hammurabi could not have been the source for Moses since the formulations and arrangement of the rules in biblical law were more 'original'.[26] On these grounds, Müller concluded that there had been no direct historical links between the two codes but that both stemmed from a common source – an original law laid down in an earlier time (*ein bereits fixiertes Urgesetz*).[27] What now appears to have been an awkward compromise became widely accepted by scholars of the time, Christian as well as Jewish.[28] A less frequent argument denied the existence of direct historical connections between the two codes and explained the similarities as being due to ideas universal in the history of law. In this vein, the German-Jewish legal historian Georg Cohn (1845–1918), president of the University of Zürich, referred to the theory of general 'elementary ideas'

*World: Archaeology and Classicism in Nineteenth-Century Culture*, ed. Richard Pearson (Cambridge, 2006), 109–23; Kevin M. McGeough, *The Ancient Near East in the Nineteenth Century: Appreciations and Appropriations*, vol. 1, *Claiming and Conquering*, Hebrew Bible Monographs 67 (Sheffield, 2015), 392–406.

24 König, 'Hammurabis Gesetzgebung und ihre religionsgeschichtliche Tragweite', 172.

25 David Heinrich von Müller, *Die Gesetze Hammurabis und ihr Verhältnis zur mosaischen Gesetzgebung sowie zu den XII Tafeln. Der Text in Umschrift, deutsche und hebräische Übersetzung, Erläuterung und vergleichende Analyse* (Vienna, 1903).

26 Ibid., 241.

27 Müller, *Die Gesetze Hammurabis*, 7.

28 See, for instance, Hubert Grimme, *Das Gesetz Chammurabis und Moses. Eine Skizze* (Cologne, 1903); David Feuchtwang, 'Moses und Hammurabi', *Monatsschrift für Geschichte und Wissenschaft des Judentums* 48 (1904), 385–99.

(*Elementargedanken*) developed by the German ethnologist Adolf Bastian (1826–1905).[29]

However, the historical relationship and the possible dependence of biblical law on Babylonian sources was not the main focus of the wider 'Moses versus Hammurabi' discourse. Much more important was the question as to the meaning and position of the two codes in the general history of human civilization, as I would like to demonstrate in the following.

## 2    Babylonian Modernity

To understand the importance of this issue, one has to consider the strange cultural competition between different pasts in European discourses on ancient history that was playing out in the nineteenth and early twentieth centuries. Intellectuals and scholars of different fields of classical and ancient studies were fiercely debating the question of 'who' (which usually meant which people or so-called race) had contributed most to the rise of modern (European) civilization. The discovery of texts and monuments in the Middle East in the late nineteenth century gave ancient Babylonia an enormous boost in this competition, and its position improved even further after the spectacular excavation of the city of Babylon by a German expedition led by Robert Koldewey (1855–1925) was underway in 1899.

It is important to emphasize, however, that Babylon did not symbolize the primitive origins of human civilization so much as its first historical peak. The result was that 'modernity' became the dominating trope in German writings on ancient Babylonia: anything regarded as Babylonian came to be seen as symbolising modernity, and references to ancient Mesopotamia abounded in the context of modern culture, art, and architecture.[30] The spectacular find of the Hammurabi stele in 1902 fit very well into the 'Babylomania' of the day. Of course, it did not matter in this context that very different Babylons were, in fact, at play here: the stele of Hammurabi was erected in the early second millennium BC, whereas the colourful Ishtar Gate and the procession street which

---

29    Georg Cohn, *Die Gesetze Hammurabis. Rektoratsrede gehalten am Stiftungsfeste der Hochschule Zürich den 29. April 1903* (Zurich, 1903), 39. On the concept of *Elementargedanken*, see Adolf Bastian, 'Ethnische Elementargedanken in der Lehre vom Menschen [1895]', repr. in idem, *Ausgewählte Werke*, vol. 7, eds Peter Bolz and Manuela Fischer, Historia Scientiarum (Hildesheim, 2007).

30    For more on the contemporary German 'Babelmania', see Andrea Polaschegg and Michael Weichenhan, eds, *Berlin – Babylon: Eine deutsche Faszination 1890–1930* (Berlin, 2017).

later became the highlights of the Berlin *Vorderasiatisches Museum* originated in the neo-Babylonian era in the mid-first millennium BC. (Imagine the historical distance: like conflating Imperial Rome and Renaissance Rome). However, scholars were the ones most fascinated by the supposed modern spirit that the majority of the paragraphs of Hammurabi's Code seemed to document – although they did not forget to mention its 'odd archaic traits' in certain fields, such as penal law.[31] Most important in this respect were the detailed regulations of the economic sphere, which had no parallels in biblical law and thus seemed to prove that Babylonia, even in this very early era, had arrived at a level of civilization much higher than that of ancient Israel. Many of these views were highly anachronistic and obviously reflected the capitalist reality of their own time, as does the passage below, taken from the Peiser and Kohler's introduction to their edition of the codex:

A flourishing agriculture, fairly unrestrained private ownership of land and soil can already be seen: the population buys and sells, rents and lets freely; there is a bustling river trade on the Euphrates, business is done between companies, loans and other monetary transactions are the order of the day.[32]

When these lines were written, Kohler had just returned from a three-month trip to the United States (which included a private audience with President Theodore Roosevelt [1858–1919]). Given his deep fascination for the emerging economic power of the former European colonies and the 'exuberant vitality' (*überschäumende Lebenskraft*) of its population, this passage suggests that his views on the future (American) world power had bled over into those of the (Babylonian) power of the ancient world.[33]

---

31    Josef Kohler, 'Hammurabis Gesetz [1903]', repr. in idem, *Aus Kultur und Leben. Gesammelte Essays* (Berlin, 1904), 58–64, at 59.

32    Kohler and Peiser, *Hammurabi's Gesetz*, 1:2; see also Josef Kohler, 'Das Recht der orientalischen Völker', in *Allgemeine Rechtsgeschichte*, Pt 1, *Orientalisches Recht und Recht der Griechen und Römer*, eds Josef Kohler and Leopold Wenger, Die Kultur der Gegenwart. Ihre Entwicklung und ihre Ziele 2/7.1 (Leipzig, 1914), 49–153, at 57. In his analysis of the neo-Babylonian law, Kohler went even further and described a full developed modern banking system: ibid., 63.

33    Josef Kohler, *Lehrbuch des bürgerlichen Rechts*, vol. 1, *Allgemeiner Teil* (Berlin, 1906), xi (taken from the preface which was written during his passage to the US); further, Josef Kohler, *Aus vier Weltteilen. Reisebilder* (Berlin, 1908), 96–146; on his view of America, Günter Spendel, *Josef Kohler. Bild eines Universaljuristen*, Heidelberger Forum 17 (Heidelberg, 1983), 36–37.

A second field of the alleged Babylonian modernity was that of administration and the legal system during Hammurabi's reign. Already in his first Babel-Bible lecture – given before the code was published – Delitzsch emphasized the advanced character of the legal system and rule of law in ancient Babylonia, identifying similarities between the Babylonian and modern European (or rather German) civilization in this respect.[34] He saw the publication of the code as vindicating his views and was thus emboldened to call Babylonia a highly developed or even exemplary *Rechtsstaat*.[35]

Most of his assyriologist colleagues agreed on this point and underlined the supposed progressive character of Babylonian law as compared to that of other ancient legal traditions – including biblical law. The historian and orientalist Lehmann-Haupt, for instance, wrote:

> In the high ethical [*sittlich*] consciousness of what is right expressed in many provisions, [in] the high and then already long-established level of commerce and business that they imply, [i]n the, at times, extremely fine casuistry, these laws far surpass anything that we can find in ancient collections of statutes from the beginnings of the history of any people.[36]

In this respect, most scholars pointed out the abolition of blood vengeance – regarded as a major problem of 'oriental' and especially 'Semitic' societies[37] – in the context of the enforcement of the rule of law under Hammurabi. Comparisons were drawn to biblical law, with its laxer regulation of blood vengeance. These were not comparisons favourable to the latter, as Kohler underlined: 'in this respect, the Babylonian civilization is superior to the Israelite civilization'.[38] Summarizing the 'penal law' in the Code of Hammurabi, Kohler and Peiser explicitly emphasized its modernity – not only in comparison to biblical law but also in comparison to any of the legal traditions in the Middle East that came after it, especially Islamic law.[39] Furthermore, they claimed that Babylonian judicial and political life as a whole was also modern in many

---

34    Delitzsch, *Babel und Bibel. Ein Vortrag*, 25.

35    Delitzsch, *Zweiter Vortrag über Babel und Bibel*, 21; Delitzsch, *Babel und Bibel*, 19.

36    Carl Friedrich Lehmann-Haupt, *Babyloniens Kulturmission einst und jetzt. Ein Wort zur Ablenkung und Aufklärung zum Babel-Bibel-Streit*, 2nd ed. (Leipzig, 1905), 6.

37    See the references on the Semitic peoples in Albert Hermann Post, *Grundriss der ethnologischen Jurisprudenz*, vol. 1, *Allgemeiner Teil* (Oldenburg, 1894), 226–61.

38    See Josef Kohler, 'Hammurabis Gesetz', in *Aus Kultur und Leben*, 59. See also Kohler and Peiser, *Hammurabi's Gesetz*, vol. 1, 126, 139; Kohler, 'Das Recht der orientalischen Völker', 57, 62.

39    Kohler and Peiser, *Hammurabi's Gesetz*, vol. 1, 139.

respects, concluding that 'Babylonia developed a legal culture that had much more in common with our culture' than with the biblical traditions.[40]

## 3      Babylonian Secularism

The main aspect of the assumed modernity of Hammurabi's reign, however, was its supposed non-religious or even secular character – again always in contrast to Mosaic law. According to Winckler, the Babylonian king had always stood in strong opposition to the priests of Marduk, the main Babylonian god. Accordingly, the supposedly non-religious dimension of Hammurabi's rule explained the 'practical' character of his code: 'His laws are responses to practical needs [...] not to spiritual or theological speculations as with the biblical legislation'.[41]

Kohler and Peiser went even further, claiming that the modern distinction between morality and law could be traced back to ancient Babylonia. In their view, its secular or profane character is what distinguished the Hammurabi Code from, and rendered it superior to, the theocratic systems of law of all other oriental civilizations. This entire section deserves attention:

> With an Oriental law, the primary question to be considered is whether the law is a purely legal act or whether it is of a theocratic-religious character, addressing the whole life of human beings. [...] The Indian law books are entirely theocratic in kind, link morality and law together; but the legal provisions of Israelite law, specifically those of the so-called Covenant Code, the Book of Deuteronomy and the priestly law, are also theocratic. Legal provisions alternate with ethical prescriptions therein [...] This theocratic type [of law] appears again, much later, in the Koran. The law of Hammurabi is quite the opposite of this. In an almost modern way, the juridical has been extracted from the prescriptions governing other aspects of life and anything to do with moral doctrine is left out entirely, particularly the debates on the moral and immoral use of law, because these should be left to the purview of religious morality.[42]

The rejection of theocracy, with its supposed conflation of morality and law, clearly mirrored the position of these scholars in the debate on the meaning

---

40    Ibid., 142–43.
41    Winckler, *Die Gesetze Hammurabis*, xxxi.
42    Kohler and Peiser, *Hammurabi's Gesetz*, vol. 1, 137–38.

and the status of (Christian) religion in contemporary German society and thus reflected the ideological formation of German secularism.[43]

In scholarship in this vein, Hammurabi's image differed fundamentally from that of a theocratic oriental despot – a common narrative pattern used for the representation of rulers in the Islamic periods.[44] It further illustrates the fact that the construction of affinities and parallels was at least as important in contemporaneous Orientalism as the construction of 'otherness'.[45] Most scholars who wrote about Hammurabi definitely admired the Babylonian king for his supposed innovations or revolutions.

The theologian and historian Paul Rohrbach (1869–1956) called him the 'first clearly outlined person' in history, while the assyriologist Bruno Meissner (1868–1947) ranked him among the 'greatest historical characters'.[46] By calling Hammurabi a great *Persönlichkeit*, both drew upon a concept that played a central role in German idealistic philosophy as well as in historiography. One need only think of Jacob Burckhardt's famous definition of the 'great man in history', who 'represent[s] the coincidence of the general and the particular, of the static and the dynamic, in one personality'.[47] Linked to the concept of *Persönlichkeit* was the idea that the great men were connected in a kind

---

43     On contemporary German secularism, see, among others, Rebekka Habermas, ed., *Negotiating the Secular and the Religious in the German Empire: Transnational Approaches,* New German Historical Perspectives 10 (New York, 2019). Kohler, for instance, held a pantheistic worldview and was a convinced supporter of the at that time still controversial cremation (see Spendel, *Josef Kohler,* 45).

44     See, among others, Dorothy M. Figueira, 'Oriental Despotism and Despotic Orientalisms', in *Bucknell Review* 38/2: 'Anthropology and the German Enlightenment: Perspectives on Humanity', ed. Katherine M. Faull (1995), 182–99; Joan-Pau Rubiés, 'Oriental Despotism and European Orientalism: Botero to Montesquieu', *Journal of Early Modern History* 9 (2005), 109–80; Michael Curtis, *Orientalism and Islam: European Thinkers on Oriental Despotism in the Middle East and India* (Cambridge, 2009). On the history of this trope, see the classic article by Richard Koebner, 'Despot and Despotism: Vicissitudes of a Political Term', *Journal of the Warburg and Courtauld Institutes* 14 (1951), 275–302; but cf. also the different perspective of Werner Kogge and Lisa Wilhelmi, 'Despot und (orientalische) Despotie – Brüche im Konzept von Aristoteles bis Montesquieu', *Saeculum* 69 (2020), 305–42.

45     On this often-neglected aspect of European orientalism, see David Cannadine, *Ornamentalism: How the British Saw Their Empire* (London, 2001), xix–xx, 3–11.

46     Paul Rohrbach, *Die Geschichte der Menschheit* (Königsstein, 1914), 37; Bruno Meissner, *Könige Babyloniens und Assyriens. Charakterbilder aus der altorientalischen Geschichte* (Leipzig, 1926), 53.

47     Jacob Burckhardt, *Force and Freedom: Reflections on History,* trans. James Hastings Nichols (New York 1943), 325; originally published as *Weltgeschichtliche Betrachtungen,* ed. Jakob Oeri (Berlin, 1905).

of historical chain, which opened the possibility to identify connections and parallels. Which historical parallels were drawn by whom was quite telling in this respect: whereas French scholars favoured comparisons between Hammurabi and Napoleon Bonaparte (1769–1821) – both 'fathers' of famous law codes[48] – the Germans compared Hammurabi to Charlemagne (ca. 747–814) or the much-admired Prussian kings of the eighteenth century, Friedrich Wilhelm I (1688–1740) and Friedrich II, also known as Frederick the Great (1712–1786).[49] (Figure 9)

What all these highly anachronistic parallels have in common is the intention to present Hammurabi's reign as a kind of 'enlightened absolutism' (*erleuchteter Absolutismus*), as Lehmann-Haupt explicitly called it.[50] In doing so, these scholars referred to a highly influential yet controversial concept in German historiography introduced in the mid-nineteenth century. Most important in this respect was the approach of the historian and economist Wilhelm Roscher (1817–1894), who distinguished three types of absolutism: the confessional form, as represented by Philipp II of Spain, or Philip the Prudent (1527–1598); the classical courtly type of Louis XIV, or Louis the Great (1638–1715); and, last but not least, the 'enlightened absolutism' of Friedrich II and of Joseph II of Austria (1741–1790). Roscher's national biases, however, were all too obvious: he made it very clear that the last type – represented by Prussian and Austrian rulers – was the most advanced one. Thus, he contrasted the (in)famous *l'état c'est moi!* ('the state is me') of the French kings with the *le roi c'est le premier serviteur de l'état!* ('the king is the foremost servant of the state') of the Prussian and Austrian kings, who were motivated, he claimed, by moral and enlightened ideals in their pursuit of legal reforms.[51]

---

48    On this analogy, see Johannes Renger, 'Noch einmal: Was war der "Kodex" Ḫammurapi – ein erlassenes Gesetz oder ein Rechtsbuch?', in *Rechtskodifizierung und soziale Normen im interkulturellen Vergleich*, ed. Hans-Joachim Gehrke, Script-Oralia 66 (Tübingen, 1994), 27.

49    See Kohler and Peiser, *Hammurabi's Gesetz*, vol. 1, 2; Lehmann-Haupt, *Babyloniens Kulturmission einst und jetzt*, 46–47; Meissner, *Könige Babyloniens und Assyriens*, 53.

50    Lehmann-Haupt, *Babyloniens Kulturmission einst und jetzt*, 46; see also Kohler, 'Hammurabis Gesetz [1903]', in *Aus Kultur und Leben*, 64.

51    Wilhelm Roscher, 'Umrisse zur Naturlehre der drei Staatsformen III', *Allgemeine Zeitschrift für Geschichte* 7 (1847), 436–473, at 451. On the concept and its problems, see H. M. Scott, 'Whatever Happened to the Enlightened Despots?', *History* 68 (1983), 245–57; Charles Igrao, 'The Problem of "Enlightened Absolutism" and the German States', *Journal of Modern History* 58 (1986), 161–80; Peter Baumgart, 'Absolutismus ein Mythos? Aufgeklärter Absolutismus ein Widerspruch? Reflexionen zu einem kontroversen Thema gegenwärtiger Frühneuzeitforschung', *Zeitschrift für historische Forschung* 27 (2000), 573–89; Helmut Reinalter, 'Der aufgeklärte Absolutismus – Geschichte und Perspektiven der Forschung', in *Der aufgeklärte Absolutismus im europäischen Vergleich*, eds Helmut Reinalter and Harm Klueting (Vienna, 2002), 11–19. On the political and scholarly background of the concept in the nineteenth century, see Reinhard Blänkner, 'Absolutismus'. *Eine begriffsgeschichtliche Studie zur politischen Theorie und zur Geschichtswissenschaft in Deutschland (1830–1870)* (Frankfurt, 2011).

FIGURE 9    Satirical cartoon in *Jugend* – an influential weekly arts magazine based in
Munich – which proposes sculptures for a "Babel-Bible Alley" in Berlin, including
Hammurabi, Abraham (with antisemitic tropes), Moses, Homer, William I, and
Chamberlain (1903, vol. 13, p.224)

Cuneiform tablets from the period of Hammurabi that had already been
excavated in the late nineteenth century bolstered this line of argument. After
all, they seemed to document the personal interest and commitment of the
king both in general legal affairs and with respected to individual cases. As
claimed by Kohler and Peiser:

> Sometimes, with these letters, we might almost be looking at official
> documents of Frederick II or Frederick William I. His [sc. Hammurabi's]
> activity is feverish; what he commands should immediately be so. [...]
> All of this presents to us the portrait of an extraordinarily circumspect,
> temperamental ruler who is active day and night, one who intervenes in
> any and all matters and who obviously brought the entire state adminis-
> tration of Babylon to an unprecedented level. And indeed, his great legis-
> lative work is consistent with this.[52]

In their praise of Hammurabi's style of government, these German scholars
making tacit allusions to the personal style of government (the so-called
*persönliches Regiment*) attributed to their own Kaiser, William II (1859–
1941), who was himself, for the very same reason, a great admirer of the
ancient Babylonian king.[53] In 1903, William II listed Hammurabi – alongside Moses, Abraham, Charlemagne, Luther, Shakespeare, Kant, and Kaiser
Wilhelm the Great (his own grandfather) – to those 'great wise men and
kings' through whose deeds God revealed himself to humankind.[54] In a
remarkable book on kingship in ancient Mesopotamia, written in his Dutch
exile in 1938, he went even further and called Hammurabi a 'Babylonian
forerunner of King Friedrich Wilhelm I, the master builder of the Prussian
State'.[55]

---

52   Kohler and Peiser, *Hammurabi's Gesetz*, vol. 1, 2–3.
53   See Isabel V. Hull, '"Persönliches Regiment",' in *Der Ort Kaiser Wilhelms II. in der Deutschen
     Geschichte*, ed. John C. G. Röhl, Schriften des Historischen Kollegs Kolloquien 17 (Munich,
     1991), 3–23.
54   [Wilhelm II], 'Babel und Bibel. Ein Handschreiben Seiner Majestät Kaiser Wilhelms des
     Zweiten an das Vorstandsmitglied der Deutschen Orientgesellschaft, Admiral Hollmann',
     *Die Grenzboten. Zeitschrift für Politik, Literatur und Kunst* 62, no.8 (1903), 493–96.
55   Wilhelm II, *Das Königtum im alten Mesopotamien* (Berlin, 1938), 27.

## 4 Superiority of Biblical Ethics

Although most scholars of the time accepted the image of Hammurabi as a modern and enlightened ruler, the preference that the 'furor orientalis' granted to the Babylonian king at the expense of Moses was far from uncontested. Theologians and biblical scholars were by no means alone in their attempts to rescue the singularity and incomparable importance of the biblical lawgiver.

The main line of argument in this respect was developed by David Heinrich von Müller in his seminal work on the relationship of the two codes. Recalling that the treatment of slaves under biblical law was far laxer than that under the Code of Hammurabi, he argued that Moses was responsible for introducing key elements that had not existed in law until then: 'Wisdom, mercy and ethical greatness'.[56] So the Hammurabi-vs.-Moses debate encompassed questions about the normative and religious foundations of modern law. In a long review of Müller's book published in the *Monatsschrift für Geschichte und Wissenschaft des Judentums*, Rabbi David Feuchtwang (1864–1936), of Vienna, expanded on this point, emphasizing the 'moral chasm' between the two codes. On these grounds, he strongly denied any continuity: 'No direct path would have led from here to the flowering of all laws, to the world-transcending Ten Commandments'.[57]

There was no difference between Christian and Jewish religious scholars in this respect: neither group had any difficulty accepting the cultural superiority of the Babylonians as this view fit very well into the biblical narrative, which equally decried the decadence of Babylon. But in contrast to the modernist and secularist perspective of scholars like Kohler, Winckler, Peiser, or Lehmann-Haupt, scholars of both religions claimed an ethical or moral superiority for the Israelites. In other words, whereas modernists and secularists regarded the unity of, or non-distinction between, law and morality in Mosaic law as a sign of its backwardness and inferiority, religious scholars pointed to that same lack of distinction as proof of its progressivity and superiority. This aspect was emphasized by Samuel Oettli (1846–1911), a Swiss professor of Old Testament at Greifswald:

> There is no question but that the civil life reflected in the Codex Hammurabi is far more developed than that reflected in the Covenant Code; but it is equally beyond doubt that a different, a truly humane spirit

---

56    Müller, *Die Gesetze Hammurabis*, 242.
57    Feuchtwang, 'Moses und Hammurabi', 393–94: 'Von hier aus hätte kein direkter Weg zur Blüte aller Gesetzgebungen, zum weltbezwingenden Dekalog geführt'.

struggles forth in this [latter] and in the later law collections of the Torah, one whose source lies in the religious faith of Israel, which is incomparably purer and more fruitful ethically.[58]

Other theologians, like the Leipzig pastor Johannes Jeremias (1865–1942), drew a sharp distinction between the revelation narratives in the Bible and on the stele of Hammurabi. Jeremias argued that unlike the revelation on Mount Sinai, which led to the further transmission of Mosaic law, the relief depicting the revelation on the Babylonian stele already betrays – through the absence of a 'free spiritual and moral acceptance of faith' – this was a revelation 'of a pagan kind'. Furthermore, Jeremias explicitly denied that Hammurabi had overcome and moved beyond 'oriental despotism'.[59] Thus, according to this theological narrative, Hammurabi remained a despot, and Babylonian law remained despotic law, whereas only Moses – and ultimately, of course, Jesus – revealed the true divine law.

Whereas Müller and Feuchtwang presented a Jewish standpoint, and Jeremias and Hommel a Protestant one, Catholic scholars participated much less in the debate. Hubert Grimme (1864–1942), an orientalist and popular biographer of Muhammed, was an exception in this respect, though there was nothing specifically Catholic in his arguments.[60] Grimme, a specialist in Arabic history, developed a new line of argument: he adopted Müller's theory that Babylonian and biblical codes had common roots and took it even further, identifying this supposed shared source in the customary law of the ancient Semitic tribes of the desert.[61] In line with widespread narratives of cultural pessimism in *fin de siècle* Europe, he contrasted the supposedly pure and noble customs of the Bedouins with the decadent Babylonian civilization. Grimme claimed that Mosaic law was closer to the original Semitic law, which, as he saw it, ruled out any direct connection between Moses and Hammurabi. According to Grimme, the Hammurabi Code, with its detailed rules for trade and commerce, was suited only to a feudal society based on a slaveholder economy, whereas the Mosaic law mirrored an egalitarian society of free nomads.[62] Hence, like the theologians, Grimme claimed an ethical or moral superiority on the part of Mosaic law. But unlike them, he found the

58    Samuel Oettli, *Das Gesetz Hammurabis und die Thora Israels. Eine religions- und rechts-geschichtliche Parallele* (Leipzig, 1903), 87.
59    Johannes Jeremias, *Moses und Hammurabi*, 2nd ed. (Leipzig, 1903), 56–57.
60    See Hubert Grimme, *Mohammed,* Weltgeschichte in Karakterbildern (Munich, 1904).
61    Grimme, *Das Gesetz Chammurabis und Moses*, 25.
62    Ibid., 29.

reason for this superiority not in divine revelation but in the purer way of life, as yet undistorted by civilization – in contrast to the hyper-civilized and thus decadent Babylonians.

## 5     Sittlichkeit and/or Morality

When contemporary German scholars wrote about morality or ethics, rather than using the German equivalents of those terms (*Moralität* or *Ethik*), they tended to use a term that is quite difficult to translate into other languages: *Sittlichkeit*.[63] Today, even native speakers of German can find it difficult to understand the meaning once conveyed by this word, because it has nearly vanished from modern German-language discourses on culture, history, and politics.[64] The complexity of the term in the language of the nineteenth and early twentieth centuries becomes evident not least when one looks at the debate on Hammurabi and Moses: *Sittlichkeit* was an important category not only for those who insisted on the moral or ethical superiority of the Mosaic law and the Israelites but also for the modernists who insisted on the *cultural* (technological, economic, secular etc.) superiority of Hammurabi and the Babylonians.

This situation, wherein the same concept is used by different factions in the rationales for opposing arguments, is characteristic for so-called key, or basic, concepts.[65] Like concepts of comparable importance like *Fortschritt* (progress) or *Kultur* (culture or civilization), *Sittlichkeit* was omnipresent in contemporaneous discourses but rarely ever explained. Again, this is typical of key concepts: everyone in a given era seems to feel that their meaning is obvious to all, yet the use of the concepts is contradictory and contested, so they do not fit into simple definitions.[66] As I intend to demonstrate in this final section, the debate on Moses and Hammurabi reflects – in a way – different

---

63     The reason that the term is almost untranslatable is that it combines the meanings of morality and custom: It is derived from (or implies) the German term *Sitte* (custom, tradition) and could thus be used, as Hegel did, in the sense of 'customary morality' as opposed to 'reflected morality' (see below).

64     The term is still present in juridical language, though. The best-known examples are *Sittlichkeitsvergehen* ('acts of indecency') or *Sittlichkeitverbrechen* ('sex crimes').

65     For more on the history of the concept in nineteenth-century Germany, see Karl-Heinz Ilting, *Naturrecht und Sittlichkeit. Begriffsgeschichtliche Studien,* Sprache und Geschichte 7 (Stuttgart, 1983), 238–82.

66     See Reinhart Koselleck, 'Introduction and Prefaces to the *Geschichtliche Grundbegriffe,'* trans. Michaela Richter, *Contributions to the History of Concepts* 6/1 (2011 [1972]), 1–37.

and even contradictory understandings of *Sittlichkeit* in the German discourse of circa 1900.

For several reasons, *Sittlichkeit* was of particular importance in discourses on (ancient) history and (ancient) law. Kohler's discussion of the supposed separation of *Recht*, or jurisprudence, from *Sittlichkeit* in the Hammurabi Code clearly reflects a more general debate on this relationship in German philosophy of law of the nineteenth and early twentieth centuries.[67] Nonetheless, the debate was certainly shaped primarily by G.W.F. Hegel's (1770–1831) famous distinction between *Moralität* and *Sittlichkeit* in his critique of Immanuel Kant's (1724–1804) moral philosophy.[68]

Kant used the two terms more or less synonymously to characterize those actions that are motivated solely by duty (*Pflicht*) to the moral law: thus contrasting duty with mere conformity with law of another kind.[69] But the most important aspect was that Kant's principles of morality are universal, unconditional, and formal, based on the idea of rational agents who autonomously impose the moral law upon themselves. Hegel, in contrast, criticized Kant's concept of morality as being too abstract and only formal – an 'empty principle of moral subjectivity'[70] – and introduced the sharp distinction between *Moralität* and *Sitichkeit* in this context. Hegel reserved the term *Sittlichkeit* for a

---

67    See, for instance, Otto von Gierke, 'Recht und Sittlichkeit', *Logos. Internationale Zeitschrift für Philosophie der Kultur* 6 (1916/17), 211–64.

68    Most famously expressed in his philosophy of right. See G. W. F. Hegel, *Grundlinien Der Philosophie des Rechts* (Berlin, 1821), parts 2 (*Moralität*) and 3 (*Sittlichkeit*). There is a huge body of literature on Hegel's distinction. See, among others, Joachim Ritter, 'Moralität und Sittlichkeit. Zu Hegels Auseinandersetzung mit der kantischen Ethik [1966]', repr. in idem, *Metaphysik und Politik. Studien zu Aristoteles und Hegel* (Frankfurt, 1977), 281–309; Gabriel Amengual, 'Der Begriff der Sittlichkeit. Überlegungen zu seiner differenzierten Bedeutung', *Hegel-Jahrbuch* 1 (2001), 197–203; furthermore, see the articles in Wolfgang Kuhlmann, ed., *Moralität und Sittlichkeit. Das Problem Hegels und die Diskursethik* (Frankfurt, 1986).

69    Immanuel Kant, 'Die Metaphysik der Sitten [1797]', repr. in *Kant's Gesammelte Schriften: Erste Abteilung: Werke*, vol. 6, ed. Preußische Akademie der Wissenschaften (Berlin, 1907), 219. Consequently, in English translations *Moralität* and *Sittlichkeit* are both translated as morality: cf. Immanuel Kant, *Groundwork of the Metaphysics of Morals: A German-English Edition* [German text from the second original edition (1786)], eds and trans Mary Gregor and Jan Timmermann (Cambridge, 2011), 13–119.

70    G. W. F. Hegel, *Elements of the Philosophy of Right*, ed. Allen W. Wood, trans. H.B. Nisbet (Cambridge, 1991), 191. See also §135 (pp. 162–63). The term *Sittlichkeit* is usually translated as 'ethics' or 'ethical life'. See the remarks of the translator (pp. 403–4); also Charles Taylor, *Hegel* (Cambridge, 1975), 376–78; others prefer the German term as a loanword. See e.g. Philip J. Kain, *Hegel and Right: A Study of the 'Philosophy of Right'* (New York, 2018), 83–137.

more objective form of ethics, referring to those moral obligations that people have to the communities of which they are part: something he then contrasts with abstract and subjective morality. Thus, Hegel took social entities like the family, the civil society, and, last but not least, the state as expressions of this highest form of moral life: 'The ethical [*das Sittliche*] is a subjective disposition, but of that right which has being in itself'.[71]

There were two aspects of Hegel's concept of *Sittlichkeit* that were of particular relevance to German discourses on law and morality in the nineteenth century: firstly, his insistence on concrete social and historical contexts; and secondly, the historization of moral beliefs and values that results from this dependence. Hegelian *Sittlichkeit* differs not only from one society to the next but also from one era to another. Moreover, as Hegel's followers were convinced, it *evolved* over the course of history. This assumption opened the possibility of identifying different stages or levels of *Sittlichkeit* in different societies and epochs.

As a result, in writings of the later nineteenth century, the differences between *Sittlichkeit* and other key concepts like *Kultur* became increasingly blurred. *Sittlichkeit* and, no less important, the law thus advanced to the status of an indicator for cultural progress and vice versa.[72] Accordingly, scholars who insisted on the modernity and superiority of Babylonian law usually also found evidence for a high level of *Sittlichkeit* in it – because they regarded Babylonia as the cradle of civilization and thus the most advanced society of the ancient Near Eastern world.[73]

However, as already mentioned, *Sittlichkeit* remained a contested concept. Furthermore, with the rise of philosophical neo-Kantianism in the late nineteenth century, the use of the term as a synonym for morality (in the sense of universal ethic) gained new currency. The most important proponent of neo-Kantian ethics at the turn of the twentieth century was certainly the German-Jewish philosopher Hermann Cohen (1842–1919). When, in his remarkable article *Religion und Sittlichkeit* (1907), he identifies the 'nature of God' with the 'nature of human *Sittlichkeit*' and calls God himself an 'archetype and

---

71    Hegel, *Elements of the Philosophy of Right*, 186.

72    For this reason, some scholars accused the philosophers, historians and jurists of the nineteenth century of moral relativism and of devaluing the ideal of universal ethics in the long run. It might be better, however, to speak of a banalization of the term *Sittlichkeit* in the nineteenth century. See especially Ilting, *Naturrecht und Sittlichkeit*, 238–47; but see also the fairly balanced perspective on the history of the concept by Wolfgang Kersting, 'Sittlichkeit, Sittenlehre', in *Historisches Wörterbuch der Philosophie*, vol. 9, ed. Joachim Ritter and Karlfried Gründer (Basel, 1995), 907–23.

73    See, specifically, Lehmann-Haupt, *Babyloniens Kulturmission einst und jetzt*, 6.

model' (*Urbild und Vorbild*) of human *Sittlichkeit*, it is clear that Cohen is using the term to denote a universal concept of 'morality' in the Kantian sense of the word.[74] Although Christian theologians like Hommel, Jeremias, or Oettli would certainly not have concurred with Cohen's insistence on the original Jewish character of ethical monotheism, from which Christianity deviated in some respects, they shared his understanding of *Sittlichkeit* as a synonym for universal ethics.[75]

Yet remarkably, all scholars agreed that the cultural superiority of the Babylonians would not have been accompanied by a superior morality (in the sense of a humane and universal ethic). This can be best demonstrated through the writings of Kohler. One of the forerunners of the so-called neo-Hegelian school of law at the turn of the twentieth century,[76] he became the leading historian of law, and his writings covered a wide range of subjects including the comparative ethnology of law.[77]

Though very interested in the customs and rules of the so-called primitive or natural peoples,[78] Kohler left no doubt that there was development and progression in the history of law (finally leading to modern Western law) that corresponded to the general social and economic development and progression.

---

74    Hermann Cohen, 'Religion und Sittlichkeit. Eine Betrachtung zur Grundlegung der Religionsphilosophie [1907]', repr. in idem, *Jüdische Schriften*, vol. 3, *Zur jüdischen Religionsphilosophie und ihrer Geschichte*, ed. Akademie für die Wissenschaft des Judentums (Berlin, 1924), 98–168, at 134–35.

75    See, for instance, Jeremias, *Moses und Hammurabi*, 25; Oettli, *Das Gesetz Hammurabis und die Thora Israels*, 88; for Jewish scholars, see, e.g., Feuchtwang, 'Moses und Hammurabi', 393. On the debate on ethics and its biblical foundations among Jewish scholars, see Kerstin von der Krone, *Wissenschaft in Öffentlichkeit. Die Wissenschaft des Judentums und ihre Zeitschriften*, Studia Judaica 65 (Berlin, 2012), 327–74.

76    On Kohler as a forerunner of the neo-Hegelian school of law, see Spendel, *Josef Kohler*, 5. On the neo-Hegelian school of law in general, see, among others, Christoph Mährlein, *Volksgeist und Recht. Hegels Philosophie der Einheit und ihre Bedeutung in der Rechtswissenschaft*, Epistemata 286 (Würzburg, 2000); Andreas Großmann, 'Recht verkehrt. Hegels Rechtsphilosophie im Neuhegelianismus', in *Recht ohne Gerechtigkeit? Hegel und die Grundlagen des Rechtsstaates*, eds Mirko Wischke and Andrzej Przyłębski (Würzburg, 2010), 191–208.

77    See, for instance, Josef Kohler, 'Rechtsgeschichte und Kulturgeschichte', *Zeitschrift für das Privat-und öffentliche Recht der Gegenwart* 12 (1885), 583–93; idem, 'Begriff und Aufgabe der Weltgeschichte [1899]', in idem, *Aus Kultur und Leben. Gesammelte Essays* (Berlin, 1904), 15–22.

78    In order to capture the customs and laws of the non-European *Naturvölker*, he designed a questionnaire for the German colonial administration. See Josef Kohler, 'Fragebogen zur Erforschung der Rechtsverhältnisse der sogenannten Naturvölker, namentlich in den deutschen Kolonialländern', *Zeitschrift für vergleichende Rechtswissenschaft* 12 (1897), 427–40.

For him, the law of Babylonia being more highly developed than that of the Bible seemed to admit of no doubt, if only in view of the modernity ascribed to Babylonia in other respects. Thus, in a highly critical review of Oettli's book on Hammurabi and Moses, Kohler explicitly linked economic progress with the rise of private property and egoism. He asserted, even further, that it would be unhistorical to consider the more altruistic and humane (in the sense of modern morality) provisions of biblical law to be an indication of its 'higher' character:

> Of course, this communism [of the Mosaic law] is associated with a lot of altruistic phrases, which one tends to regard as more humane, and which the author [Oettli] also draws attention to. But it is not correct to say that such philanthropic institutions would prove an increased higher culture. On the contrary: the progress of culture initially pushes towards a well defined form of private property, and, as a result, towards the egoism of property and commercial transactions. This decisive egoism in the use of property is characteristic for a more advanced stage of civilization. [...] It is therefore unhistorical to expect the so-called philanthropism, i.e. the communist features, of the Torah from the developed Babylonian law.[79]

## 6 Conclusion

As a historian of modern history, I have little to say about the historical relationship between Babylonian and biblical law. From today's perspective, however, questions such as whether Mosaic law represents an 'advance' or 'regression' from Hammurabi's seem outdated and misleading. The same applies to historical parallels that scholars drew in the early twentieth century: it is difficult for us to see those similarities between ancient Babylonia and modern Prussia that were so obvious to them. What interests me, however, are not these questions and parallels themselves but why scholars discussed them so intensively in early twentieth-century Germany. As I hope to have made clear, the debate on Moses and Hammurabi was not primarily about different approaches to, and interpretations of, certain ancient sources. Like so many other discourses on ancient civilizations of the time, this debate raised central issues and questions of the day: the relationship between ethics (or morality), religion, and

---

79    Josef Kohler, review of *Das Gesetz Hammurabis und die Thora Israels* by Samuel Oettli, *Deutsche Literaturzeitung* 24 (1903), 1543–49, at 1547.

law; the normative sources and underpinnings of law; and, most immediately, the Near Eastern roots of human civilization. Further still, the debate clearly reflects contemporary perceptions of – and attitudes to – the 'modern' and capitalist society they lived in. Thus, these scholars recognised in the historical material the same questions and problems that were central for understanding their own present – and, not least, they used this historical backdrop to take stands in these debates.

There is no reason to look down on these scholars and their strange, anachronistic approaches to the past. To be sure, Hammurabi has completely – and Moses nearly – disappeared from today's debates on the normative sources and foundations of law and on the relationship between ethics, religion, and law. When we think today about modern society and its historical roots, we certainly do not associate them with ancient Babylonia. But the issues themselves are all still raised and debated today: coupled with historical references and mythologies, which, though different from those invoked by our forerunners, will certainly seem strange to future historians.

### Acknowledgements

The article is based on research made possible by a fellowship of the DFG Center for Advanced Studies (2615) 'Rethinking Oriental Despotism' at the Free University of Berlin in 2018. Apart from quotations from Kant, Hegel, and Burckhardt, all translations from German sources are mine. Where expressions or phrases posed translation difficulties, I have included the original German wording in parenthesis or in a footnote.

## PART 2

### *Transformations in the Present*

∵

# CHAPTER 4

# Gesetz als Gegensatz

## The Modern Halachic Language Game

*Irene Zwiep*

Law is not law which alters when it alteration finds. This, I admit, is a rather cheap pun on lines two and three of Sonnet 116, Shakespeare's famous ode to constancy in love. Still, the phrase kept going through my head, with and without a question mark, while I was working on this paper. In William Shakespeare's (1564–1616) rhyme, love is a rock-solid foundation, celebrated as 'not Time's fool' but 'an ever-fixed mark' that is never shaken. Jewish law, by contrast, though in many respects ever fixed, was shaken by time or, better perhaps, by its historical collision with modernity. By the end of the eighteenth century, the immobile history (as French micro-historian Le Roy Ladurie would call it) of the Jews in Western Europe had come to an end, and with it a relatively sovereign period in the development of its communal legal corpus, known as the halacha. But in the fray of the modern experiment, was there any room left for that hoary mix of divine statutes and human jurisprudence, which had demarcated Jewish life for so many centuries?

The socio-political changes of the eighteenth and nineteenth centuries did not fundamentally alter the rules and principles of traditional Jewish legal practice. They did, however, affect the ways in which Jewish law was perceived, practiced, and obeyed, triggering a wealth of – more often than not conflicting – Jewish reactions. On the one hand, the (gradual) dismantling of the pre-modern Jewish semi-autonomy in favour of individual citizenship limited its span of control, sparking reflection on the status and validity of Jewish law *as law*. In parallel, processes of integration and assimilation called for a restyling of the pre-modern Jewish habitus – of which halacha supplemented by *minhag*, or Jewish (legal) custom, had been a central ingredient. In the course of that procedure, the reigning Protestant paradigm of inward piety put pressure on Jewish 'legalism', with its aura of collective obedience and accountability, as a suitable vehicle for modern personal *religiosity*. The critical historicism that dominated nineteenth-century scholarship did little to ease the pain. The *Wissenschaft des Judentums*, especially the classical branch as initiated by Leopold Zunz (1794–1886), came close to excluding the law from its model for Jewish *national* belonging altogether. Turning its lens on the ancient sources,

Christian higher criticism relativized, and thus undermined, the authority of Mosaic law as a divinely inspired *system of belief.* Part of the groundwork for this high-brow demolition act had been laid by Kantian philosophy and its rejection of revealed religion as a source of *moral inspiration.* In the main part of this chapter I will return to these issues in somewhat more detail.

All things considered, it seems fair to say that in its confrontation with modernity Jewish law, both as a concept and as a practice, did find alteration. The question how this confrontation affected the Jewish framing of 'the law' is the starting point of this paper. Needless to say, the modern rethinking of that law as a defining aspect of the Jewish *modus vivendi* has been the subject of intense research. Much effort has gone into reconstructing its role as a catalyst in the fragmentation of Judaism into the various denominations we know today.[1] The aim of this chapter is not to revisit the case of Zacharias Frankel (1801–1875) versus S.R. Hirsch (1808–1888) versus Abraham Geiger (1810–1874) and remap its communal and cultural implications.[2] The reflection on halacha's divine or human nature, its aptitude for governance or guidance, and its role in modern public and private life and religion have been thoroughly explored. Instead, I propose a much more modest conceptual analysis of the nineteenth-century discourse on law-based Judaism as an inherently problematic combination of legal procedure and normative religion. As we shall see, it was this ancient combination that distinguished Judaism from its Christian counterpart but simultaneously reduced it to an anomaly, a legal and religious relic in a new order that was built on the separation of church and state and, accordingly, of (private, moral) religion and national legislation.

In her 2011 monograph *How Judaism Became a Religion*, Leora Batnizky described the transformation of the pre-modern Jewish polity into a religious system known as Judaism, juxtaposing the new religion to the competing

---

1   Major milestones are Jay M. Harris, *How Do We Know This. Midrash and the Fragmentation of Modern Judaism.* SUNY Series in Judaica: Hermeneutics, Mysticism, and Religion (New York, 1995); Andreas Gotzmann, *Jüdisches Recht im Kulturellen Prozeß. Die Wahrnehmung der Halachah im Deutschland des 19. Jahrhunderts,* Schriftenreihe wissenschaftlicher Abhandlungen des Leo Baeck Instituts 55 (Tübingen, 1997); David Ellenson, 'Antinomianism and its Responses in the Nineteenth Century', in *The Cambridge Companion to Judaism and Law,* ed. Christine Hayes (Cambridge, 2017), 260–86; Michael Meyer, *Response to Modernity. A History of the Reform Movement in Judaism,* Studies in Jewish History (Oxford, 1988); Mordechai Breuer, *Modernity within Tradition: The Social History of Orthodox Jewry in Imperial Germany* (New York, 1992); Andreas Brämer, *Rabbiner Zacharias Frankel. Wissenschaft des Judentums und konservative Reform im 19. Jahrhundert,* Netiva 3 (Hildesheim, 2000).
2   See further Judith Frishman's chapter in this volume.

conceptual domains of culture, nation, reason, and secularity.[3] Seven years before, Simone Lässig documented the process of cultural embourgeoisement that accompanied this religious transformation, facilitated by school, synagogue, and community and broadcast via a new – emphatically Jewish – public sphere, in which traditional halacha played a less than prominent role.[4] The present chapter hopes to unravel a small additional thread by examining how the modern Jewish mind tackled the problem of Judaism's inner contradiction, *viz.* its legal core which, in nineteenth-century terms at least, constituted at once the essence and the embarrassment of Jewish religiosity, besides giving ground to legal disqualification. More specifically, we shall have a closer look at what, for want of a better term, I have dubbed 'the halachic language game', i.e. the ways in which authors tried to affect current perceptions of Jewish law by absorbing its old vocabulary into novel contexts and usages, thus creating new types of meaning. Before we turn to the actual analysis, I will briefly explain this – inevitably somewhat vulgarized – use of Wittgenstein's idea of linguistic usage as *Sprachspiel*, as language game.

The contested convergence of law and religion seems epitomized in the German term *Gesetzesreligion*, an originally Pauline notion which gained renewed import in nineteenth-century religious polemics. Initially the term was coined to denote a difference in theological outlook, with Christianity seeking release from sin through grace and Judaism seeking justice before God by performing his commandments. In the nineteenth century this opposition was increasingly framed as an opposition of religious group mentalities, with the Jews as the embodiment of calculated judgement and Christians representing love and mercy. In addition, the post-Kantian primacy of (non-enforceable) private moral conscience clashed with Judaism's reliance on (enforceable) halachic justice. It was, then, Jewish *Gesetz als Gegensatz*, in short, the antithesis of the Christian doctrine of redemption and of the processes of *Verinnerlichung* and *Versittlichung* (internalization and civilizing/ ethicizing) that shaped modern bourgeois piety and morality, if often only on paper.[5]

One might add that, from the Enlightenment onwards, Jewish educators readily embraced this ideal of internalization and moralization, with the

---

3 Leora Batnitzky, *How Judaism Became a Religion. An Introduction to Modern Jewish Thought* (Princeton, 2011).

4 Simone Lässig, *Jüdische Wege ins Bürgertum. Kulturelles Kapital und sozialer Aufstieg im 19. Jahrhundert* (Göttingen, 2004).

5 Cp. Lucian Hölscher, 'Die Religion des Bürgers. Bürgerliche Frömmigkeit und protestantische Kirche im 19. Jahrhundert', *Historische Zeitschrift* 250 (1990), 595–630.

aesthetization of Jewish religious practice as one natural corollary.[6] Across the
board, i.e. from radical Reform to restorative Orthodoxy, they turned to the
Hebrew Bible as the *summa* of Jewish ethical edification, passing over the esoteric legal corpus as a source of Jewish morality and devotion.[7] In a pioneering article on the place, or rather on the absence, of halacha in nineteenth-
century children's textbooks, Andreas Gotzmann defined this development as
a 'process of delegalization', during which definitions of halacha faded while
its juridical contours became increasingly blurred. In the formulation of modern Jewish belief, the divine will as laid down traditional halacha was almost
unanimously discarded as a viable normative system.[8] It took a thinker of the
devotional and philosophical calibre of Rabbi Joseph Soloveitchik (1903–1993)
to eventually reappropriate Jewish religiosity through law and to rehabilitate,
with the help of scientific reasoning, halachic observance as a source of Jewish
spirituality.[9]

Unlike the field of elementary education, Jewish theological-political thinking could not afford the luxury of tacitly relinquishing the halachic heritage.
In an era of religious transformation and secular state formation, the lingering
presence of an ancient law-based religion raised multiple questions. Though
originating from new debates, these questions often tied in with old clichés.
Some touched upon Judaism as a superseded, redundant religion (building
on Friedrich Schleiermacher's [1768–1834] famous image of Judaism as *eine
unverwesliche Mumie*, an imperishable mummy), while others targeted the
Jews' empty legalism as opposed to Christian piety, or their habit of keeping
double, in-group and out-group, standards (as in Weber's no less famous allusion to Jewish *Binnen-und Außenmoral*, to internal and external morality). The

---

6  For the latter aspect, see Uta Lohmann, 'Das bürgerliche Leben als humanistisches
   Kunstwerk. Reflexionen zum universal-ästhetischen Selbst-und Gesellschaftsbild des
   jüdischen Kaufmann's David Friedländer und zur Ikonographie der Haskala', *Trumah* 22
   (2014), 39–68.
7  Andreas Gotzmann, 'The Dissociation of Religion and Law in Nineteenth-Century
   German-Jewish Education', *Leo Baeck Institute Year Book* 43 (1998), 103–26; Uta Lohmann,
   'Wissensspeicher, Lehrbuch, Erkenntnisquelle. Zur Rolle der Hebräischen Bibel im
   Bildungskonzept der Berliner Haskala', in *Deutsch-jüdische Bibelwissenschaft. Historische,
   exegetische und theologische Perspektiven*, eds Daniel Vorpahl, Sophia Kähler, and Shani
   Tzoref, Europäisch-jüdische Studien 40 (Berlin, 2019), 77–92; Dorothea Salzer, *Mit der Bibel
   in die Moderne. Entstehung und Entwicklung der Gattung jüdische Kinderbibel*, Studia Judaica
   122 (Berlin, 2023).
8  Gotzmann, 'The Dissociation of Religion and Law', 11off.
9  Joseph B. Soloveitchik, *Halakhic Man*, trans. Lawrence J. Kaplan (Philadelphia, 1983). For
   a thorough exposé, see Dov Schwartz, *Religion or Halakha: The Philosophy of Rabbi Joseph
   B. Soloveitchik* (Leiden, 2007).

need to respond to these old-new biases prompted a wealth of speculation on the meaning of law for Jewish civil and religious life. In this chapter I will look at some of the main stakes in the debate. Ignoring obvious nodes and overlap, I have lumped together these concerns into three rubrics, which roughly cover the arena of the modern halachic language game, each with its own approach of the triad religion-law-morality: (1) actual legal practice; (2) the juncture between moral and religious philosophy; and (3) legal history and its interface with theological-political thinking. In the remainder of this chapter, we will take a closer look at the dynamic of that language game by revisiting three classic sample texts[10] and looking at their framing of Jewish law in relation to each of these – again, never mutually exclusive – categories. But first an ultra-short note on Wittgenstein's theory of language and meaning, and its political implications, is in place.

## 1    Meaning beyond the Lexicon

He is a man, but not a man. It was phrases like this that convinced Ludwig Wittgenstein (1889–1951) of the shortcomings of logic in determining the meaning of any given utterance. To be a man and not to be a man is a logical impossibility. Within the social setting of human communication, however, the statement is not only possible but can also be effective, because it forces us to rethink precisely how we define 'man'. Language, in other words, is not a closed circuit; words do not have one single denotation but acquire meaning by being used, by being 'woven into actions', as Wittgenstein himself put it. Outside this active setting words are meaningless. It is their context or, more precisely, it is the particular language game that is being played – each with its own grammatical rules – that determines whether a word, or a string of words, makes sense and, if so, what sense it makes. When cited in a logic textbook, the phrase 'law is not law' will be a contradiction. At the beginning of an academic paper, however, it may become the starting point of historical reflection.

It is here that the political implications of the language-game come in. Whether or not we call somebody 'a man' follows from our perception of the person in question, but that perception is shaped by our expectations regarding 'man-hood'. Our definitions thus to a large extent determine what we see and what we see is what we name. Still, there is always room for reconsideration, a

---

10    I take Heymann Steinthal's *Allgemeine Ethik* (Berlin, 1885) and Moritz Lazarus's closely related *Die Ethik des Judentums*, 2 vols (Frankfurt, 1898–1911) as one representative text.

fact which Wittgenstein illustrated by referring to the 'rabbit-duck illusion': an ambiguous figure taken from an 1892 German cartoon and later used in experimental psychology (Figure 10). Either you identify the animal drawing as a rabbit and you name it a rabbit, or you see a duck and you name it a duck. If someone tells you the duck is a rabbit, you will see the rabbit, but it does take some additional mental work. In a similar vein the concept of 'man-hood' is ambiguous, and this ambiguity leaves room not just for perception (seeing a man) but also for interpretation (seeing a person *as* a man, which involves reflection on why we use the word 'man'). It is in this interpretative space that definitions are challenged, perceptions alter, and words can change. As the current shift in discourse on sex and gender shows, it is possible to rebel successfully. While the phrase 'he is a man, but not a man' will always be a logical impossibility, its meaning – being culturally contingent – is transient and language-game dependent.

## Welche Thiere gleichen einander am meisten?

## Kaninchen und Ente.

FIGURE 10    Earliest known illustration of the 'rabbit-duck illusion'
First printed in 1892 in the German magazine *Fliegende Blätter*.
ARTIST UNKNOWN; IMAGE IN PUBLIC DOMAIN, AVAILABLE ON
WIKIMEDIA COMMONS

When Jewish scholars joined German-language scholarship, they faced an unprecedented opportunity to observe and tweak the rules of the game. Reconceptualizing Judaism was an obvious priority, with the word *Gesetzesreligion* as perhaps the ultimate test. Throughout the exercise, the need for cross-cultural translation proved a bonus. New disciplines, media, and styles of prose forced them to adopt a relative outsider perspective to their own tradition, from where they could locate new ambiguities and try to reverse established connotations. As Wittgenstein teaches us, the meaning of words like *Mosaismus* and *Rabbinismus* – Mosaism and Rabbinism – is not a matter of lexicography. It resides in the lapse between perception and interpretation, where the rabbit is the duck, religion equals law, and, with the right argumentation, the two become morality.

## 2    Legal Pluralism and Religious Tolerance

In the early days of emancipation, one domain in which Judaism had to reassert itself was that of the everyday reality of legal practice and its intersections and overlap with other judicial systems. New forms of interaction with other legal systems and corpora prompted Jewish reflection on the individuality and limits of halacha, especially (but, as we shall see, not exclusively) in the field of family law and private law and on the nature and scope of Jewish legal expertise. No less importantly, it provided an impetus for Jewish thought on the possibility of legal pluralism and on halacha's place between church and state, on the hand, and within Jewish life in civil society, on the other.

An early prototype of the genre was the brochure *Ritualgesetze der Juden* (*The Ritual Legislation of the Jews*), completed in 1776 and published in Berlin in 1778.[11] It was the result of a collaboration between chief rabbi Hirschel Lewin (1721–1800), who supplied the impetus and halachic approbation, and Moses Mendelssohn (1729–1786), who had been asked to execute the project because of his German competence and fluent prose. The book, in later editions often ascribed to Mendelssohn only, was a belated answer to a request by the Prussian justice department, issued in 1770, for information on Jewish legal

---

11    Moses Mendelssohn, and Hirschel Lewin, *Ritualgesetze der Juden, betreffend Erbschaften, Vormundschaften, Testamente, und Ehesachen, in so weit sie das Mein und Dein angehen* (Berlin, 1778). On Lewin and his intellectual contacts with Mendelssohn, see the *Biographisches Handbuch der Rabbiner*, vol. 1, *Die Rabbiner der Emanzipationszeit in den deutschen, böhmischen und großpolnischen Ländern, 1781–1871*, ed. Carsten Wilke (Munich, 2004).

procedures, more specifically on rabbinic jurisprudence regarding marriage, tutelage, inheritance, donations, and legacies in as far they touched upon property issues: *in so weit sie das Mein und Dein angehen*, the title page read ('in so far as they concern what is mine and what is yours'). Seven years after its publication, Mendelssohn reported in a letter to his relative Elkan Herz (d. 1816) that the book was now for sale in virtually all Berlin bookshops. Despite its technical, esoteric nature, it apparently enjoyed considerable exposure.[12]

As an early prototype in German, *Ritualgesetze* showed due sensitivity to matters of language, especially in relation to genre and audience. On multiple occasions the authors emphatically positioned themselves outside the living legal tradition. This book is not a code (*Gesetzbuch*), they warned their readers, let alone a manual compiled to streamline the act of dispensing justice. Rather, it was meant as a first introduction to the Jewish legal system, written for the gentile professional who wished to learn about Jewish legal principles (*Urteilsgründe*) rather than how to apply them. In cases of doubt regarding the text and validity of the law in question (*dubium iuris*), they cautioned, a judge should always have access to the sources and spirit of his legal material. Accordingly, those who wished to engage in Jewish jurisdiction should know Hebrew, in order to consult independently the Pentateuch and Talmud (i.e. the agreed core of the legal canon) as well as the rabbinic jurisprudence that provided additional argumentation through individual case law rulings. For this type of active legal engagement, the current book simply offered too slim a base. In fact, the authors confessed, they were not sure if their German rendition always did full justice to the scope and meaning of the technical Hebrew idiom.[13]

Ironically, it was the act of translation that facilitated the language game that allowed Mendelssohn to reposition the Jewish legal apparatus *vis-à-vis* current perceptions of law and society. In his German recapitulation, the process of *psiqah*, or 'doing halacha', was translated as *Rechtsgelehrsamkeit* (jurisprudence), the term *Rechtswissenschaft* (legal science) having yet to become common property. The entanglement of written and oral Torah became straightforward *Gesetz*, while post-Mosaic additions were presented as *Satzungen* (normative regulations), and the *takkanot* – rabbinic repeals of obsolete or impracticable biblical laws – were introduced as *Schonungsgesetze*, as laws of preservation. Together with the nationally accepted *Religionsgebräuche*

---

12   M. Wishnitzer, 'Moses Mendelssohn', *Jewish Quarterly Review* 25 (1935), 307–10, esp. 308f.

13   '[*Aber*] *wie wir aufrichtig gestehen müssen, selbst nicht versichert sind, daß die Worte in der deutschen Sprache genau von derselben Bedeutung und von demselben Umfange sind, als die hebräischen*' (*Ritualgesetze*, xix).

(*minhagim*) and '*übrigen Schriftsteller*' (other authors, notably the medieval codices and early modern *posqim*), they made up the corpus of Jewish *Ritualgesetz*, a container concept that failed to become a household word in subsequent legal speak, despite the book's attested circulation.[14] As Daniel Krochmalnik has pointed out, the choice of the term was by no means coincidental. It had been politically motivated, he writes, to preclude any association of Jewish law with political autonomy, i.e. with the Jewish polity as a sovereign, self-ruling state within a state. As subjects of a strictly *religious* law, the term signalled, the Jews were fully deserving of enlightened *religious* tolerance.[15]

Choosing the right German equivalents for the Hebrew terminology obviously meant highlighting certain properties and, possibly, shifting a priori perceptions in a certain direction. In the eye of the eighteenth-century gentile beholder, the picture that emerged from Mendelssohn's description of Jewish *Ritualgesetz* will have been that of a Hebrew *ius canonicum*, a body of laws and regulations developed and enforced by the ecclesiastical authorities. Like canon law, Mendelssohn's halacha was a combination of divine law and human, positive law: the latter formulated by legal scholars in under constant reference to the *ius divinum*.[16] Like its Catholic counterpart, halacha consisted of a universally accepted central code, supplemented by a continuous and varied commentary tradition.[17] Both traditions included customary law (*ius non scripture*, Hebrew *minhag*) alongside written, expert law, and both were subject to periodization (compare *ius antiquum, novum, novissum* with the division of Jewish legal scholars into *tannaim, amoraim, geonim, rishonim* [early] and *acharonim* [late authorities]). The judicial hierarchy, too, resembled that

---

14     As Gotzmann's studies suggest, nineteenth-century (popular) authors rather employed the term *Zeremonialgesetz*, i.e. ceremonial law; Gotzmann, *Das jüdische Recht*, 25ff., and idem, 'The Dissociation of Religion and Law', 111.

15     Daniel Krochmalnik, 'Mendelssohn's Begriff "Zeremonialgesetz" und der europäische Antizeremonialismus. Eine begriffsgeschichtliche Untersuchung', in *Recht und Sprache in der deutschen Aufklärung*, eds Ulrich Kronauer and Jörn Garber (Tübingen, 2001), 129–60, at 155f.

16     Witness the emphasis on the apostolic nature of the oral Torah, its ordinances as well as its exegetical rules; *Ritualgesetze*, esp. p. iii.

17     Importantly, Mendelssohn's differentiation of legal doubt into *dubium iuris* (see above) and *dubium facti* (doubt of fact) also points at canon law as his source of inspiration. Finally, the injunction to consult the original sources mirrors the importance of *fontes iuris cognoscendi*, to know the material and formal sources of canon law in the Latin tradition. For the relevant terminology, see Charles G. Herbermann et al., eds, *The Catholic Encyclopedia: An International Work of Reference on the Constitution, Doctrine, Discipline, and History of the Catholic Church*, 16 vols (New York, 1907–14), entries 'Canon Law' and 'Doubt'.

of canon law, where old law prevailed over new (no *lex posterior* principle here) and legal commentary would never acquire the status and authority of law. Likewise, in Mendelssohn's summary the combination of (revealed) written and (apostolic) oral Torah was presented as the legally binding core, while the authors of later codices such as the Shulchan Arukh, highly esteemed and holy though they were, lacked actual legislative power.[18]

Despite these final reservations, the *Ritualgesetze* was essentially an introduction to those parts of the Shulchan Arukh – notably the orders *Even ha-'Ezer* (*Stone of Help*) and *Choshen Mishpat* (*Breastplate of Judgement*) – that overlapped with gentile private law.[19] Public law was obviously left out of the discussion. By framing the halacha in question as the Jewish equivalent of canon law, Mendelssohn managed to present it not as the competing legal system of a parallel society but as a *ius vigens*, the complex of ecclesiastical laws in force alongside state legislation. If there was any overlap, as inevitably there must be between non-state and state legislation when it comes to everyday civil matters, that overlap was thus embedded in different zones of competence and rooted in distinct, yet equally solid, legal traditions. In thus stressing the authenticity of the Jewish legal system, the book was more than a simple halacha for dummies. Within the context of the enlightened *Toleranz*-debate it became a plea for legal pluralism and, in its wake, for religious toleration.

To be sure, as a manual intended for professional outsiders the *Ritualgesetze* was a novum, and the format offered by canon law will have created a fruitful patch of common ground between Mendelssohn and his gentile readership. Still, although the choice of a Christian clerical template may have been partly pragmatic, its implications were of a more fundamental nature. If endorsing legal pluralism had been the book's implicit sub-goal, portraying the Jewish polity as a positive, law-based religion alongside Christianity was its implicit, yet by no means accidental, result. Precisely how to define that positive, revealed religion, and how to harmonize it with the demands of enlightened natural theology, was the subject of much of Mendelssohn's oeuvre, first and foremost of his *Jerusalem oder über religiose Macht und Judentum* of 1783 (*Jerusalem, or On Religious Power and Judaism*).[20] For Mendelssohn halacha

---

18     'Da indessen die Verfasser des Schulchan Aruch zwar als Männer von sehr erleuchteten Einsichten und heiligen Sitten in großen Ansehen bey der Nation stehen, aber doch keine gesetzgebende Gewalt haben' (*Ritualgesetze*, x).

19     Ibid., ix–x, xiii.

20     The literature on Mendelssohn's reconceptualization of Judaism is endless. For varying assessments of its embeddedness in Jewish thought, Spinozism and Wolffian-Leibnitzian thinking, see Alexander Altmann, *Die trostvolle Aufklärung. Studien zur Metaphysik und politischen Theorie Moses Mendelssohns* (Stuttgart-Bad Cannstatt, 1982); Allen Arkush,

would always remain a key ingredient, both for the preservation of the eternal truths of rational universalism (Judaism as revealed legislation was, if anything, conveniently praxis-oriented and blissfully undogmatic) and for keeping the Jewish nation together in times of civic integration.[21] On the pages of the practical *Ritualgesetze* there was no room for explaining this curious merger of natural theology and national religion as encapsulated in Jewish law. In subsequent post-Kantian philosophy, however, the obvious tension between the two would prove one of the toughest nuts to crack.

## 3  Lex as ius: Jewish Law as the Principle of Humanism

'*Vernunft, Humanität, Idee, Sittlichkeit – es sind Synonyma*'. Reason, humanism, transcendent ideas, morality: these were, in the words of philosopher and *Völkerspychologe* Heymann Steinthal (1823–1899), the central ingredients of moral philosophy according to Immanuel Kant (1724–1804).[22] In modern ethics, morality encompassed the totality of humankind; it was motivated by human reason and proceeded from the – non-enforceable, private – *Wille zum Guten*, the benevolent resolve to do the right thing by one's neighbour. Moral humanity was, literally, autonomous, obeying only to self-imposed laws that lacked both ulterior motive and hidden purpose.[23] Its moral compass was self-evident, in the sense that it was not validated by a higher order, *in casu* by revealed religion, nor steered by a holy writ that conveyed the will of God. This lack of divine sanction was what distinguished the *Vernunftreligion* (the natural – and

---

    *Moses Mendelssohn and the Enlightenment* (Albany, 1994); David Sorkin, *The Religious Enlightenment: Protestants, Jews, and Catholics from London to Vienna* (Princeton, 2008); Michah Gottlieb, *Faith and Freedom: Moses Mendelssohn's Theological-Political Thought* (Oxford, 2011).

21    For this reconstruction of Mendelssohn's argumentation and its inherent tensions, see, e.g., Robert Erlewine, *Monotheism and Tolerance: Recovering a Religion of Reason* (Bloomington, IN, 2010), esp. ch. 3, 'Mendelssohn and the Repudiation of Divine Tyranny', 43–68. For an analysis of Mendelssohn's stance on halachic observance and ceremonial practice (*Uebung*) as forces of historical preservation, see Elias Sacks, *Mendelssohn's Living Script: Philosophy, Practice, History, Judaism* (Bloomington, IN, 2016).

22    Heymann Steinthal, *Allgemeine Ethik* (Berlin, 1885), 412 (§ 262). For an extensive analysis and contextualization of the work, see Ingrid Belke, 'Steinthals *Allgemeine Ethik*', in *Chajim H. Steinthal. Sprachwissenschaftler und Philosoph im 19. Jahrhundert*, eds Hartwig Wiedebach and Annette Winkelmann, Studies in European Judaism 4 (Leiden, 2002), 189–236.

23    See, e.g., Jerome B. Schneewind, *The Invention of Autonomy: A History of Modern Moral Philosophy* (Cambridge, 1997) part IV 'Autonomy and the Divine Order', 429–530.

thus universal – moral religion) from the various *Offenbarungsreligionen* (the particular belief systems built on historical experience, such as the revelation of a set of – enforceable – laws on Mount Sinai). Being grounded in the empirical rather than in the rational, such revealed laws lacked moral substance in the Kantian sense. Being obeyed not for their own sake but from fear of divine retribution, they also lacked the selflessness of true morality. For Kant, one should add, revealed religions, first and foremost Christianity, had not been completely pointless. As long as humankind continued working towards the bliss of natural religion, he believed, their mutable forms could play a role in the transmission of the abstract, immutable truths of reason by tying them to concrete experience. Their ultimate destination, however, was to collectively dissolve into the one moral religion during the enlightened eschaton.[24]

It takes no genius to see that the stress on moral self-governance put considerable pressure on law-based Judaism as a source of moral thought and action. Its central corpus of divinely revealed legislation, which in the Lewin-Mendelssohn brochure had been the binding core of Jewish canon law, was rendered not just morally meaningless but became an actual obstacle on the road to moral perfection. After Kant, Jewish practical philosophy had to reconcile moral autonomy with Jewish 'heteronomy' ('law from the other side') and reflect on the mediating role of Judaism. Given the primacy of ethics, this invariably meant to drastically redefine (often synonymous with reduce) its legal substance to meet the criteria of modern morality. It should not surprise us that neither God nor Moses, the divine lawgiver and his human legislator, emerged from this exercise quite unscathed. Nor did the law. In this corner of the modern halachic language-game, the interest shifted from law as *lex* to law as *ius*, i.e. from halacha as binding practice to Jewish law as the embodiment of justice and a suitable base for Jewish religion.

Heymann Steinthal devoted a mere three paragraphs of his *Allgemeine Ethik* to religion, in the section in which he outlined the socio-political order in which human moral life takes shape.[25] In a distinctly utopian mode, he pictured society as the place where the ideal intelligible realm was put into practice. For Steinthal, religion was but a minor part of the required equipment, a mere moment – important but fleeting – in the totality of morality, as he put it. All good religions, he wrote, were concretizations of the *Liebe zum Guten*, i.e. of the enthusiasm for goodness, truth, and beauty that inspired morality and

---

24    A lucid, user-friendly summary of the relevant passages in Immanuel Kant's *Die Religion innerhalb der Grenzen der bloßen Vernunft* (1793) and *Der Streit der Facultäten* (1798) is Saskia Wendel, 'Religionsphilosophie nach Kant', *Colloquia Theologica* 2 (2001), 203–14.

25    Steinthal, *Allgemeine Ethik*, 225–29 (§§ 147–149).

expressed itself in empathy and altruism. In that capacity, he conceded, religion was a form of ethical life. And while Judaism and Christianity each served as particular 'personifications' of 'the intelligible realm of humanism', atheism (for Steinthal literally 'religion without a God') cultivated the same humane benevolence without the need for such mediating personification.

As Ingrid Belke suggests, it was in all likelihood the atheist Steinthal who added this final observation.[26] In his discussion of the central concept of benevolence (*Wohlwollen*), however, it was the Jew Steinthal who unexpectedly spoke out. When trying to grasp the essence of this most noble form of altruism, he found well-tried terms such as 'love' and 'sympathy' fell short. In two lengthy *Anmerkungen* he identified the Hebrew word *chesed* (perhaps best translated as 'unconditional affinity') as the archetypal equivalent of German *Wohlwollen*, invoking the biblical injunction to love thy neighbour (Lev 19:18) and the inclusive legislation regarding the stranger and the enemy, together representing the ultimate 'neighbour', by way of illustration.[27] 'Equality before the law for the native and the stranger', he concluded, 'is what Leviticus 24:22 asserts with such emphasis that today's legislators would do well to marvel. Monotheism not only managed to add love to justice; it also apprehended justice much more thoroughly than polytheism ever did'.[28] Or German legislature, for that matter. Steinthal's utopian *Ethik* had been triggered by the loss of his two children, by the epistemological shock of Darwinism, and by recent academic antisemitism and its disqualification of Jewish citizenship.[29] It is the latter challenge that echoes through his short paean of Torah's strict egalitarianism, which Steinthal – far from accidentally – reduced to the Golden Rule of moral cosmopolitanism. It was a rare partisan moment in what was otherwise a testimony to pure practical reason.

The serial, partly posthumous publication *Die Ethik des Judentums* by Steinthal's fellow *Völkerpsychologe* and former brother-in-law, the Reform

---

26 Belke, 'Steinthals *Allgemeine Ethik*', 207.
27 Steinthal, *Allgemeine Ethik*, 122–24 (§§ 82–83).
28 Ibid., 123.
29 Belke, 'Steinthals *Allgemeine Ethik*', 225. For Jewish responses to Darwinism, see Daniel Langton, *Reform Judaism and Darwin: How Engaging with Evolutionary Theory Shaped American Jewish Religion*, Studia Judaica 111 (Berlin, 2019), who mentions Steinthal in passing (p. 76) when discussing US rabbi Emil Hirsch, the son of German Reform thinker Samuel Hirsch. For the legal background of contested citizenship in Germany, see Dieter Gosewinkel, 'Citizenship in Germany and France at the Turn of the Twentieth Century: Some New Observations on an Old Comparison', in *Citizenship and National Identity in Twentieth-Century Germany*, eds Geoff Eley and Jan Palmowski (Stanford, 2008), 27–39.

rabbi Moritz Lazarus (1824–1903), can be read as a Jewish supplement to the *Allgemeine Ethik*.[30] Lazarus' choice to approach the universal via the particular was a brave one, and it involved hard work and a heady dose of rhetoric, sometimes bordering on sophistry. Where Steinthal had glossed over Jewish law and reduced religion to a simple personification of moral principle, Lazarus knew that in adding Judaism to the equation he had to thematize the combination. In doing so he would have to walk, in the words of Alan Mittleman, 'a narrow line between Judaism as a form of autonomous moral conscience and Judaism as a heteronymous [*sic*] religious system'.[31] Walking that line meant addressing a few obvious stumbling blocks, including the law, the people of Israel who were its primary addressees, and the respective roles of God and Moses, each of which were subjected to a thorough revision. As Mittleman writes, Lazarus's strategy of merging old Jewish texts with modern insights did not make for a coherent philosophical system.[32] Yet if his method of preserving the traditional idiom and charging it with new meaning was flawed, it did produce a new perception of Judaism as a robust moral religion that could be experienced collectively, in the presence of God and the law, and with open arms towards the rest of ethical humanity.

The 'Jewish essence' of Lazarus's argumentation can be recapped in a short but far from simple formula: *Heiligung ist Versittlichung ist Gesetzlichkeit*, i.e. the hallowing of life equals its ethical realization, which in turn implies compliance with the law.[33] It is in this (in modern terms problematic) equation that the gap between Jew and Judaism – i.e. between private moral conscience and public moral involvement – was bridged. In Lazarus's system, the collective public dimension was crucial. As he reminded us elsewhere, all commandments were given in the second person singular, but the *mitzvah* to be holy – issued five times in the book of Leviticus – was always cast in the second

30    Moritz Lazarus, *Die Ethik des Judentums*, 2 vols (Frankfurt, 1898–1911). For a technical introduction, see David Baumgardt, 'The Ethics of Lazarus and Steinthal', *Leo Baeck Institute Year Book* 2 (1957), 205–17. For its place within Lazarus's life and thought, see Hanoch Ben-Pazi, 'Moritz Lazarus and the Ethics of Judaism', *Daat: A Journal of Jewish Philosophy and Kabbalah* 88, Special Issue: 'Wissenschaft des Judentums: Judaism and the Science of Judaism. 200 Years of Academic Thought on Religion' (2019), 91–104. For an excellent summary of its central ideas and argumentation, see Alan L. Mittleman, *A Short History of Jewish Ethics: Conduct and Character in the Context of Covenant* (Chichester, 2012), 181–84. On Steinthal and Lazarus, cf. also the chapter by Carlotta Santini in this volume.

31    Mittleman, *A Short History of Jewish Ethics*, 182.

32    Ibid; Baumgardt, 'The Ethics', uses the word 'eclectic'.

33    Lazarus, *Ethik* vol. 2 (1904), ch. 5 ('Versittlichung is Gesetzlichkeit'), 220–240 (§§ 203–209).

person plural, a sure sign of its being a communal assignment.[34] It was the law that facilitated this shared mission to try to be a holy nation, Lazarus wrote, by absorbing the individual into an overarching totality (*Gesammtheit*). And it was in this moral-legal *Gesammtgeist* – the Jewish equivalent of Steinthal's more neutral *Wir* (We) – that *die wahre Heiligung des Lebens* took place, the true hallowing of life.

At this point, various elements in Lazarus's argumentation need qualification. First of all, there is his notion of enforceable *Gesetz* and its relation to non-enforceable morality. Jewish law, Lazarus argued, should not be seen as a source of mandatory moral teaching. It was a goal in itself, in the sense that its only purpose was to instill *Gesetzlichkeit*, i.e. a mentality of compliance with the law. Through that single purpose it became a source of freedom (*Quelle der Freiheit*) rather than control.[35] Being obliged by law in all aspects of life, he explained, the Jews were prevented from ever acting out of base self-interest.[36] As an agent of gallant selflessness, their *gesetzliches Handeln* (legal dealings) thus closely resembled moral autonomy. In a paragraph devoted to the head-on confrontation between the apparent opposites *Gesetzlichkeit und Autonomie* ('Legality and Autonomy') Lazarus found their compatibility confirmed by the longstanding rabbinic preference for obedience over voluntarism. Among the Sages, he wrote, the *metzuvveh ve-'oseh* (the one who is commanded and performs the deed for no other reason than having been ordered to do so) ranked higher than the one who performed it voluntarily (read: out of hope for benefit or reward). Legal obedience as the moral superior of (potentially unreliable) free will: what at first sight must have looked like a rabbinic paradox turned out to be a higher form of truth. It was in the shared selflessness of Jew and law, of the commanded and the commandment, that law begot freedom and compliance acted like autonomous will. Lazarus had achieved the near-impossible: in his summary of Jewish ethics there was no longer any either/or. He had created

---

34 Ibid., 312.

35 In a rare reference to revelation, Lazarus stressed that Israel had been delivered not on the shores of the Red Sea, but 'only at Sinai, and with Israel the totality of humanity' (ibid., 25).

36 In ch. 6, on 'Naturgesetz und Sittengesetz', Lazarus brought in his psychological expertise when speaking of *das vom Naturtrieb erlösende Gesetz*, the law that helps us channel our natural urge (§ 239). Judaism's ritual laws (*Zeremonialgesetze*) in particular fulfilled the mediating, didactic role which Kant had allotted to all revealed religions. By hallowing daily routine, they reminded the Jews of their dual state, i.e. of their being part of this world *and* of the realm of transcendent ideas; Mittleman, *A Short History of Jewish Ethics*, 184.

a linguistic territory where religion equalled law equalled morality: where the rabbit, one might say, *was* the duck.[37]

Although he had succeeded in shifting the perception of Jewish law as ethics, Lazarus conceded that those ethics came 'in a completely different format than Kant's absolute morality'.[38] One such difference was the presence of the divine. In Kant's practical reason, the primacy of the autonomous individual had evolved out of the failure to prove God's existence in metaphysical terms. In Lazarus's adaptation, by contrast, that divine existence was an indisputable, if awkward, axiom. In a book on the ethics of Jewish religion there was no way of avoiding the 'ultimacy language' of religious philosophy.[39] In the paragraph in which he defended his Jewish understanding of moral autonomy, he therefore reintroduced God as a generic framework, a décor almost, for Jewish ethical behaviour. God, he explained, was the archetype (not the source, mind you) of holiness and morality, created humankind in his likeness, and had encoded them to imitate his holiness.[40] Simultaneously, as an ethical being, each individual was his own creator (*Selbstschöpfer*), spurred by his own will and tapping into his own natural resources. Yet for all his independence, the *Grund*, i.e. the efficient cause for his moral actions, resided elsewhere, in the opaque ultimacy of the divine creator in whose image humankind had been created.[41]

In the process of moral auto-creation, the letter of the law as received by Moses on Sinai was of course irrelevant, a conviction Lazarus shared with countless Reform thinkers before him. What mattered, if one chose to uphold the law in the functional mode of *Gesetzlichkeit*, was its spirit, which scripture itself had managed to condense into a single legal principle. Unlike Steinthal, Lazarus did not refer to the universal Golden Rule but instead identified the prophet Habakkuk's adage that 'the righteous shall live by his faith' as the

---

37    Lazarus, *Ethik* vol. 2, 228 (§ 208).
38    Ibid., 102.
39    For the expression, see Robert Neville, 'Philosophy of Religion and the Big Questions', *Palgrave Communications* 4/126 (2018), https://doi.org/10.1057/s41599-018-0182-9 (accessed 14 February 2024).
40    In Lazarus's representation, God's holiness was ontological, i.e. part of his divine essence. Israel's pursued holiness, by contrast, was relational. Through the commandment to be holy, they could aspire to a holiness that was greater than, and thus *relative to*, their own original state and that of other nations. For the distinction, see Alan L. Mittleman, 'Introduction: Holiness and Jewish Thought', in *Holiness in Jewish Thought*, ed. idem (Oxford, 2018), 4.
41    Lazarus, *Ethik* vol. 2, 103–111 (§ 101).

one dominant norm behind the 613 commandments.[42] *Tzaddik be-munato yichyeh*: justice through faith and, again, Jewish law as *ius* rather than *lex*. For Lazarus too, biblical righteousness, not halachic tradition, was the key to the marriage of religion, law, and morality.

With the law thus transformed into a moral instrument and God into a holy exemplum, what remained was to redefine the role of his earthly legislator. In a passage devoted to the importance, or rather to the irrelevance, of genealogy for Jewish morality, Lazarus introduced Moses as little more than a dialectical moment in a much broader argumentation.[43] The Sages, he argued, taught that spiritual and moral achievement were unrelated to lineage and should be admired over birth, descent (*Stamm*), and race (*Blut*). The highest moral post to be held, by gentile and Israelite alike, was prophethood, with Moses as the undisputed paragon of prophets. Therefore, if scripture says that no prophet arose *in Israel* after Moses (Deut 34:10, emphasis mine), this must necessarily mean that gentile Balaam, known only for his ambiguous appearance in Num 22–24, should be identified as his rightful heir and successor. Following this surprising argument, Lazarus continued with an equally programmatic paragraph on proselytes as the spiritual heroes of Judaism who, by voluntarily embracing Jewish *Gesetzlichkeit*, became the pinnacle of Jewish ethics – moral autonomy squared, one might say.[44] This generous stance on conversion was but one expression of Judaism's inherent humanism, he concluded. The other was its natural inclination towards universalism, which had set it apart from the other nations to such a degree that Israel had been driven into particularism to protect its early humanist creed.[45]

Heteronomy as autonomy, Balaam as Joshua, particularism as the protective husk of universalism – at first sight there is every reason, with Baumgardt and Mittleman, to charge Lazarus with dilettante eclecticism. In his defense,

---

42   Ibid., 105–08. Lazarus's argumentation follows the gist of bMakkot24a, where R. Simlai reduced the 613 Mosaic commandments first to eleven (Psalm 15), then to six (Isaiah 33:15), three (Micah 6:8), two (Isaiah 56:1) and finally one legal principle (Habakuk 2:4).

43   Ibid., 158–59 (§ 155), based on a tendentious reading of *Ba-Midbar Rabba*, ch. 14. On the dialectics of universalism in early rabbinic exegesis, see Marc Hirschfeld, 'Rabbinic Universalism in the Second and Third Centuries', *Harvard Theological Review* 93 (2000), 101–15.

44   Lazarus, *Ethik* vol. 2, 159–61 (§ 156). NB: in 1895, Lazarus had entered a second marriage with the convert Nahida Remy. For a portrait, with a focus on philo-semitism and feminism, see Alan T. Levenson, 'An Adventure in Otherness: Nahida Remy-Ruth Lazarus (1849–1928)', in *Gender and Judaism: The Transformation of Tradition*, ed. Tamar M. Rudavsky (New York, 1995), 99–111.

45   '[U]m diesen Vorzug zu pflegen ... musste Israel sich absondern' (Lazarus, *Ethik* vol. 2, 164 [§ 159]).

however, it could be said that he was trying to serve (too) many masters at once. In ranking religiosity above halachic practice, he joined a long line of Reform thinkers who had attacked Mendelssohn's idea of Judaism as revealed legislation and had tried to replace it with a more dynamic Jewish *Geist*.[46] In choosing to define that spirit through a mixture of rabbinic idiom and modern philosophical ideas, he appealed to an educated class whose knowledge of Judaism was waning, as was its appetite for a socially demanding *Gesetzesreligion*. In explaining Jewish particularism as a retreat on behalf of universalism, he endorsed the spirit that had enlightened his lecture 'Was heißt und zu welchem Ende studirt man jüdische Geschichte und Litteratur' ('What Is Jewish History and Literature, and Why Does One Study It'), deliberately named after Schiller's *Antrittsvorlesung* of 1789.[47] And finally, in stressing the importance of the Jewish collective but dismissing the tribal and the racial, he reissued the call for ethnic pluralism (*Mannigfaltigkeit*) that had echoed through his address 'Was heißt national?' ('What is the National?'), delivered and published in 1880 at the height of the Berlin *Antisemitismusstreit*, or Antisemitism Controversy.[48]

Much had changed in German politics since 1778, when Mendelssohn had promoted straightforward legal pluralism in order to excite straightforward religious tolerance. Writing in a different climate, Moritz Lazarus knew he had to play a more subtle, less exclusively legal card. So he traded the Mosaic authority of canon law for the prophetic principle of righteousness, proposed rabbinic *Gesetzlichkeit* as a tool for private moral conscience, and succeeded in upgrading law-based Judaism into an open, egalitarian humanistic enterprise. If ever a circle had been squared, it was on the 400-plus pages of Lazarus's theoretically flawed, perhaps, but at the same time bracingly creative reconstruction of the *Ethics of Judaism*.

---

46    Daniel Schwartz traces the origins of this critique back to Saul Asher (1767–1822), whose 1792 *Leviathan* renounced Mendelssohnian legalism and located the core of Judaism in fourteen (sic) principles of faith; Daniel B. Schwartz, *The First Modern Jew: Spinoza and the History of an Image* (Princeton, 2012), 60.

47    Published as Moritz Lazarus, *Was heißt und zu welchem Ende studirt man jüdische Geschichte und Litteratur*, Populär-wissenschaftliche Vorträge über Juden und Judentum 1 (Leipzig, 1900).

48    The address was reprinted in Lazarus's collected essays, *Treu und Frei. Gesammelte Reden und Vorträge über Juden und Judentum* (Leipzig, 1887), 53–113. I have given a short characterization in 'Nation and Translation: Steinschneider's *Hebräische Übersetzungen* and the End of Jewish Cultural Nationalism', in *Latin-into-Hebrew*, vol. 1, *Studies*, eds Resianne Fontaine and Gad Freudenthal, Studies in Jewish History and Culture 39/1 (Leiden, 2013), 421–45, esp. 432–34.

## 4 Law as Right. Legal History and the Perfect Synthesis

Though ultimately sharing the objective of reconciling Jewish nomism with moral cosmopolitanism, Mendelssohn and Lazarus found themselves at opposite ends of the scale in their evaluation of the law. Mendelssohn ended up with a Jewish version of canon law, Lazarus with a liberating legal principle, each in its own way compatible with the tenets of natural religion. Needless to say, there were acres of middle ground between them, in the heart of which we find Rabbi Zacharias Frankel (1801–1875), architect of Conservative Judaism and founder of its first theological seminary in Breslau (Figure 11). Navigating between the extremes of Reform and Orthodoxy Frankel was, as Andreas Brämer phrased it, a man of the *juste milieu*.[49] In Frankel's case middle ground did not, however, mean compromise. On the contrary, it became a centre space where all possible perspectives converged and where Jewish law, not as *Gesetz* but as *Recht*, was transformed into an academic subject in its own right. As we shall see, Frankel's Jewish *Rechtswissenschaft* revealed a legal corpus that could stand the comparison with model Roman law, enlightened *Vernunftgesetz*, *and* superior revealed religion, all at once. A whole set of circles to be squared indeed, which Frankel managed to do with the help of his newly devised positive-historical methodology.

As indicated above, the early *Wissenschaft des Judentums* had come close to excluding the study of the halachic archive from its agenda. In *Etwas über die rabbinische Literatur*, published in 1818, Leopold Zunz had built his new research programme around a universal knowledge order that consisted of three clearly demarcated domains: the realm of (human, social) culture; the realm of (divinely created) nature, which included the material worlds of commerce, technology, and the arts; and finally, the transcendent, immaterial realm of the divine itself. In the latter section he placed, besides such disciplines as theology, dogmatics, and liturgy, the categories of legislation and ethics. In grouping them under the heading of the divine sciences, Zunz effectively eliminated law and morality from the definition and study of Jewish national culture that was at the heart of his academic programme. *Wissenschaft des Judentums* was to be all about '*Literatur und Bürgerleben* (literature and civil life)', and philology and statistics were perfectly adequate for tackling the combination.[50]

---

49    Andreas Brämer, 'The Dilemmas of Moderate Reform: Some Reflections on the Development of Conservative Judaism in Germany 1840–1880', *Jewish Studies Quarterly* 10 (2003), 73–87, at 75, 84. See idem, *Rabbiner Zacharias Frankel,* for an exhaustive intellectual biography.

50    Thus Immanuel Wolf, 'Über den Begriff einer Wissenschaft des Judenthums', *Zeitschrift für die Wissenschaft des Judentums* 1 (1822), 1–24, 23.

FIGURE 11    Photograph of the Jüdisch-theologisches Seminar zu Breslau
(Fraenckel'sche Stiftung). Originally published in Marcus Brann,
*Geschichte des jüdisch-theologischen Seminars (Fraenckel'sche Stiftung)
in Breslau, Festschrift zum fünfzigjährigen Jubiläum der Anstalt* (Breslau:
Schatzky, 1904), p. 74.
CREATOR UNKNOWN; IMAGE NOW IN THE PUBLIC DOMAIN AND
AVAILABLE ON ARCHIVE.ORG.

In reaction to this omission Zacharias Frankel founded the so-called
historical-positivist school, which advocated the integration of critical and
theological methods and introduced oral law and halacha as viable topics of

academic research.[51] In response to Reform modernization, he adopted the term 'positive' as a 'defiant reassertion of Judaism's fundamental legal character'. The adjective 'historical', in its turn, was added in order to stress the importance of 'the past as a source of values, inspiration, and commitment', i.e. of living, human history, as the second defining aspect of Jewish religious life.[52]

Scattered over many publications, Frankel's treatment of Jewish law was rich and erudite, and much can be, and has been, said about it – especially in relation to the religious breakup of modern Judaism.[53] In the final section of this chapter, I will briefly look at one text to explore what happens when a rabbinical scholar applies the apparatus of German secular *Rechtswissenschaft* to Jewish religious law. As several scholars have pointed out, the Romantic *Historische Rechtsschule*, or Historical School of Jurisprudence, founded by Berlin professor Friedrich Carl von Savigny (1779–1861), was an obvious source of inspiration.[54] Its central notion of law as an organic part of national life, its attention to historical evolution, and its stress on the importance of early 'folk' sources for current legislation were useful strategies in the battle against the Reform's rejection of oral law as bad exegesis and Orthodoxy's insistence on Sinaitic origins.[55] But they did not suffice to solve the other issues that had nagged Jewish scholars from Mendelssohn to Lazarus, especially the tension between divine and natural law and morality and between Jewish legislature, statehood, and citizenship. It is the conceptual-linguistic framing of these two topics that we will briefly look into here. For a better appreciation, it is worth sketching the contemporary German context first.

---

51 See, e.g., Zacharias Frankel's address 'Das Talmudstudium', *Monatsschrift für Geschichte und Wissenschaft des Judentums* 18 (1869), 347–57, esp. 355–56. In addition to Bräme, *Rabbiner Zacharias Frankel*, see Ismar Schorsch, 'Zacharias Frankel and the European Origins of Conservative Judaism', *Judaism* 30 (1981), 344–54; Roland Goetschel, 'Aux origines de la modernité juive: Zacharias Frankel (1801–1875) et l'école historico-critique', *Pardès* 19–20 (1994), 107–32; and Andreas Brämer, 'Jüdische "Glaubenswissenschaft" – Zacharias Frankels rechtshistorische Forschung als Herausforderung der Orthodoxie', in *Die Wissenschaft des Judentums. Eine Bestandsaufnahme*, eds Thomas Meyer and Andreas Kilcher (Paderborn, 2015), 79–94.

52 Schorsch, 'Zacharias Frankel', 346–47.

53 See the literature mentioned in fn. 2, 49 and 51.

54 Most recently Theodor Dunkelgrün, 'The Philology of Judaism: Zacharias Frankel, the Septuagint, and the Jewish Study of Ancient Greek in the Nineteenth Century', in *Classical Philology and Theology: Entanglement, Disavowal, and the Godlike* Scholar, eds Catherine Conybeare and Simon Goldhill (Cambridge, 2020), 63–85, 75–76, and the publications listed in fn. 53.

55 For an engaging introduction to Von Savigny's work, see Susan Gaylord Gayle, 'A Very German Legal Science: Savigny and the Historical School', *Stanford Journal of International Law* 18 (1982), 123–46.

German legal thinking after 1814 had been dominated by the
*Kodifikationsstreit*, a debate on the nature and future of German civil law
between Anthon Friedrich Justus Thibaut (1772–1840) and Friedrich Carl von
Savigny.[56] In this controversy, Thibaut stressed the need for one general civil
law for Germany, as the title of his 1814 publication indicates.[57] Against this
French-style national codification, von Savigny pleaded for a deferral and for a
German legislation based on the careful reconstruction of its historical foundations. On one level it was a debate about top-down statutory law imposed by
a legislative body versus 'bottom-up' customary law sanctioned by a common
consciousness or *Volksgeist*. Simultaneously, it was a confrontation between
the idea of natural law (*Vernunftrecht* as grounded in the permanent, universal
natural order) and positive law, human-made, organically grown and therefore
historically variable. Proponents of the former found themselves united in the
philosophical school, the 'armchair version' of legal scholarship,[58] where the-
orists postulated rational axioms and built a system *more geometrico*, i.e. with
the help of logical deduction. Their opponents formed the historical school
that concentrated on tracking down historical sources and extrapolating their
legal principles. Always in the background loomed Roman law, no longer normative but converted into a methodological prism for all who engaged in contemporary *Rechtswissenschaft*.[59]

Echoes from these developments trickled into Frankel's analysis of Jewish
law, with one important proviso. For all their differences, the German scholars were united in their secular approach to the law. Frankel, by contrast,
examined the Jewish legal corpus through the combined lenses of history and
theology,[60] two conflicting disciplines with incompatible truth claims. An
example of how he managed to reconcile those claims can be found in the
section on *mosaisch-talmudisches Recht* ('Mosaic-Talmudic Law') that served

---

56   See the entry 'Kodifikationsstreit' by Joachim Rückert in *Handwörterbuch zur deutschen
     Rechtsgeschichte*, 2nd ed., 16th instalment (2012), https://www.hrgdigital.de/HRG.kodi
     fikationsstreit (accessed 14 February 2024).

57   Anthon Friedrich Justus Thibaut, *Ueber die Nothwendigkeit eines allgemeinen bürger-
     lichen Rechts für Deutschland* (Heidelberg, 1814). In response Savigny wrote his famous
     *Vom Beruf unserer Zeit für Gesetzgebung und Rechtswissenschaft* (Heidelberg, 1814). On
     its reception and influence, see Benjamin Lahusen, *Alles Recht geht vom Volksgeist aus.
     Friedrich Carl von Savigny und die moderne Rechtswissenschaft* (Berlin, 2012).

58   Gayle, 'A Very German Legal Science', 125–27.

59   Von Savigny extolled paradigmatic Roman law in his 1814 *Vom Beruf* (*passim*) and devel-
     oped the paradigm in his *System des heutigen Römischen Rechts* (Berlin, 1840); Gayle, 'A
     Very German Legal Science', 133–34.

60   Witness the title of Zacharias Frankel, *Die Eidesleistung der Juden in theologischer und
     historischer Beziehung* (Dresden, 1840), on the Jewish taking of the oath before a court.

as an introduction to his 1846 study of Jewish procedural law. The book, which dealt with the weighing of legal evidence, was written as a proof of competence in Jewish *Rechtswissenschaft* and as an attempt to put Jewish legal expertise on a par with its German equivalent. Written towards the end of emancipation, it was a plea for full equality before the law.[61] As a work of scholarship, it was an explicit emulation of the work of John Selden (1584–1654), Johann David Michaelis (1717–1791), and Moses Mendelssohn, whose *Ritualgesetze* Frankel dismissed as inadequate and makeshift.[62] Its bottom-line argument was that in pairing principle to equity, and strictness to moderation, Jewish law was the perfect synthesis of Roman and German law and, as such, worthy of academic attention and legal esteem.[63] The question remained, of course, how to blend God into the mixture.

Frankel began his exposé by saying the Mosaic and Talmudic legal systems had become so entwined (compare his persistent, deliberate use of the composite '*mosaisch-talmudisch*') that it was hard to define the idea of *rein mosaisch* ('purely Mosaic'). As we have seen, in Mendelssohn's 'makeshift' resumé the difference between the two had been that of revealed versus apostolic tradition, together constituting one *ius divinum*. In Frankel's system the hierarchy was reversed: the folkish, historical positive law of the Sages became the prism through which the entire law was perceived, in simultaneous refutation of Geiger's accusations of artificiality and of Hirsch's claim to revealed origins.[64]

---

61    In this aspiration the book continued the political aims of *Die Eidesleistung der Juden* of 1840, which had been part of a current debate on the humiliating tradition of the oath *more judaico*. For a survey of contemporary Jewish publications on the topic, see David Philipson, 'The Rabbinical Conferences, 1844–6', *Jewish Quarterly Review* 17 (1905), 656–89, 674 fn. 2.

62    '[N]ur ein zu unzulänglicher Nothbedarf' (Frankel, *Der gerichtliche Beweis nach mosaisch-talmudischem Rechte. Ein Beitrag zur Kenntnis des mosaisch-talmudischen Criminal-und Civilrechts. Nebst eine Untersuchung über die Preussische Gesetzgebung hinsichtlich des Zeugnisses der Juden* [Berlin, 1846], v [Vorwort]. A more exhaustive, and critical, list of previous studies ended up in a footnote on p. 112). Further to Michaelis, see the essays by Ofry Ilany, Carlotta Santini, and Michael Ledger-Lomas in this volume.

63    '[A]ber auch in ihnen schon gewahrt man eine eigenthümliche Verschmelzung des Characters des römischen und des deutschen Rechts: die mit strenger Consequenz durchgeführte Folgerung aus einem Grundsatz, und die die Lebensverhältnisse würdigende Billigkeit, die vermittelnd zwischen das strenge Recht und die Forderungen der Menschlichkeit tritt' (ibid., v-vi, quoting from Carl Joseph Anton Mittermaier's *Grundsätze des gemeinen deutschen Privatrechts*, 3rd ed., 2 vols [Landshut, 1827]). In 1834, Mittermaier had written an influential study on evidence and the examination of witnesses in criminal procedural law. For Frankel's contact with Mittermaier, see Andreas Brämer, '"Wissenschaft des Judentums" und "Historische Rechtsschule": Zwei Briefe Zacharias Frankels an Carl Josef Anton Mittermaier', *Aschkenas. Zeitschrift für Geschichte und Kultur der Juden* 7 (1997), 173–79.

64    See esp. Harris, *How Do We Know This*, chs. 6–8.

In legal philosophical terms, the search for the 'purely Mosaic' amounted to a reconstruction of the earliest principles on which that incremental legal system had been built. Given the trends in contemporary discourse, the revealed nature of those principles needed vindication, a task which Frankel readily took to hand.

One way or the other, he stated, the Mosaic constitution was divine legislation or, at least, announced itself as such. Through its many details, the will of God penetrated every part of the human – public as well as private – realm. This had led people to conclude that Mosaic law was not the product of internal reason but had been externally motivated, and hence did not qualify as *Vernunftgesetz*. Upon closer inspection, however, the divine will as laid down in Mosaism appeared to operate in a much more subtle, far less absolutist manner. Rather than demanding legal submission, its aim was to alert humanity to its innate ideas of truth and justice and bring these ideas from potentiality to actuality. Revealed yes, dictated no, in other words. If we are to believe Frankel, the realization of Mosaic justice was a process of gentle anamnesis, of, literally, re-cognizing true ideas and stirring them into consciousness. What at first sight may have looked like external motivation was thus in fact double immanence. On the one hand God represented, or better, *was* the *höchste Idealität des Rechts* ('highest ideality of law'). Simultaneously, this supreme Idea of Right was the innate substrate of all worldly affairs and social relations in the Mosaic republic.[65] It should be added here that, in *Der gerichtliche Beweis* (*The Judicial Evidence*), Frankel never explicitly identified Mosaic law with natural law. We can nevertheless deduce this identification, inter alia, from his use of the adjective '*natürlich*', natural, for example in his assertation that Mosaic law introduces itself as divine legislation but 'adheres to natural ground' when it comes to law and justice.[66]

As this summary shows, Frankel's identification of *mosaisches Recht* with *Vernunftgesetz* owed more to the idiom of Hegelian Idealism than to Romantic legal historicism.[67] Following the gist of G.W.F. Hegel's (1770–1831) *Phenomenology*, Frankel situated the highest idea (c.q. the objective *Geist*) of *Recht* in an abstract outer world, from where it emanated into the subjective, this-worldly *Geist* of human society. There it found its particular, concrete

---

65    Frankel, *Der gerichtliche Beweis nach mosaisch-talmudischem Rechte*, 8–10.

66    '*Sie kündigt sich, wie oben bemerkt wurde, als eine göttliche an, und wenn sie auch hinsichtlich des Rechts auf natürlichem Boden bleibt*' (ibid., 52).

67    For the early *Wissenschaft des Judentum*'s encounter with Hegelian philosophy, see Sven Erik Rose, *Jewish Philosophical Politics in Germany, 1789–1848* (Waltham, MA, 2014), ch. 3, 'Locating Themselves in History: Hegel in Key Texts of the Verein', 90–145, esp. 70–71.

expression in the *principia* of Mosaic law. In Frankel's summary, God appears – again, never explicitly but between the lines – not as the vengeful God of the Hebrews but in the form of Hegel's *absoluter Geist*, or Absolute Spirit. In this absolute, highest manifestation of the spirit, natural law did not just coincide with the divine will, but in fact constituted its very being.[68] For Hegel, religion, alongside art and philosophy, had been one way of accessing the absolute *Geist* and bringing it to actualization. With the Mosaic God equaling (natural) law, Frankel saw the study and practice of that law as the highest, most effective form of Jewish religion, ranking it higher than the 'so-called purely religious deed', i.e. than the rites and rituals of cultic worship.[69] By thus putting the commitment to law and justice at the heart of the Jewish religion, Frankel managed to elevate the idea of Judaism as an ancient *Gesetzesreligion* (a word which, I should add, does not occur in the pages of *Der gerichtliche Beweis*) to a whole new philosophical level.

If Mosaic law had its ultimate origins in a God who was the absolute *Geist*, its legal content and procedures were very much of this world. Likewise, although this treasure was best accessed via the Jewish religion, its norm for what was good and right did not derive from religious values but remained within the confines of the legal. Religion, in other words, was the instrument – not the substance – of Mosaism. 'Its laws', Frankel summarized, 'are thus grounded in legal principle itself, in the ethics (*Sittlichkeit*) and free esteem for the True and the Good, they dwell on legal ground, deriving their legitimacy from the Idea of Right, without ever crossing over into religious territory'.[70] Mosaic law, in short, was rooted in morality; its practice constituted the core business of Jewish religion. Like Lazarus, we may conclude, Frankel managed to reconcile the three. Only this time it was not the law but religion that was forced into an instrumental role.

*Free esteem* for the True and the Good: almost by way of an afterthought did Frankel make room for the modern moral subject within the collectivity of the ancient Mosaic polity. Again following Hegel, he claimed that the only type of government suitable for this combination of civic legislation and morality had been the republic, with God as its invisible ruler (*das unsichtbare Oberhaupt*)

---

68 '[*I*]*hm inhärirend und sein eigenes Wesen bildend*' (Frankel, *Der gerichtliche Beweis nach mosaisch-talmudischem Rechte*, 10).

69 Ibid., 9.

70 '*Die Rechtsgesetze sind also im Rechtsprincipe selbst, in der Sittlichkeit und der freien Achtung vor dem Wahren und Guten begründet, sie verbleiben auf dem Rechtsboden, leiten ohne auf das Religionsgebiet hinüberzuschlagen ihre Legitimität aus der Idee des Rechts ab*' (ibid., 10).

from whose divine being all right proceeded.[71] Michaelis, among others, had been wrong to mistake this kind of republic for a theocracy, Frankel argued.[72] More than anything, Mosaic law had been anthropocentric, tailored to the here and now, helping humanity to move through this world, not prepare for the next. Christian doctrine, by comparison, was tuned towards the afterlife and closed its eyes to earthly realities. Accordingly, it advocated passive virtue, not active jurisdiction (*nicht Rechtsgesetze, sondern Tugendvorschriften*) and told its followers to turn the other cheek (read: to surrender justice, in anticipation of better times to come). In political terms, this was a choice which, in Frankel's view, totally disqualified church law as modern state legislation. What with Christianity putting itself above state and legal principle, he concluded, a Christian state was a contradiction in terms.[73] In the Mosaic republic, however, virtue and right, i.e. moral conduct and law, had been balanced in perfect harmony, ensuring a superior form of justice in which stringent right would always be mitigated by ethical considerations.[74]

The central principle upon which this superior form of justice was built was that of retributive justice (*die Wiedervergeltungstheorie*), a theory which, Frankel was happy to announce, had often been misunderstood but had recently been reestablished.[75] In Mosaic law, he explained, crime was considered an injury both to the person who was injured and to the principle of justice that had been violated. Therefore, the dual goal of punishment was to compensate the individual damage as well as restore the collective respect for the law, which in turn would strengthen society's disinclination from evil.[76] By

---

71    Ibid., 10–11. Frankel's argumentation seems to mirror, mutatis mutandis, Hegel's position that in the modern state representative legislative bodies were to formulate the law, which the constitutional monarch should then ratify by issuing them as his 'will': in G.W.F. Hegel, *Elements of the Philosophy of Right: A Critical Guide*, ed. Allen W. Wood, trans. H.B. Nisbet (Cambridge, 1991), 275–380 (the section on 'the state'). For a characterization of Hegel's republicanism in relation to civic virtue and the autonomous personality, see Andrew Buchwalter, 'Hegel, Modernity, and Civic Republicanism', *Public Affairs Quarterly* 7 (1993), 1–12.
72    In the footnote on page 12–13, Frankel referred to Michaelis, *Mosaisches Recht*, vol. 1, § 35.
73    Ibid., 13–14.
74    Ibid., 15–16.
75    Ibid., 23–24. In a footnote Frankel referred to the relevant paragraphs (§§ 96–104) of Hegel's *Grundlinien der Philosophie des Rechts* (1821) as the basis of his interpretation.
76    Frankel explicitly related this disinclination to society's religious sense: '*So lange also der religiöse Sinn im Volke lebt und die Achtung vor diesen geoffenbarten Vorschriften aufrecht erhält, wird das Verbrechen geflohen: die Religion lehrt es verabscheuen, und was der innere Rechtssinn nicht vermag, wird an der Hand des als göttlich und heilig geachteten Gesetzes erlangt werden*' (ibid., 53).

and large, Frankel's summary of Jewish legal principle followed Hegel's theory of retribution-as-annulment, in which punishment was believed to somehow cancel the crime and, in doing so, to serve as an affirmation of right. Hegel being Hegel, modern commentators have struggled to understand the precise nature of his 'opaque' theory of annulment through punishment.[77] Frankel's version certainly was a lot more straightforward. In his adaptation, punishment was portrayed as a psychological instrument that enabled the state to deter crime and boost the people's moral resolve not to transgress the law. At the risk of overinterpreting (Frankel was never one to share his underlying theories), one could say that this final allusion to the people's collective moral sensitivity reflected Hegel's definition of *Sittlichkeit* as 'conventional ethical order' (from *Sitte* meaning 'custom'), as opposed to rational, individual morality *à la* Kant.[78] This might explain Frankel's minimal, cursory reference to the Kantian 'free esteem for the True and the Good'. In a similar way, Frankel's unapologetic identification of 'an eye for an eye' as the central principle of *das Mosaische* is a far cry from the centrality of the 'cosmopolitan' Golden Rule in the philosophical ethics of Steinthal and Lazarus.

Talmudic law, Frankel stated at the end of his introduction, related to these Mosaic foundations as positive law did to canon law. By using exegesis, it hoped to capitalize on Mosaism's revealed status. Simultaneously, in filling the gaps and addressing new realities it enjoyed a validity of its own, independent from the divine will: comparable, in a way, to natural law in the strict sense of the word. Mosaic and Talmudic law were inextricably linked, however, in their religious colouring (*religiose Färbung*) and in the *moralische Wirkung* described above, which relied on the sanctity of the law and the public endorsement of its holiness. It would be wrong, Frankel cautioned in good historical-positivist fashion, to mistake the Talmud for law in decline, running contrary to the will and spirit of the modern nation.[79] Such anachronistic thinking should be avoided. Deeply rooted in its own particular time and place, it had developed a unique character that was best described not in terms of rise and fall but through a synchronic comparison with model, exemplary Roman law. For all

---

77    See, among many others, Thom Brooks, 'Is Hegel a Retributivist?' *Hegel Bulletin* 25 (2004), 113–26, and the vast corpus of recent studies listed in footnote 1.

78    For this dichotomy as a critique of the ethics of Kant and Fichte, see Wood, *Elements of the Philosophy of Right*, 58–76, ch. 3, 'Hegel on Morality'.

79    Cp. Kristiane Gerhardt, 'Frühneuzeitliches Judentum und "Rabbinismus". Zur Wahrnehmung des jüdischen Rechts in den Zivilisierungsdebatten der Aufklärung', *Trajectoires* 4, 'Postkolonial' (2010), https://journals.openedition.org/trajectoires/473 (accessed 14 February 2024).

the material parallels, we learn, it was an encounter of two completely different legal worlds, where Roman principle was balanced by Jewish creativity, Latin *Tiefsinn* by Semitic *Scharfsinn*, and synoptic standardization by casuistic fragmentation. A meeting, in short, of the systematic, methodical Occident and the inventive, adventurous Orient that was the ultimate home of the Jewish legal heritage.[80]

## 5    Afterword

In the words of the British writer George Orwell (1903–1950), prose should resemble a freshly cleaned windowpane: transparent, near-invisible, allowing the beholder an honest, unspoilt view of what is behind the glass. For philosopher Ludwig Wittgenstein, honest prose was but a language game, a moment of action during which words acquire meaning and their referents – be they concrete or abstract – are subject to interpretation. In the sample texts discussed above, we have watched an intricate language game enfold, when the ancient Hebrew concept 'halacha' became the object of inter-cultural translation.[81] On the one hand, the process enabled authors to reposition and universalize this traditional Jewish 'walk through life' by framing its contents in terms of law, religion, and morality. By the same token, the effort to match halachic properties with these Western categories, and to find all three united in modern Judaism, proved a tour de force. At a time when religion was turning inward, when laws were based on the moral codes of nations, and when morality was an effect of personal will, a divinely inspired law-based religion was an oxymoron.

Fortunately, the idiom of both the source and the target language was conveniently open and ambiguous. The act of cultural translation was a language game in which the interplay of lexicon, discipline, genre, and audience transformed meanings on both sides of the language divide. In different ways, all texts discussed in this chapter were about legal and religious pluralism and equality before the law. Within that framework, Judaism's legal component was singled out and defined either as legislation (Mendelssohn), righteousness (Lazarus), or – ideal and actual – *Recht* (Frankel). Its moral foundations, the absolute principles of the Good behind the Right, were located in God's

---

80    Frankel, *Der gerichtliche Beweis nach mosaisch-talmudischem Rechte*, 55–62.
81    The notion of the *Wissenschaft des Judentums* as a 'translation act' was first articulated in Ismar Schorsch, *From Text to Context: The Turn to History in Modern Judaism*, Tauber Institute for the Study of European Jewry Series (Hanover, NH, 1994), 151–76.

essence, in the human faculty of reason, or, preferably, in both. Patently contra-dicting modern conceptions of law and morality, the religious dimension was hard to accommodate. In the context of the *Ritualgesetze*, Mendelssohn iden-tified it as the ecclesiastical zone of competence in which Jewish canon law had evolved. For Lazarus, it was the open Jewish spirit under which humanis-tic morality blossomed, in collective imitation of God's holiness. In Frankel's Jewish legal science, it served as the locus of transcendentally sourced legal practice and as a prop for society's legal sensitivity. What emerged from each of these efforts was more than the sum of law, ethics and religion; more, too, than arcane halacha or biased *Gesetzesreligion*. What emerged was Judaism, staunchly defying secularization and delegalization, inviting us to a never-ending language game and, never-fixed, altering when it alteration finds.

CHAPTER 5

# The Truth Shall Abide

*Samson Raphael Hirsch and Abraham Geiger on the Binding Nature of Torah*

*Judith Frishman*

Outside the winter storm blows through the bare treetop, breaking off branch after branch and casting them down into the shiny, luminous snow below. Do you hear how the branches rejoice in the merry dance and the surroundings glow, and jeer at the dark, old motionless trunk? 'We pay homage to progress! Are borne by modern times (*Zeitgeist*)! Bear illumination and clarity! Old useless trunk, with your unmoving rigidity; with your insensitivity to movement and light! Will you never move from your spot? Do you think that light and life will come to you? Must you not seek them and dance the dance of the times?' But the trunk doesn't answer. It is not yet time for it to answer. How untouched it lets the gusts of wind blow in its top; allows the lightly clothed twigs to swirl in the bright snow. But when the snow's glow has finally faded, when the merry branches have long withered, new life stirs in the trunk that has not budged from its place. The spring sun returns and with its rays the trunk bears buds and twigs and offers shade and freshness. Then the time for its answer has arrived:

> And though a tenth remains in the land,
>     it will again be laid waste.
> But as the terebinth and oak
>     leave stumps when they are cut down,
> so the holy seed will be the stump in the land.
> Is 6.13[1]

∴

---

1 The title of this article refers to the motto from bShabbat 104a – 'Truth will abide, lies will not' (*qushta qa'e shuqra la qa'e*) – found on the title page of Samson Raphael Hirsch, *Naftuli Naftali. Erste Mittheilungen aus Naphtali's Briefwechsel* (Altona, 1838).

FIGURE 12    Title page of נפתולי נפתלי Erste Mittheilungen aus Naphtali's Briefwechsel, ed. Ben
Usiel (1838)

These are Naphtali's opening words in Chapter 4 of Samson Raphael Hirsch's
(1808–1888) *Naftuli Naftali. Erste Mittheilungen aus Naphtali's Briefwechsel*
(Altona, 1838), or *The Wrestlings of Naphtali: First Communications from
Naphtali's Correspondence* (Figure 12). With them, he answers Simeon, who
wonders whether things will ever change in Israel and if ever in the past
things were (in so bad a state) as they are in his own times.[2] This fictitious
literary exchange between Naphtali and Simeon is the sequel to Hirsch's
1836 *Neunzehn Briefe über Judenthum* (*Nineteen Letters on Judaism*), in which
Naphtali convinces the doubting Benjamin of the value of Judaism, leading to
the latter's revaluation of his ancestral faith after his near dismissal of the same.

---

2   Ibid., 66–67. The cover of this work attributes authorship to Ben Uziel, S.R. Hirsch's pseudo-
nym used earlier for *Igrot Tsafon. Neunzehn Briefe über Judenthum* (Altona, 1836).

Whereas the *Nineteen Letters* address those members of the German-Jewish bourgeoisie who were rapidly assimilating into their non-Jewish surroundings, these preliminary notifications are aimed primarily at the – at least in Hirsch's eyes – defecting group of rabbis urging for the reform of Judaism. The *Mittheilungen* are more specifically a reaction to (a) the synod convened by Rabbi Abraham Geiger (1810–1874) in Wiesbaden in 1836, (b) Geiger's declarations concerning the reform of Judaism, and (c) his three-part review of Hirsch's *Nineteen Letters*, published in the *Wissenschaftliche Zeitschrift für jüdische Theologie*, or *Scientific Journal for Jewish Theology*.[3]

Rabbi Samson Raphael Hirsch and Rabbi Abraham Geiger were contemporaries, fellow students, and friends (Figure 13). Hirsch, the son of a merchant, was born in 1808 and grew up in Hamburg, where he attended both the Talmud Torah founded by his grandfather as well as the gymnasium. As a student of Chacham Isaac Bernays (1792–1849) and, later, Rabbi Jacob Ettlinger (1798–1871) in Mannheim, Hirsch was well acquainted with the Talmud. The combination of university training and traditional Jewish education made him a true *Jissroel-Mensch*: the type of Jew he himself regarded as best fit for modernity. That type embodied the ideal union of *Torah* and *Derekh Erets*, namely being true to the Torah while at the same time adhering to 'the ways of the world', i.e. (ethical) behavior appropriate in modern society.[4] Opposed to the Reformers, he broke away from the Jewish community of Frankfurt and began his own *Austritt-Gemeinde* (seceded community).[5] He is known as the founder

---

3  Geiger was the founder and editor of this journal that first appeared in 1835. The first article in the first volume is Geiger's evaluation of the present state of Judaism: Abraham Geiger, 'Das Judenthum unsrer Zeit und die Bestrebungen in ihm', *Wissenschaftliche Zeitschrift für jüdische Theologie* 1 (1835), 1–12. For his reviews of Hirsch, see Abraham Geiger, 'Neunzehn Briefe über Judenthum, von Ben Uziel (Recension)', *Wissenschaftliche Zeitschrift für jüdische Theologie* 2 (1836), 351–359, 518–548; 3 (1837), 74–91. Hereafter, the journal is abbreviated as *WZJT*.

4  For biographies of Hirsch, see Noah H. Rosenbloom, *Tradition in an Age of Reform: The Religious Philosophy of Samson Raphael Hirsch* (Philadelphia, 1976) and, more recently, Roland Tasch, *Samson Raphael Hirsch. Jüdische Erfahrungswelten im historischen Kontext* (Berlin, 2011). On Hirsch's concept, see, for example, Mordechai Breuer, *The 'Torah-Im-Derekh-Eretz' of Samson Raphael Hirsch* (Jerusalem, 1970).

5  What is still probably the best discussion of the circumstances surrounding Hirsch's secession may be found in Robert Liberles, *Religious Conflict in Social Context: The Resurgence of Orthodox Judaism in Frankfurt am Main, 1838–1877* (Westport, 1985). For a more recent discussion, see Matthias Morgenstern, 'Rabbi S. R. Hirsch and his Perception of Germany and German Jewry', in *The German-Jewish Experience Revisited*, eds Steven E. Aschheim and Vivian Liska (Berlin, 2015), 207–230.

FIGURE 13    Portraits of A. Geiger between 1838 and 1843 (left) and S.R. Hirsch between 1830
             and 1841 (right)
             HELD BY THE NATIONAL LIBRARY OF ISRAEL, ABRAHAM SCHWADRON
             COLLECTION; IMAGES IN THE PUBLIC DOMAIN AND COURTESY OF
             WIKIMEDIA COMMONS

of neo-Orthodoxy, today called Modern Orthodoxy and mainly located in the USA and Israel.

Geiger was a leader of the *Wissenschaft des Judenthums* and one of the most influential founders of Liberal Judaism.[6] Born into a traditional Jewish home, in Frankfurt in 1810, he received a good Jewish education. Geiger acquainted himself with secular works that deeply influenced him. A book on the political use of myths in Greece and Rome led him to discover that the Bible, too, included myths. So inspired, he questioned the historicity of the Bible and the interpretation of Jewish history. Despite his lack of formal education, Geiger enrolled at the University of Heidelberg and later at the universities of Bonn and Marburg. Having first studied classical and oriental languages, he moved from philology to philosophy and history, receiving his doctoral degree with his essay *Was hat Mohammed aus dem Judenthume aufgenommen* (*What Did*

---

6  For a biography of Geiger, see Susannah Heschel, *Abraham Geiger and the Jewish Jesus*,
   Chicago Studies in the History of Judaism (Chicago, 1998). On Geiger and the scientific study
   of Judaism, see Christian Wiese, Walter Homolka, and Thomas Brechenmacher, eds, *Jüdische
   Existenz in der Moderne. Abraham Geiger und die Wissenschaft des Judentums* (Berlin, 2016).

*Mohammad Borrow from Judaism*).[7] After lay leaders in Hamburg had created a new liturgy, Geiger called for a reform of Judaism based on scholarly methods, as opposed to what he considered the arbitrary approach initiated by laymen.[8] Both men sought answers to the pressing matters of their day. These included accusations arising from Christian theological supersessionism as well as from philosophical views of morality. For Christian theologians, Judaism represented the dead letter rather than spirit, while the rabbinic literary corpus represented a post-Mosaic *Entartung*, or degeneration. For philosophers like Immanuel Kant (1724–1804), Judaism's law was 'heteronomous', and its God demanded slavish behaviour, rather than the development of free moral consciousness. Both groups, supercessionists and philosophers alike, criticized the Jews' reclusive nature, labelling them misanthropic and therefore unable to integrate in society. Moreover, the Jews themselves were said to be in need of regeneration and *Bildung*, although the cause of their lowly state was a matter of debate.

Hirsch and Geiger countered this critique yet simultaneously internalized it. Adopting and adapting Enlightenment ideals, they formed their own understandings of Judaism suited for modernity. The works discussed in what follows represent an early phase in the crystallization of Hirsch and Geiger's ideas about what it meant to be a Jew in modernity. Over the course of time, the views they held on subjects ranging from the Torah and the Talmud to the mitzvoth and the role of *Wissenschaft* became increasingly disparate, causing a rift between the two erstwhile friends.

1    Geiger's Criticism of Hirsch's Nineteen Letters

The first two parts of Geiger's fifty-five-page review of Hirsch's *Neunzehn Briefe über Judenthum* are rather mild, with Geiger referring to Hirsch as 'my friend' throughout. It is only in part three that Geiger's critique grows sharper and where he offers us some brief insight into his own hopes for the future of Judaism. Geiger feels attracted to Hirsch's warm religiosity and morality (*Sittlichkeit*)[9] and shares the deep pain Hirsch feels due to the degeneration of Judaism. Like Hirsch, he would like to 'recognize pure Judaism, regenerate

---

7   Abraham Geiger, *Was hat Mohammed aus dem Judenthume aufgenommen* (Bonn, 1833).
8   Geiger contrasts the arbitrary reform of the prayer book by laymen with his idea of true reform. Geiger, 'Neunzehn Briefe über Judenthum, von Ben Uziel', WZJT 3 (1837), 89.
9   *Sittlichkeit* here as an innate sense of the good and what is right. For more on this concept, see the contribution by Felix Wiedeman in this volume.

the faith, present the truth clearly, and heal the present times that are sick and wounded by the reign of confusion and contradictions'.[10]

Geiger notes, however, that there are many weaknesses in the direction Hirsch has taken. The solutions he offers the doubting Benjamin would only cause the times to sink even deeper in morbidity. For Benjamin, the role of religion is to bring people closer to their destiny, which is nothing other than 'happiness and perfection' (*Glückseligkeit und Vollkommenheit*). He does not see this in Judaism, which is dejected by its fate and stagnated by its own rules. Judaism isolates its followers, so Benjamin, and closes off the path to free research and artistic creation. The Talmud distorts the mind, moving neither the heart nor life. To the contrary, it focuses on all sorts of inconsequential details and relates them all to God.

Geiger wonders why Hirsch has not asked Benjamin about the nature of his own relationship to Judaism. Even more surprising to Geiger is Hirsch's failure to address the evident contradictions that may be found in the Bible, Talmud, and later rabbinic works. These should have led him to question whether they were, in fact, written in one spirit, forming an unambiguous and indivisible whole. How, he asks, could such a gifted and highly intelligent man offer such vacuous instruction on so many points?

### 1.1    Teleological Pedagogy

The main points of Geiger's criticism relate to Hirsch's teleological pedagogy, his ahistorical approach to the Bible and other major Jewish sources, and his philological methodology. Hirsch, says Geiger, learns the Jewish way to live from the Jewish sources, from Torah. It is, for him, the only source of knowledge about Judaism and the purpose of Judaism in the course of its history and in the history of the world and all of humanity. According to Torah, the world was created to serve God by fulfilling the divine task: to act justly and lovingly, to submit to God both consciously and out of free will, and to overcome pride, greed, and hedonism. Hirsch does recognize that humans are able to discern the norms for their own behaviour, and he points to consciousness as that which distinguishes between animals and humans. Yet he turns to an outside cause: God's commandments – a cause that is *purposely* unfounded. Hirsch then claims that Judaism demands obedience. Why, Geiger asks, should humans not obey God by way of their insight? Is recognizing God's will not precisely what consciousness is? And does obeying God's will not represent

---

10    Abraham Geiger, 'Neunzehn Briefe über Judenthum, von Ben Uziel: eine Recension', *WZJT* 2 (1836), 352.

the exercise of free will, choosing fulfilment of the commandments rather than our own desires, such as pleasure and lust.

The question for Geiger concerns the definition of obedience to God. Is obedience only obedience when the mitzvoth are fulfilled, simply because one is commanded to do so, as Hirsch avers? Or does obedience entail the recognition by insight that God's commandments are just and good? Hirsch implies the latter is obedience to oneself rather than to God, leading Geiger to conclude that Hirsch thereby nullifies the very gift of consciousness with which humans are endowed – to which he himself had pointed – and therefore reduces them to the level of animals.[11]

From the superiority of humans above animals, Geiger then moves on to the specific role Jews are to play in history and the related question of election. Hirsch and Geiger assign similar roles, or a mission, to the Jews (*Israeliten*) in history. For them, Jews are supposed to serve as a silent example for others: as a warning, as a teacher, as a model priest of justice and love, of true acknowledgement of God and of a moral-divine (*sittlich-göttlich*) life. This function contrasts with *Jissroeilthum*, i.e. an inward facing stand towards the Jewish community.

However, they part ways when Hirsch introduces a special category of commandments – the *Edoth* – in his sixfold system of classification.[12] The *Edoth* are symbolic acts and seasons that 'all give expression to ideas. [...] They present themselves with all the force of a single, undivided and indivisible appeal to the soul. [...] They are, all of them, reminders, or vivid expressions of sentiment by means of the significant language of action. The greatest and the least among them ... are all symbols that teach important lessons.'[13]

As opposed to Hirsch, Geiger cannot find meaning in many of the ceremonial laws. He remains unconvinced that they all still function as reminders of Israel's role. However, his main objection to the *Edoth* is the notion that the Jews have received these as a sign of their superiority, insinuating a differentiation based on race or class. Geiger exercises a form of relativism: Judaism, he says, is the way for Jews to achieve their destination as true humans and worshippers of God so that Jews and non-Jews – each on *their own historical*

---

11    Ibid., 538–539.

12    In the *Nineteen Letters*, Hirsch divides the commandments into six categories. The *Edoth* are discussed in the thirteenth letter.

13    Samson Raphael Hirsch, *The Nineteen Letters on Judaism*, ed. Jacob Breuer, trans. Bernard Drachman (Jerusalem, 1969), 85.

*paths* – will grow ever closer and, finally in brotherhood and complete equality, pursue life's goals. The peoples and their decline do not simply serve as a warning against trifling existence or as a contrast to the human greatness to be achieved by Israel, as Hirsch would have it. Could this be the only purpose for the existence of millions of people endowed with freedom and reason? All, including the weak and feebleminded, have the obligation to strive for a religiously virtuous life at their own level, Geiger argues, without any additional obligation simply because one belongs to a specific tribe. Israel is neither God's favourite, nor does it have a spiritual spark that allows it to immediately discern what is true and good. Were this the case, then the additional obligations would have been a logical consequence. Geiger labels this understanding of chosenness and mission as fanaticism, not true Judaism.[14]

### 1.2    The Bible and History

Hirsch refrains from proving the truth and eternal validity of the Jewish documents in their unchanged form, something Geiger considers essential for the instruction of someone in doubt. Rather, he argues that the Torah is a fact, like heaven and earth: something that is unquestionable and to be taken at face value, needing no historical proof of its validity. To the Torah, Hirsch simply adds the Talmud, treating the latter's detailed halachic discussions as if they arose directly from the biblical concepts.

In doing so, Hirsch rejects historical methods, where the biblical texts are compared with other ancient works and where questions of origin, dating, and composition are raised. He adjures his readers, instead, to approach the text without preconceived ideas or insights from elsewhere – i.e. without comparative history or comparative philology – in order to understand the Torah's foundations and its aims. Hirsch claims his method is comparable to those used in the study of natural history. What he seems to have in mind is the theory of preformationism. In this view, organisms develop from pre-existing miniature versions of themselves, stemming from the beginning of creation.

Challenged by the epigenesists, this theory was on the wane by the end of the nineteenth century. Geiger, then, ridicules Hirsch's comparison between the investigation of Torah and natural history at length. If a biologist extrapolates laws about the earth when viewed solely from one moment in time, then he would falsify the entire history of nature. If a biblical scholar likewise insists that everything must be derived from and imbued by the same spirit, then he must relegate everything to one period, thus failing to discern a beginning,

---

14    Geiger, 'Neunzehn Briefe über Judenthum, von Ben Uziel', *WZJT* 2 (1836), 545–546.

middle, and end (i.e. generation, development, and decay) and denying any external influence as the cause of the varieties of form. 'Why', Geiger asks, 'should we reject every historical question and see a sealed object before us, a work that is lofty and impenetrable, that is beyond our power?' The Torah has developed over the course of time, as has the Talmud. Therefore, it is important to discern and trace this development.[15]

### 1.3 Philological Method

Hirsch's philological method, together with its origins, have been the subject of several book chapters and articles.[16] Among the writings of historians of religion said to have influenced Hirsch are those of Friedrich Schleiermacher (1768–1834), Georg Friedrich Creuzer (1771–1858), and particularly the speculative etymology of Johann Arnold Kanne (1773–1824).[17] Geiger is averse to acknowledging Hirsch's expressive theory of language and its application to the biblical text as 'method'. He characterizes it, instead, as *Spielerei*, as disingenuous: meaning is sought in the Hebrew language by way of etymology. In this, Hirsch ignores the rule of all languages that states the plain meaning is always primary. The important meanings he derives from the Hebrew language are precisely those that contradict human logic. In Geiger's view, he does not seek coherence or clear meanings. Rather, he freely plucks words out of their context and forces artificial meaning upon them so that a higher, divine wisdom flows from each word. Hirsch's approach is wholly subjective, Geiger scoffs, concluding, 'Whoever can discern exegesis in this method will also find great religious value in the egg hatched on Yom Tov.'[18]

### 1.4 Geiger's Approach

Despite his sympathy for Hirsch, Geiger concludes that his colleague and his laboured division of the mitzvoth into categories are far from convincing. But Geiger goes much further: he asks how a commandment that must be

---

15   Geiger, 'Neunzehn Briefe über Judenthum, von Ben Uziel', *WZJT* 3 (1837), 74–77.

16   David Sorkin, *The Transformation of German Jewry, 1780–1840* (New York, 1987), 156–171; Arnold Eisen, *Rethinking Modern Judaism* (Chicago, 1998), 135–155; Roland Tasch, *Samson Raphael Hirsch*, 99–107; Michah Gottlieb, 'Oral Letter and Written Trace: Samson Raphael Hirsch's Defense of the Bible and Talmud', *Jewish Quarterly Review* 106/3 (2016), 316–351.

17   David Sorkin, *The Transformation of German Jewry*, 164–166.

18   Ibid., 75. Geiger's reference to the egg hatched on Yom Tov (one of the pilgrim festivals during which work is prohibited) is derived from Hirsch's thirteenth letter where he claims that every detail, even the one 'that is the target of so much ridicule, the prohibition of the use of an egg laid on a Sabbath or holiday' teaches an important lesson. See Hirsch, *The Nineteen Letters*, ed. Breuer, trans. Drachman, 85.

performed in unchanged and meticulous fashion, without any room for free spirituality, can lead to a lofty idea. 'If I must attend to a myriad of details on Shabbat yet, in failing to fulfil one of these I have failed to fulfil everything, then how will I come to think about the correct way to use possessions, recognize the limitations of my power, and realize that the world does not belong to me?', Geiger queries.[19] Hirsch's insistence on adhering to every letter implies it would otherwise be impossible to derive the meaning of the commandments. Hence, spiritual development is not required, stimulated, or expected – but rather spiritual enslavement and mechanical observance.

How, Geiger wonders, would such an approach convince someone like Hirsch's own Benjamin, who is in search of spiritual uplifting in Judaism? Geiger's own solution is reform: not the kind offered by Hirsch or some superficial reform, as in the case of earlier liturgical innovation.[20] While in his later writings he would anchor reform in prophetic Judaism and ethical monotheism, in 1837 Geiger was still searching for the central idea, or ideas, of Judaism. In his review of Hirsch, he identifies these as sanctification of life at home and in business alike: by recognizing humanity's calling as service to God in every aspect of life, exercising justice and mildness through stimulating, unaffected actions that uplift the heart and spirit – as opposed to spiritless form and thoughtless practice. 'To this end', Geiger says,

> we will listen to the voice of the ancients in the Talmud, that documents the great and long history of Judaism. We will seek that which is excellent in it and learn how the teachers of the Talmud questioned their times when it came to institutions and changes. We will, like them, look towards the spirit and not the letter. [...] Truth and insight rather than unfounded faith; justice and mildness rather than sickly love; trust rather

---

19    Abraham Geiger, 'Neunzehn Briefe über Judenthum, von Ben Uziel', *WZJT* 3 (1837), 86–87.
20    Geiger would criticize the liturgical reforms in Hamburg extensively in *Der Hamburger Tempelstreit, eine Zeitfrage* (Breslau, 1842). For a discussion of the liturgical reforms in Hamburg see Jacob Petuchowski, *Prayerbook Reform in Europe* (New York, 1968), 49–58; Michael Meyer, *Response to Modernity: A History of the Reform Movement in Judaism*, Studies in Jewish History (Oxford, 1988), 53–61; Andreas Brämer, *Judentum und religiöse Reform. Der Hamburger Israelitische Tempel 1817–1938*, Studien zur jüdischen Geschichte 8 (Hamburg, 2000), 45–56. Geiger was not the only reformer to search for a leading principle upon which reform was to be based. Cf. Samuel Hirsch, *Die Reform im Judenthum und dessen Beruf in der gegenwärtigen Welt* (Leipzig, 1844); see further Judith Frishman, 'True Mosaic Religion: Samuel Hirsch, Samuel Holdheim and the Reform of Judaism', in *Religious Identity and the Problem of Historical Foundation*, eds Judith Frishman, Willemien Otten, Gerard Rouwhorst (Leiden, 2004), 195–222.

than the hope that comes from weak nostalgia: these are the pillars upon which Judaism rests and upon which contemporary forms will find their directives.[21]

## 2    Hirsch on Geiger and Reform

### 2.1    *Wissenschaft as a Scenario of Doom*

Hirsch concludes his *Erste Mittheilungen aus Naphtali's Briefwechsel* by correctly noting that – despite their twisting of the meaning of the Talmud – the reformers do, indeed, want to justify the new directions they are taking by referring to the Talmud. They thereby recognize the Talmudic principles. Yet in blindness and in ignorance, they attempt to change a millennia-old Judaism into a Judaism of 1837.[22] Whereas Geiger criticized Hirsch for failing to look beyond the text and for resorting forcibly to contrivances, Hirsch accuses the reformers of introducing foreign elements into Judaism. They make use of preconceived notions derived from the study of classical paganism, dogmatic philosophy, historical criticism, theology, and New Testament studies, instead of understanding the text within its own context.[23] These reformers, Hirsch exclaims, are not 'Jews in life' (i.e. Jews by living Jewish lives) but 'Jews in *Wissenschaft*'.[24] Their Judaism more closely resembles rational forms of Christianity even as they denigrate Torah-true Jews, considering them blind slaves and condemning them as idolaters and heathens.[25]

Hirsch opposes *Wissenschaft* as exercised by the reformers. At the same time, he tries to find an alternative version of *Wissenschaft*, one that will solve his greatest problem: the attack on rabbinic Judaism and substitution of progress and *Zeitgeist* in its stead. His solution, only hinted at in the *Nineteen Letters*, is most fully developed in 1838, in *Horeb* – a masterly twist on the general understanding of the relationship between the written and oral Torah.

---

21    Geiger, 'Neunzehn Briefe über Judenthum, von Ben Uziel', *WZJT* 3 (1837), 91.

22    Andreas Gotzmann discusses the irony of the reformers use of the Talmud to justify their rejection of Talmudic rulings, noting that only Samuel Holdheim and Samuel Hirsch developed an alternative system for the halakhic one. See Andreas Gotzmann, *Jüdisches Recht im kulturellen Prozeß. Die Wahrnehmung der Halachah im Deutschland des 19. Jahrhundert*, Schriftenreihe wissenschaftlicher Abhandlungen des Leo Baeck Instituts 55 (Tübingen, 1997), 278.

23    Hirsch, *Naftuli Naftali*, 66.

24    Ibid., 75.

25    Ibid., 12.

There, he argues the biblical text represents merely the short-hand notes of the complete revelation, as represented by the oral Torah.[26]

The preliminary communications (*Erste Mittheilungen*), in the guise of correspondence, constitute mainly a review of, and an attack on, several reformers. They thus function less as a proper defence of his alternative outlook. Those assailed include Michael Creizenach (1789–1842), Leopold Stein (1810–1882), Albert Cohn (1814–1877), Joseph Aub (1804–1880), and Moses Brück (1812–1849), whose articles had been published in the early volumes of Geiger's *Wissenschaftliche Zeitschrift für jüdische Theologie*.[27] Hirsch's critique targets, in particular, their lack of familiarity with rabbinic texts, incorrect interpretations of these sources, uses of statements out of context, and comparisons drawn between Judaism, the Torah, or rabbinic texts, and laws and myths from the ancient Near East.

The first few pages refer explicitly to Geiger, although Hirsch's criticism of his friend – unlike his critique of the others – remains rather temperate, as was Geiger's critique of Hirsch at the outset of his review.[28] However, in this introduction, the metaphoric description of the reformers as Hellenists (or even the Seleucids) in the story of the Maccabees is scathing. In a dream, Naphtali envisions the learned on the Temple Mount being attacked suddenly by a madding crowd:

> The latter bore torches, which they used to singe the beards and side locks of the Torah scholars and burned the texts. Although they would have liked to have spared the Temple, the fire got out of control and the Temple Mount turned into a sea of fire: Temple, altar, holy table and ark, even the tablets of the law went up in flame. The fire spread throughout Zion, travelled over land and sea and burnt everything that was holy right

---

26    Samson Raphael Hirsch, *Horeb. Versuche über Jissroels Pflichten in der Zerstreuung* (Altona, 1837). Jay Harris explains that Hirsch's literal understanding of both the written and oral law as *Torah le'moshe mi'sinai* (divine revelation) deviates from the general understanding of the relationship between the oral and the written in rabbinic literature: Jay Harris, *How Do We Know This? Midrash and the Fragmentation of Modern Judaism*, SUNY Series in Judaica: Hermeneutics, Mysticism, and Religion (Albany, 1995), 223–227. See further David Ellenson, 'Antinomianism and Its Responses in the Nineteenth Century', in *The Cambridge Companion to Judaism and Law*, ed. Christine Hayes (Cambridge, 2017), 260–86; Michah Gottlieb, 'Oral Letter and Written Trace'; idem, 'Scripture and Separatism: Politics and the Bible Translations of Ludwig Philippson and Samson Raphael Hirsch', in *Deutsch-jüdische Bibelwissenschaft*, eds Daniel Vorpahl, Sophia Kähler, and Shani Tzoref, Europäisch-jüdische Studien 40 (Berlin, 2019), 57–73.

27    The entirety of the third chapter is a review of the reformers: Hirsch, *Horeb*, 13–66.

28    Hirsch, *Naftuli Naftali*, 6–13.

down to the ground, until the earth was nothing but a steaming desert. Perets, who was bearing the torch, did not notice that its flame was dying out until it burnt his hand then wholly consumed him ... But then a group of men rose up, prophets, wise men, men of the Great Assembly, led by Moses. And they bore the only remaining light that illuminated everything. The Temple stood once more, altar, table and veil, and the divine law rested again in the tabernacle and the earth was filled with blessing and joy.'[29]

## 2.2    *Free Moral Conviction versus Obedience*

To those who consider themselves enlightened, Hirsch objects they are not *Jewishly* enlightened. Nor are the so-called scholarly systems according to which they attempt to describe and model Judaism. Geiger wants to offer a system of Judaism grounded in Judaism, one that contains the kernel and essence of Judaism – an essence from which Judaism is estranged and therefor needs reform. The core is the unfolding of moral strength according to free moral conviction, i.e. following self-recognized goals without coercion. All the rest is but a shell that accrued in the course of time and needs to be peeled off.

However, the emphasis on free development of inner moral strength and free moral conviction is faulty, Hirsch retorts. He contends that moral strength in Judaism is not a matter of personal conviction but pertains to him who performs his duty despite his urges – and does so because he fulfils the commandment of a Higher Being, a commandment he must fulfil with even the smallest part of his being.[30] 'Where does the Torah call for conscience as a necessary condition for an obligation? And when does it make a distinction between the significance of the various types of mitzvah?', Hirsch queries.[31] For Geiger, nearly the entirety of the 613 commandments is nothing but shell, to be cast aside as historically determined. Erstwhile expedient for invoking the correct sentiment rather than obedience, none but a handful remains relevant according to Geiger's criteria. Yet Geiger fails to account for the disproportionate attention the Torah devotes to offerings, dietary laws, and festivals, all of which Geiger deems to be chaff. Moreover, the punishment for transgressing them is far greater than for those he considers moral-philosophical mitzvoth. As Hirsch cleverly notes, the wood gatherer was condemned to death for transgressing the Sabbath (Num 15:32–36), but had he transgressed by telling a lie,

---

29    Ibid., 3–4. The name Perets means one who causes a breach, i.e. those who broke with tradition and burned down the entire edifice of Judaism.
30    Ibid., 7–8.
31    Ibid., 6–9.

then he would not have received this punishment, despite the negative view the Torah maintains on lying.[32] In sum, Hirsch remains unconvinced by the distinction made between rational and non-rational commandments. He also disagrees with the notion that reason can teach one duty. Should the acts of duty emanate from one's sentiment, as Geiger claims, then it would not only be ridiculous but dangerous as this would lead us to perform that which is a free moral sentiment as required obedience. This would simply cloud our free moral consciousness.

## 3    Conclusion

In this initial, rather neglected phase of their debate, both Hirsch and Geiger ask themselves why Jewish teaching, whose goal is so lofty, seems so unattractive, consisting of inexplicable demands and spiritless exercises. Hirsch attributes this to a *misunderstanding*: hundreds of years of oppression that resulted in a mummified form of Judaism; the kabbalah, which reduced pedagogical spiritual practice into a 'thing about amulets' (*Amulettenwesen*); and the *Orach Chayyim* (a part of the *Shulchan Arukh*), originally intended as an exercise for scholars, fell into the hands of laymen so that it seemed as if praying and holidays were central and everything else irrelevant.[33] For Geiger things went wrong far earlier, initially beginning with the rabbis; in his later works, he points to the periods after the Pharisees, particularly the Middle Ages.[34]

Hirsch's remedy was to turn to the sources, to Torah, Mishnah, and Midrash – and particularly to Tanach, its language, and the disposition of its spirit – so as to understand concepts pertaining to God, the world, human beings, Israel, and Torah. When finding the rabbinic tradition in the line of fire, Hirsch addressed the Bible to illuminate its brief notes. Perhaps he did so to compete with the emphasis placed on the Bible by many of the Reformers; or because of the popularity of the Bible in the version published by Ludwig Philippson (1811–1889),

---

32    Ibid., 10–11.
33    Geiger quotes Hirsch as offering these excuses for the devaluation of Judaism: WZJT 2 (1836), 525.
34    Jay M. Harris traces Geiger's shifting opinions on rabbinic Judaism, indicating that his initial total rejection of the rabbis' methods was mitigated in the course of time, specifically as a result of his debate with Christian scholars: Harris, *How Do We Know This?*, 157–172. For Geiger on the Pharisees, see Heschel, *Abraham Geiger and the Jewish Jesus*, 77–105 and Judith Frishman, *Wat heeft het christendom van het jodendom overgenomen? Abraham Geiger en de geschiedschrijving van het rabbijns jodendom*, Inaugural lecture, Katholieke Theologische Universiteit (Utrecht, 1999).

whose physical form his own commentary imitated; or for fear of higher biblical criticism?[35] For Hirsch, despite his so-called modern, scientific approach to symbols, nothing in the end was to change as far as halacha was concerned. However, the background of the rabbi, his training, and the role he was to play as an intermediary between modern society and traditional Judaism did undergo radical change. Surely Hirsch's own more open upbringing outside of the yeshivah world combined with his deep appreciation of tradition served as a model for the *Mensch-Jissroel* he propagated.

Geiger was very impressed by what Judaism would gain from comparative religion. However, his remedy of reform, which included the study of the historical development of the biblical and rabbinic texts, preserved little of rabbinic Judaism. In discerning between the wheat and the chaff, any and all of the commandments for which he could find no rational explanation fell at the wayside. Even though Geiger questioned the validity of a great deal of rabbinic Judaism, he did not expect, nor did he insist, that his congregants adopt his own radical stance on matters such as kashrut and circumcision. Moreover, he still had recourse to rabbinic texts and the halachic system when delivering proof and justification for his ideas.

Hirsch's Pentateuch was read for generations, and the notion of *Torah im Derekh Erets* – if not his symbolic system – is still at the heart of Modern Orthodoxy today.[36] In Frankfurt, Hirsch seceded from the *Einheitsgemeinde* (the unitary Jewish community), and the number of adherents to neo-Orthodoxy who joined him was limited. In its transposition to the United States, the movement underwent a slow process of development from the traditional to the less traditional.[37] The notion of *Torah uMaddah* (the combination of Torah and secular knowledge), embodied by the Yeshivah University (established in 1886) and serving as its motto, was no longer understood as simply two entities existing side by side without mutual influence. By the second half of the twentieth century, prominent American/Israeli rabbinical scholars such as Eliezer Berkovits (1908–1992), Emanual Rackman (1910–2008) and David Hartman

---

35    In his article 'Scripture and Separatism. Politics and the Bible Translations of Ludwig Philippson and Samson Raphael Hirsch', Michah Gottlieb offers alternative reasons for Hirsch's focus on the Bible (see note 26).

36    Samson Raphael Hirsch, *Der Pentateuch, übersetzt und erläutert*, 5 vols (Frankfurt am Main, 1867–78).

37    Steven M. Lowenstein traces the developments in the neo-Orthodox community in New York in his book *Frankfurt on the Hudson: The German-Jewish Community of Washington Heights, 1933–1983, Its Structure and Culture* (Detroit, 1989).

(1931–2013) appealed to sociological and historical arguments in arguing for a modern halacha.[38]

In this same period, Geiger's liberal Judaism became a transatlantic success, evolving into the largest movement in Judaism worldwide. Geiger's lack of appreciation for ceremonies and symbols was reflected in Reform services for more than a century. More recent Reform prayer books and platforms demonstrate that the understanding and appreciation of the role of ritual and even halacha has now radically changed.[39] While the Liberal/Reform movement is still the largest, its membership has sharply declined, and many of its synagogues worldwide are closing.

The questions raised by modernity and the challenges it posed to traditional understandings of torah gave impetus to the founding of both Liberal/Reform Judaism and neo-Orthodoxy as alternatives to secularism and fundamentalism. However, the questions these nineteenth-century movements sought to answer have become less pressing or even irrelevant for most post-modern Jews, who are now facing a variety of new challenges brought about by the rise of populism, fascism, and the digital revolution.[40]

---

38   For relevant works by Eliezer Berkovits, see *God, Man, and History: A Jewish Interpretation* (New York, 1959), *Not in Heaven: The Nature and Function of Jewish Law* (New York, 1983) and *Jewish Women in Time and Torah* (Hoboken, 1990). Among Emanuel Rackman's most recent works are *Modern Halakhah for Our Time* (Hoboken, 1995) and *One Man's Judaism: Renewing the Old and Sanctifying the New* (Jerusalem, 2000). From David Hartman, see *A Living Covenant: The Innovative Spirit in Traditional Judaism* (New York, 1985), *A Heart of Many Room: Celebrating the Many Voices Within Judaism* (Woodstock, Vermont, 1999), and *The God Who Hates Lies: Confronting and Rethinking Jewish Tradition* (Woodstock, Vermont, 2011).

39   For the platforms adopted by the Central Conference of American Rabbis from 1885 until the present, see https://www.ccarnet.org/rabbinic-voice/platforms/.

40   Steven Kepnes, ed., *Interpreting Judaism in a Postmodern Age, New Perspectives on Jewish Studies* (New York, 1995); Arthur Green, *Judaism for the Post-Modern Era*, Samuel H. Goldenson Lecture Delivered December 12, 1994, at the Hebrew Union College–Jewish Institute of Religion, Cincinnati, Ohio (Cincinnati, 1995); S. Daniel Breslauer, *Creating a Judaism Without Religion: A Postmodern Jewish Possibility*, Studies in Judaism (Lanham, 2001); Jack J. Cohen, *Judaism in a Post-Halakhic Age*, Reference Library of Jewish Intellectual History (Boston, 2010); Danny Schiff, *Judaism in a Digital Age: An Ancient Tradition Confronts a Transformative Era* (Basingstoke, 2023); Jerome A. Chanes and Mark Silk, eds., *The Future of Judaism in America*, Studies of Jews in Society 5 (Cham, Switzerland, 2023).

# CHAPTER 6

# 'A Law for Jews and Not for Christians'?

## Mosaic Law and the Deceased Wife's Sister Debate in Victorian Britain

*Michael Ledger-Lomas*

In the summer of 1882, the Hebrew professors of Europe received a curious circular from James Ramsay (1847–1887), the Earl of Dalhousie. He explained that he had just introduced a bill into the House of Lords 'with the object of legalizing marriage with a deceased wife's sister' and requested them to do him the 'favour of furnishing me with your opinion, together with the permission to quote it, as to whether such marriages are or are not prohibited in the Mosaic writings'. Dalhousie promised that any replies would be 'understood to be limited to an interpretation of the Levitical Law, and not as committing you to any approval, or the reverse' of his bill.[1]

From Aberdeen to Zagreb, the replies poured in. They made gratifying reading for Dalhousie and the Marriage Law Reform Association, which published them to sway opinion in their struggle to legalize marriage with a deceased wife's sister. From Königsberg, Professor Johannes Georg Sommer (1810–1900) reported that German commentators from Johann David Michaelis (1717–1791) to August Wilhelm Knobel (1807–1863), Carl Friedrich Keil (1807–1888), August Dillmann (1823–1894), and Joseph Lewin Saalschütz (1801–1863) – the Jewish commentator on Michaelis – concurred that Lev 18:18, the text central to the debate, prevented men from marrying their wives' sisters only during the latter's lifetimes.[2] Not all, though, were supportive. From Greifswald, Julius Wellhausen (1844–1918) dryly commented that as the ban on marriage with a deceased wife's sister was 'not unreasonable, I have no inclination to repeat the oft-repeated assertion that it cannot be supported by Jewish law, as it is not contained in it'.[3] Anglicans were still more disobliging. Edward Bouverie Pusey (1800–1882), the elderly professor of Hebrew at Oxford, not only repeated his

---

1  T. Paynter Allen, ed., *Opinions of the Hebrew and Greek Professors of the European Universities, of Bible Revisers, and of Other Eminent Scholars and Commentators on the Scriptural Aspect of the Question Regarding the Legalisation of Marriage with a Deceased Wife's Sister*, 2nd ed. (London, 1884), 2.

2  Ibid., 53–54.

3  Ibid., 38.

long-standing belief that Leviticus forbade such marriages but even warned that the 'terrible evil' involved in 'any relaxation of the sacredness of the law of marriage' was evident in Protestant Germany, where such marriages had long been permitted.[4] H.W. Watkins (1844–1922), professor of Hebrew at Durham, did a better job of retaining his professorial cool, tersely noting that the 'letter of the Levitical law does not forbid' such marriages, but sincerely hoped that 'your Lordship's Bill will never become the law of England'.[5] (Figure 14)

The debate over marriage with a deceased wife's sister was 'ideologically crucial' to Victorian Britain.[6] It began with the passage of Lord Lyndhurst's Marriage Act (1835), which established that sororate marriages would henceforth be void because they went against the canon law of the Church and in particular Archbishop Parker's Table of Kindred and Affinity (1563), which put them within a forbidden degree of affinity.[7] Although people could ignore Lyndhurst's Act or dodge it by marrying abroad, a growing body of activists protested against what they regarded as the reactionary incorporation of canon law into statute law – which they felt rested in any event on a superseded reading of the Mosaic writings. A Royal Commission on the operation of Lyndhurst's Act (1847) publicized their arguments, while bills for legalization passed the Commons on nineteen occasions before being defeated in the Lords, generally due to concerted action by the bishops who sat there.

The debates occasioned by this Parliamentary war of attrition, which only ended with the Deceased Wife's Sister Act (1907), have long interested historians. For historians of gender, they reveal the instability of the family unit, which was central to conceptions of property and inheritance. Marriage law reformers claimed to speak for working-class widowers who needed to hold their hardscrabble households together by marrying the deceased wife's sister or middle-class men seeking to preserve the family as a sentimental unit.[8] Yet their opponents scorned this concern as a cover for wealthy libertines who wished to size up younger sisters of ailing wives as potential bedmates. The question of whether a wife's sister was as close a kin relation as a man's blood

---

4  Ibid., 72.
5  Ibid., 21–22.
6  Margaret Morganroth Gullette, 'The Puzzling Case of the Deceased Wife's Sister: Nineteenth-century England Deals with a Second-Chance Plot', *Representations* 31 (1990), 142–66.
7  See C.F. Behrman, 'The Annual Blister: A Sidelight on Victorian Social and Parliamentary History', *Victorian Studies* 11 (1968), 483–502 and David Barrie, *Sin, Sanctity, and the Sister-in-Law: Marriage with a Deceased Wife's Sister in the Nineteenth Century* (London, 2018), ch. 1 for excellent surveys.
8  Karen Chase and Michael Levenson, *The Spectacle of Intimacy: A Public Life for the Middle-Class Family* (Princeton, 2001), ch. 5.

FIGURE 14   The Holman Hunts: Fanny, William, and Edith, all painted by William, 1866–68.
After the death of Fanny (née Waugh), he married her younger sister, Edith – but
had to travel to Switzerland to do so. Portraits held in the Toledo Museum of Art,
Ohio; Uffizi Gallery, Florence; and de Young Museum, San Francisco
IMAGES IN THE PUBLIC DOMAIN AND COURTESY OF WIKIMEDIA COMMONS

relative raised fears of incest, the shadow side of Victorian Britain's intense
domesticity.[9] The controversy even generated novels, which dramatized this
gender trouble by presenting contrasting images of the sister in the house as
rival, angelic understudy or frustrated individual struggling to express desires
of her own.[10] Political and imperial historians have explored how British colonies legalized marriage with the deceased wife's sister decades before Britain,
offering a space for secularizing change which in due course provoked reform
in the metropole as a bill to recognise colonial marriages of this kind as valid
in Britain (1906) preceded their full legalization here (1907).[11]

The economic, social, and sexual anxieties that saturated the deceased wife's
sister debate explain why it took up so much Parliamentary time. Yet this essay
concentrates not on the thoroughly mapped subtexts of the debate but on the
scriptural text that disciplined it: the book of Leviticus. This debate involved

9      Nancy Anderson, 'The "Marriage with a Deceased Wife's Sister Bill" Controversy: Incest
Anxiety and the Defense of Family Purity in Victorian England', *Journal of British Studies*
21 (1982), 67–86; Adam Kuper, 'Incest, Cousin Marriage, and the Origin of the Human
Sciences in Nineteenth-Century England', *Past and Present* 174 (2002), 158–83.

10     See Anne D. Wallace, *Sisters and the English Household: Domesticity and Women's
Autonomy in Nineteenth-century English Literature* (London, 2018), ch. 3.

11     Charlotte Frew, 'Sister-in-Law Marriage in the Empire: Religious Politics and Legislative
Reform in the Australian Colonies 1850–1900', *Journal of Imperial and Commonwealth
History* 41 (2013), 194–210.

not just competing ideas of the family but different scholarly and theological approaches to the textual foundations and current application of the Mosaic Law. Arguments about what kinds of marriage a modern state should allow had long shaped how biblical critics thought about the provisions of Leviticus and were, in turn, affected by their conclusions. As Ofri Ilany has argued, the seminal publications of Johann David Michaelis on Mosaic law were interventions in debates in mid-eighteenth-century Protestant Germany about whether the Levitical code should survive in modern state law.[12]

Yet historians have often downplayed the role of biblical scholarship in the British debate which followed the German one, in which Michaelis became a much-quoted authority. They argue that by the time Dalhousie's book appeared, reformers had moved the conversation away from scripture to more promising themes, such as the respectability of the parties who wished to engage in such marriages, the importance of public opinion, or the need to roll back ecclesiastical monopolies.[13] Historians have also cast opponents of legalization as antediluvian holdouts, clinging to ecclesiastical taboos even as biblical criticism discredited them and public opinion turned against them.[14] There was undoubtedly truth in this perspective: in introducing his bill in the summer of 1882, Dalhousie confidently emphasised the 'social aspect of the question' rather than theological considerations.[15] Yet Anglicans and Presbyterians stuck to Parker's Table or the Westminster Confession because they felt they accurately rendered the letter and spirit of Leviticus 18, a text which they regarded not as legislation binding only on the Hebrews but as a 'Marriage Code promulgated to *all nations* by God Himself, Who exterminated the Canaanites (who knew nothing of the Levitical Law) for violating that Code'.[16] Though an embattled position, theirs was hardly a doomed one. The episcopate scotched every bill for sororate marriage put before the Lords while Queen Victoria – who was exasperated by its intransigence – lived and reigned.

The Deceased Wife's Sister debate thus reveals both the continued centrality of Mosaic Law to Victorian Britain and deep disagreements about how to interpret it. This essay explains what Dalhousie thought he was doing in

---

12  Ofri Ilany, *In Search of the Hebrew People: Bible and Nation in the German Enlightenment*, trans. Ishai Mishroy, German Jewish Cultures (Bloomington, 2018), ch. 2.
13  Barrie, *Sin, Sanctity, and the Sister-in-Law*, chs 4–6.
14  Ibid., 100.
15  *Hansard*, 3rd Series, 270 (1882), col. 775.
16  Christopher Wordsworth, 'What the Bishop of Lincoln says; As addressed to the Clergy and Laity of his Diocese at his Triennial Visitation, Oct. 1882', in *Tracts Issued by the Marriage Law Defence Union*, vol. 1, *Scriptural* (1889), tract 1, pp. 7–14, at 9–10.

assembling the Hebraists of Europe to back his bill. It starts by showing that reformers had long thought it profitable to try and understand Leviticus as a Jewish text, because they felt that the 'national interpretation' of Lev 18:18 by Jewish scholars would establish an impeccably scriptural loophole for sororate marriage. Over time, though, as Dalhousie's collection will show, reformers turned to a different argument: by presenting Leviticus as an ancient Jewish text, they could argue that modern legislators could disregard it. In making this argument in print and in Parliament, and in drawing upon German orientalists to do so, they embraced the new variant of anti-Judaism introduced into pentateuchal scholarship by Michaelis, who regarded the Books of Moses not as living scripture but as an intricate but dead text.[17]

The essay then turns to explore why conservatives could bat away the scholarship that Dalhousie marshalled against them. If the strength of their position was institutional, residing in the episcopal bloc vote in the Lords, then it was also intellectual. Although conservatives engaged in grammatical haggling about the original meaning of crucial verses in the book of Leviticus, their position on the text was cushioned by an equally fierce but different form of anti-Judaism. Leviticus to them remained no mere text, but rather Christian scripture. God, not Moses, was the author of Leviticus, and they urged that Jesus Christ was the best guide to what he had meant to say in its disputed chapter 18, which they regarded as arguing that marrying a husband or wife's kin violated the mysterious sanctity of marriage. They condemned the literal interpretation of detached verses to authorise such marriages as a Talmudic perversion of its spirit. The defenders of the status quo were therefore not so much uncritical about Leviticus as hostile to the foundational assumption of orientalist critics from Michaelis onwards, namely that the meaning of the text was bounded by the time and place of its production. The deceased wife's sister debate thus illustrates that religious conflict in Victorian Britain could both activate and neutralise the findings of scholarship.

1    'Wondrous Unanmity': Scriptural Arguments for Marriage Reform

Leviticus 18:18 had long been the locus classicus for reformers like Dalhousie because they believed it authorised marriage with a deceased wife's sister. 'Neither shalt thou take a wife to her sister, to vex *her*, to uncover her nakedness,

---

17    See David Nirenberg, *Anti-Judaism: The Western Tradition* (New York, 2013), ch. 13, and passim; Michael Legaspi, *The Death of Scripture and the Rise of Biblical Studies*, Oxford Studies in Historical Theology (Oxford, 2010).

beside the other in her life *time*' reads this verse in the Authorized Version. Centuries of misapprehension by churchmen, who had drawn out from this chapter a comprehensive table of impediments to marriage, hid the fact that this verse did not expressly forbid and so by implication permitted men to marry a wife's sister after her death.

One of the fullest statements of this case came from Alexander McCaul (1799–1863), the first professor of Hebrew at King's College London. In an 1859 tract, McCaul had welcomed the willingness of high church controversialists to ask what scripture said about the deceased wife's sister question, only to revel in their defeat. To begin with, the 'national interpretation' was against them. Jews had always understood that Lev 18:18 forbade marrying a wife's sister only in the former's lifetime, implying a permission to do so thereafter, a view which the first Jewish Christians clearly shared.[18] The New Testament said nothing against this 'received Jewish interpretation', for while Jesus had criticised Jewish teaching on divorce, he was silent on marriages with the deceased wife's sister.[19]

Furthermore, McCaul could not find down to the Council of Trent any record of Christian theologians understanding the Hebrew of Lev 18:18 differently than the Authorized Version rendered it. He enthused at the 'wondrous unanimity of all ages, countries, and climes – of Jews before the coming of Christ and after the coming of Christ – of Eastern Christians and Western Christians – of Romanists and Protestants' at the meaning of these words. Anyone who disputed that unanimity suggested that the Jews did not know their language or that the 'gigantic scholars of the age of the Reformation were unable to learn Hebrew' properly.[20] This was a suggestive variation on McCaul's usual mode of argument. An evangelical Protestant conversionist, McCaul had gone to Warsaw to learn Hebrew from rabbinic students of the Talmud, the better to convince Jews that the Old Testament proclaimed Jesus as the Messiah. Yet although McCaul's controversial writings tirelessly assailed rabbis for spiritual authoritarianism and perversion of the true meaning of the Hebrew scriptures, he was on this occasion content to argue that rabbinical

---

18    Alexander McCaul, *The Ancient Interpretation of Leviticus XVIII.18, as Received in the Church for more than 1500 Years, A Sufficient Apology for holding that According to the Word of God, Marriage with a Deceased Wife's Sister is Lawful: A Letter to the Rev. W.H. Lyall* (London, 1859), 5.
19    Ibid., 10.
20    Ibid., 24.

Judaism and the 'ancient interpretation' of Leviticus spoke with the same voice.[21]

This unanimity was important because one of the talking points of conservatives was to appeal to a marginal reading of Lev 18:18 contained in the Authorized Version, which read, 'Neither shalt thou take one wife to another to vex her, to uncover her nakedness beside the other in her life-time'. The sixteenth-century Christian Hebraists Franciscus Junius (1545–1602) and Immanuel Tremellius (1510–1580) – a Jewish convert – had introduced this reading in their Latin translation of the Old Testament from 1579. They cited the argument of Karaite Jews that as the Hebrew word for sister could also be an idiomatic expression for one more of a thing, so that it was possible to speak of a curtain having a sister. If their argument were right, Lev 18:18 was not a ban on marrying a wife's sister in her lifetime, which lapsed with her death, but a sweeping condemnation of polygamy. Junius and Tremellius had produced what was in their day 'the preeminent Protestant Latin Bible, an icon of Reformed learning' and part of a new attempt to recover the literal sense of scripture.[22] McCaul, though, dealt with this problem briskly, commenting that theirs had been not a scholarly advance but a cul-de-sac, with the vast majority of subsequent authorities returning to the 'judgment of antiquity' which was also Jewish practice.[23] He was insistent that the word for sister meant just that in the context of this passage. Nor did it make sense to read Lev 18:18 as a condemnation of polygamy when 'the law of Moses presupposes the existence of polygamy'. The 'literal grammatical sense of the Hebrew words' supported the translation in the Authorized Version and therefore did imply that widowers could marry their wife's sister.[24]

Having defused this linguistic mine, McCaul dealt with a deeper question: the kind of inferences about the moral law that one could draw from Leviticus. For opponents of legal change, Lev 18 promulgated a sexual code, with Lev 18:6 as its foundation: 'None of you shall approach to any that is near of kin to him, to uncover *their* nakedness: I *am* the Lord'. The following verses

---

21    See David Ruderman, 'Towards a Preliminary Portrait of an Evangelical Missionary to the Jews: The Many Faces of Alexander McCaul (1799–1863)', *Jewish Historical Studies* 47 (2017), 48–69; idem, *Missionaries, Converts, and Rabbis: The Evangelical Alexander McCaul and Jewish-Christian Debate in the Nineteenth Century* (Philadelphia, 2020).

22    Bruce Gordon, 'Creating a Reformed Book of Knowledge: Immanuel Tremellius, Franciscus Junius, and their Latin Bible, 1580–1590', in *Calvin and the Book: The Evolution of the Printed Word in Reformed Protestantism*, ed. Karen E. Spierling, Refo500 Academic Studies 25 (Göttingen, 2015), 95–122, at 96.

23    McCaul, *The Ancient Interpretation of Leviticus XVIII.18*, 24.

24    Ibid., 60.

showed that 'near of kin' meant bonds not just of consanguinity but also of affinity by providing a host of examples to illustrate the principle. Particularly crucial in this regard was Lev 18:16, 'Thou shalt not uncover the nakedness of thy brother's wife: it is *thy* brother's nakedness'. If men were not – with an exception to be discussed later – to marry the wives of their deceased brothers, then by 'parity of reasoning' that women should not marry the husbands of their deceased sisters.

Defenders of the ban on sororate marriage argued that readers must approach Lev 18:18 with this inference in mind. 'Inference is their stronghold', snapped McCaul, but he insisted that readers draw the right kind of inference from scripture. And here conservatives had gone awry, because there was no good scriptural foundation for the argument that marriage between a man and a woman collapsed the distinction between consanguinity and affinity. Having punctured this inference from silence, McCaul wished to defend the validity of 'inference from limitations'. If Lev 18:18 said that one could not marry a wife's sister during the former's lifetime, then it was legitimate to infer that one could do so after her death.[25] McCaul was not worried that on his reading Lev 18:18 had merely imposed an imitation on an institution – polygamy – which Christ had abrogated. 'No Christian would think of appealing from the Law of Christ to the Law of Moses', he conceded. But in teaching us that polygamy was 'contrary to the original purpose of the Creator', Christ had said nothing about the negative permission which McCaul had extracted from the Levitical limitation on polygamy. It would be a 'strange argument indeed' to say that Christians who availed themselves of the right to marry a deceased wife's sister were obliged to approve of taking more than one wife at once.[26]

For McCaul, the right to marry a deceased wife's sister was a justified inference from Lev 18:18, because it respected what both Jews and Christians had always understood to be the text's literal meaning. His camp made concurrent appeal to Jewish practice and to the long history of Christian Hebraism. The testimony of the Chief Rabbi Nathan Adler (1803–1890) to the 1847 Commission had been a great and much cited coup for them, for he had stated that sororate marriages were 'not only not considered as prohibited' but were 'distinctly understood to be permitted', with 'neither the Divine Law, nor the Rabbis, nor historical Judaism' leaving 'room for the least doubt', citing authorities from the Mishnah to the Shulchan Arukh. The claim that Lev 18:16 established a '*degree*' of forbidden affinity which also encompassed sororate marriage could

---

25    Ibid., 39–40.
26    Ibid., 56.

not outweigh the 'clear and explicit words' of Lev 18:18. And he ruled out the Karaite marginal reading of 18:18 as 'destitute of all authority, and discordant with the spirit of the sacred language'.[27] 'Can words be plainer?' asked the Earl of St Germans in introducing a deceased wife's sister bill into the Lords in 1851. Not only had Hebraists from Michaelis to McCaul endorsed the received translation of the verse, in which the ban on sororate marriage was limited to the wife's lifetime, but so too had 'all the Jewish writers' down to Nathan Adler.[28] In Dalhousie's volume, the Grand Rabbi of Belgium, who answered his query on behalf of the Free University of Brussels, argued that nothing in the Septuagint or the Talmud ruled out such marriages.[29] With the professorship of Hebrew vacant at the university of Graz, its professor of Sanskrit enlisted the town's rabbi Samuel Mühsam (1837–1907), a noted Talmudist who reported that there was no problem with marrying a deceased wife's sister. Other Jewish authorities in Dalhousie's volume were closer to home. From Manchester, the Reform rabbi Tobias Theodores (1808–1886) drew on the sixteenth-century digest of the Talmud Shulhan Arukh and the writings of Maimonides to argue that most Jews had always understood Lev 18:18 to authorise such marriages.[30] From Regent's Park, the eminent Reform rabbi David Woolf Marks (1811–1909) insisted that

> this law has been understood, and by those to whom Hebrew was a living and familiar language, simply as a prohibition to marry a second wife, who might be sister to a first wife then living. But after the first wife's death there is not a particle in the Mosaic code that prevents marriage with her sister. In fact, Jews have rather leaned to such marriages, and I can bear testimony to couples frequently going abroad to get married because the Jewish minister is prevented by the law of the land from solemnizing such marriages in England.

Driving his point home, he related that he had lately been at the death bed of a woman whose last request was that her husband marry her sister to assure

---

27　N. Adler to the Secretary of the Marriage Commission, 13 March 1848, in *First Report of the Commissioners Appointed to Inquire into the State and Operation of the Law of Marriage, as Relating to the Prohibited Degrees of Affinity, and to Marriages Solemnized Abroad or in the British Colonies; with Minutes of Evidence, Appendix, and Index* (1847–48), appendix no. 35, pp. 151–52.

28　Earl of St Germans (Edward Eliot [1798–1877]), *Hansard*, 3rd Series, 114 (1851), cols. 902–03.

29　While the volume gives the name as Very Rev. T. H. Dreyfus, the office was held by Jacques-Henri Dreyfuss (1844–1933) at this time.

30　Allen, ed., *Opinions of the Hebrew and Greek Professors of the European Universities*, 67.

their children's welfare.[31] Marks was a suggestive witness because he was a protégé of McCaul who had founded the West London Reform Synagogue to elevate the scriptures above the Talmud.[32] In an 1854 course of lectures, he had condemned 'traditional' or 'historic' Judaism for its claim that the five books of Moses would be a sealed book were it not for the Talmud, likening this position to Roman Catholicism in its contempt for scripture.[33]

Until the very end of the nineteenth century, Protestant reformers in Parliament enlisted Jews as their allies. 'The text was interpreted by the Jews themselves in the way he contended for', huffed Lord Bramwell in the debate on the second reading of a bill in 1886. 'Who would say that he could interpret the books of the Jews better than the Jews themselves?' Added to the consensus of 'all the best authorities out of England-in the Colonies, and in general in all foreign countries', the argument for change became irresistible, especially to a swashbuckling lawyer like Bramwell who was voluble in his contempt for theological obfuscation.[34]

## 2      'A Law for Jews and Not for Christians': Orientalizing Leviticus

Reformers did not, though, have to start from the position that the law of Moses should be in force today. German Hebraists had long promised freedom from the Mosaic yoke. In his *Abhandlung von den Ehegesetzen Mosis welche die Heyrathen in die nahe Freundschaft untersagen* of 1755 (*Treatise on the Marriage Laws of Moses Forbidding Marriages in Close Affinity*), Johann David Michaelis sought to understand how far the proscriptions of Moses should remain in force by reconstructing their original logic. He represented Moses as a cautious reformer of his people's customs. The bans on marriages between near of kin were not timeless divine commandments but his attempts to safeguard the cohesion of the household, which would be destroyed by indiscriminate promiscuity. Because the logic of these proscriptions was social rather than divine, there was no reason to think that they translated immediately into modern Europe.

---

31    Ibid., 18, 60.
32    See David Feldman, *Englishmen and Jews: Social Relations and Political Culture, 1840–1914* (New Haven, 1994).
33    David Woolf Marks, *'The Law is Light': A Course of Four Lectures on the Sufficiency of the Law of Moses as the Guide of Israel* (London, 1854), 5–6.
34    *Hansard*, 3rd series, 305 (1886), col. 1812.

His arguments were sparked by, and contributed to, initiatives by Protestant German governments to strip out references to the Levitical code from marriage law.[35] Although the *Abhandlung* had little impact on Britain, Michaelis summarised its findings in his elaborate study of *Mosaisches Recht* (1770–75), which reached a wide audience in Alexander Smith's fluent translation (1813). Here Michaelis portrayed Moses as an ancient Montesquieu (1689–1755), a legislator whose regulations were tailored to his nomadic Near Eastern people. The very terms in which Michaelis praised Mosaic prudence raised questions about the relevance of his laws to modern Europe. A ban on lighting fires on the Sabbath might not be onerous in the Arabian desert but would be unfortunate in chilly Norway. Similarly, tolerance of polygamy need not unbalance the population structure of ancient Israel – because its men could raid adjoining tribes for more wives – but would be an evil in modern Germany, which enjoyed no such sexual outlet.[36] Most of Mosaic law had then to be understood as responses to the temporally and geographically distant world in which Moses lived – part of what now appeared to be an 'alien Old Testament'.[37] Even if his laws offered lessons to modern legislators, who, in their efforts to repress duelling, for instance, likewise had to tangle with obdurate popular customs, they did not constitute a model.[38] Michaelis therefore understood Lev 18 as the fragmentary record of Moses's attempt to firefight problems with ancient polygamy as they arose. If Moses had not explicitly forbidden marriage with a deceased wife's sister, there was no reason to think that it was forbidden, then or now.[39]

Notwithstanding the widespread nervousness about German higher criticism in early nineteenth-century England, which led Michaelis's translator Smith (1830–1867) to apologise pre-emptively for his speculations, English advocates for marriage reform made good use of his historically bound Moses. In case the claim that Leviticus authorised marriage with a deceased wife's sister failed to stick, it was handy to be able to suggest that its provisions were

---

35    Ilany, *Hebrew People*, ch. 2.
36    Johann David Michaelis, *Commentaries on the Laws of Moses*, trans. Alexander Smith, 4 vols (London, 1814), 2:21, 24. On Michaelis, see also the chapters by Ofry Ilany, Carlotta Santini, and Irene Zwiep in this volume.
37    For this phrase see Jonathan Sheehan, *The Enlightenment Bible: Translation, Scholarship, Culture* (Princeton, 2005), ch. 5; see also Legaspi, *The Death of Scripture and the Rise of Biblical Studies*, chs 4 and 6.
38    Michaelis, *Commentaries on the Laws of Moses*, vol. 2, 17. Suzanne Marchand, *German Orientalism in the Age of Empire: Religion, Race, and Scholarship*, Publications of the German Historical Institute, Washington D.C. (Cambridge, 2009), 39–41.
39    Michaelis, *Commentaries on the Laws of Moses*, 2: 118–19.

no longer relevant anyway. Having aired the seventeenth-century theologian Jeremy Taylor's (1613–1667) claim that Christians were no longer bound by the Mosaic law, the barrister Henry Revell Reynolds (1775–1854) thus invoked Michaelis in a much discussed pamphlet to claim that the efforts by Moses to regulate kinship relations among ancient Hebrews were not relevant to or binding on Christians now.[40] Yet he also cited the 'elaborate work' of Michaelis in making the argument that Jews had always rightly understood that Lev 18:18 banned sister marriage only during the wife's lifetime.[41] In 1851, the Earl of St Germans had also cited Michaelis as part of his argument that Lev 18's restrictions belonged to the vanished life of 'oriental nations'. All that mattered about 18:18 was that it was a 'prohibitory verse' and that what it did not prohibit was therefore permitted. Decades later, Dalhousie's respondents continued to point to the *Abhandlung* and the *Mosaisches Recht* as convincing explanations of why Lev 18:18 was no more nor less than a limited restriction on the now vanished institution of polygamy.[42]

The irrelevance of the polygamous Mosaic law to modern Christians continued to be a leitmotif of German orientalist scholarship. In the substantial essay he sent to Dalhousie, Paul de Lagarde (1827–1892) developed his thesis that Leviticus was the product of a thoroughly polygamous people, arguing the practice had persisted among Jews until the 'Christian Germans' had put a stop to it in the eleventh century.[43] (Figure 15) Lagarde carried out a leisurely investigation of the word commonly translated as 'sister' in Lev 18:18, diverting himself with waspish asides on the blunders of his predecessors. Yet although he convinced himself that it did indeed mean 'sister', the question was academic, for

> It is impossible to consider Leviticus 18:18 as being in force in Christian times, because polygamy is by this verse supposed to be in general use: if no one is allowed with us to have two *wives* at the same time it is not necessary to forbid two *sisters* to be married to the same man at the same time. *Qui genus negat, species negat.* But if any one should feel inclined

---

40  Henry Revell Reynolds, *Considerations on the State of the Law Regarding Marriages with a Deceased Wife's Sister* (London, 1840), 27.

41  Reynolds, *Considerations on the State of the Law*, 30.

42  See, e.g., Victor Chauvin of Liege (1844–1913), August Ferdinand Michael van Mehren (1822–1907) of Copenhagen, and Count Wolf von Baudissin of Marburg (1847–1926) in Allen, ed., *Opinions of the Hebrew and Greek Professors of the European Universities*, 18, 59, 70. The name of van Mehren was evidently misread and/or misprinted as 'Nehren' in the work: kind thanks go to Jesper Høgenhaven (Copenhagen) for clarifying this matter.

43  Ibid., 27.

to make use of this law as far as use can be made of it in Christian times, it must be acknowledged that its effect is, to allow marriage with a Deceased Wife's Sister.[44]

Lagarde handed back Dalhousie his scriptural precedent, having snapped it in half. This was in keeping with his scholarly preoccupations, which put positivist methodology into the service of liberating modern religion from Semitic hang-ups.[45] The major project of his scholarly life was an edition of the Septuagint, designed to free the Old Testament from what he regarded as the misleading additions to the Masoretic text. In settling the exact nature of Judaism's textual influence on Christianity, this project would allow modern believers to escape from it.[46] His essay for Dalhousie was a work of just this kind: a fastidious plotting of religious customs he regarded as irrelevant.

Other respondents to Dalhousie shared Lagarde's insistence that the authority of Leviticus to restrict or authorise marriage with a deceased wife's sister had passed away with the social forms that occasioned it. Adalbert Merx (1838–1909), the professor of Hebrew at the university of Heidelberg, also sent a long essay to Dalhousie. A champion of scriptural interpretation as a check on the theological misappropriation of the Old Testament, Merx defined this discipline as a determined effort, as much psychological as linguistic, to think oneself into the position of the original author of a text. Although this could lead to a clearer understanding of the Mosaic law, it could not resolve the question of which parts were binding on Christians. Here Christians needed to look back to Jesus, who had come neither to bind nor to loose but to fulfil the Mosaic law. In seizing on the central moral ideas of the Law, he taught Christians an approach which was far superior to the empty formalism [*leerer Formalismus*] of the rabbis.[47] What bearing then did the study of Leviticus have on marriage reform in the present? Merx echoed Lagarde when he commented that the

whole of this law of matrimony is based on a social condition which in our time has been entirely suspended in Christian society. It proceeds on the general supposition that a man may at the same time have more than

---

44      Ibid., 35.
45      Marchand, *German Orientalism in the Age of Empire*, 169; Ulrich Sieg, *Deutschlands Prophet. Paul de Lagarde und die Ursprünge des modernen Antisemitismus* (Munich, 2007).
46      Marchand, *German Orientalism in the Age of Empire*, 170–72.
47      Adalbert Merx, *Eine Rede vom Auslegen ins besondere des Alten Testaments. Vortrag gehalten zu Heidelberg im wissenschaftlichen Predigerverein Badens und der Pfalz am 3 Juli 1878* (Halle, 1879), 16–17, 40–41.

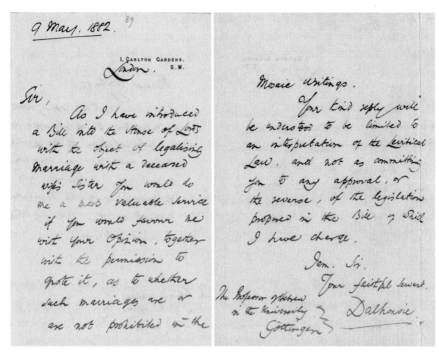

FIGURE 15    Letter from Earl of Dalhousie to Paul de Lagarde (1882)
HELD BY THE STATE AND UNIVERSITY LIBRARY OF THE UNIVERSITY OF
GÖTTINGEN, FILED UNDER COD. MS. LAGARDE 150:238, WITH COURTESY

one wife; it allows polygamy, and takes it as a regular and lawful form of matrimony.

Merx regarded Lev 18 as a complex text, because it is 'self-evident that in polygamic matrimony the degrees of affinity are more complicated than where matrimony prevails'. A close investigation of Lev 18:18 established that it had indeed only sought to prevent simultaneous sexual relations with two sisters, but we needed to remember that this problem only arose in a polygamous society. At the same time, 'whoever transfers this Jewish limitation of the permission to have more than one wife at the same time, to a Christian state of society, can draw but one conclusion, that is: *The Mosaic law clearly allows marriage with the deceased wife's sister*'.[48]

---

48    Allen, ed., *Opinions of the Hebrew and Greek Professors of the European Universities*, 41–6.

The British orientalists whose testimonials appeared in the second and expanded edition of Dalhousie's book were no less insistent than Lagarde in distancing Leviticus from the present. The assyriologist Archibald Sayce (1845–1933), who had served as a reviser of the Old Testament, wrote that whether it is forbidden or not 'in the Mosaic Law seems to me a matter of little moment, as far regards the obligations of Christians'. George Vance Smith (1816–1902), a Unitarian theologian who was, like Sayce, a member of the Old Testament Revision company, was still more outspoken on the irrelevance of Leviticus, for

> we should strive to remember that the Mosaic law is a law for Jews and not for Christians. It is obviously unfit to apply the laws and customs of an ignorant and semi-barbarous age to the regulations of modern life. To attempt this seems repugnant to common sense, and if it were done on any large scale, would only lead to disastrous consequences.

Where George Vance Smith waxed polemical, Lagarde's correspondent William Robertson Smith (1846–1894), who just years earlier had been deposed from his professorship at New College, Edinburgh for questioning the Mosaic authorship of the Pentateuch, exemplified this intellectual distancing. 'Among the Hebrews, as among the Semitic peoples', he wrote,

> the laws restricting marriage within certain degrees of consanguinity or affinity appear to have undergone changes at various stages of the progress of society. ... But at no stage in the development represented by the Pentateuch, with its various laws of different dates, do we find an ordinance which, with any fairness, can be held to prove that marriage with a Deceased Wife's Sister was forbidden.[49]

Even as Smith conceded Dalhousie's point, which was that Leviticus did not positively forbid sororate marriage, he hinted at the oddity of basing modern legislation on the mores of 'Semitic peoples'.

---

49      Ibid., 147–48.

3    'The Broken Cisterns of the Talmud': Opposition to Marriage Law
      Reform and Anti-Judaism

While reformers slowly shifted from invoking to sidelining Lev 18 in their cam-
paigning, defenders of Lyndhurst's Act and their ecclesiastical allies insisted
until the very end of the nineteenth century that the prohibition of marriage
with a deceased wife's sister was religious – and not just religious, but scriptural. After Dalhousie had suggested, in moving the first reading of his bill, in
June 1882, that the Lords should concentrate on the 'social aspect of the question', Lord Balfour of Burleigh restated his belief in 'a Scriptural prohibition
binding upon us at this present day against legalizing the marriage which this
Bill proposes to legalize'.[50] When, having lost his bill, Dalhousie tried his luck a
year later, Lord Cairns restated Balfour's case. Cairns swatted aside Dalhousie's
recently published volume, tutting that the question was not how to interpret
this or that clause in Leviticus. 'The objection is simply this – There is undoubtedly a Code of Law with regard to marriages contained in the Old Testament. Is
that Code a Code which applied only to the Jews, and which did not apply further?' Because Lev 18:3 prefaced this code by reproving the Egyptians and the
Canaanites for their 'doings', it appeared that the ban it placed on marriages
between the near of kin applied to all nations and thus for all time.[51] Balfour
was a Presbyterian Scot, and Cairns an evangelical Anglican, but what united
them was this ownership of Leviticus. Because both they and the scholars who
stood behind them were convinced that Leviticus contained a moral law in
force today, they were ready to move aggressively against suggestions that it
was merely an ancient Jewish text or that Jews rather than Christians were best
qualified to interpret it.

   Edward Bouverie Pusey exemplifies the anti-Judaism which pervaded conservative argument. Pusey was a lynchpin of Anglican opposition to changing
the marriage laws. Having published a *Letter on the Proposed Change in the Laws
Prohibiting Marriage Between Those Near of Kin* (1842), he appeared as a hostile
witness before the Royal Commission of 1847 and returned to the charge with
*God's Prohibition of the Marriage with a Deceased Wife's Sister, Leviticus XVIII.
6., Not to Be Set Aside by an Inference from a Restriction of Polygamy among the
Jews, Leviticus XVIII. 18* (1860). As Regius Professor of Hebrew at Oxford, he
had sent in one of the few dissents to Dalhousie's circular. Pusey warned that

---

50    *Hansard*, 3rd Series, 270 (1882), col. 782.
51    *Hansard*, 3rd Series, 280 (1883), col. 152.

legalisation of sororate marriage would destroy the English family and English society.

The justification for his turbid passion was scriptural. As a leading light of Tractarianism – often known as 'Puseyism' – Pusey belonged to a movement attacked for elevating church tradition over scripture and often faulted since for its ignorance of German biblical criticism. Yet as Timothy Larsen has argued, Pusey saw no tension between scripture and Catholic tradition, venerating the latter because correct interpretations of scripture were trapped within it like flies in amber. Having become Regius Professor in 1828, Pusey spent half a century writing gargantuan commentaries on the Hebrew scriptures which minutely examined and rejected what he regarded as the errors of German higher critics – notably their refusal to acknowledge that the Pentateuch and the prophets had anticipated the Christian dispensation.[52]

Pusey's commentaries on the deceased wife's sister question took a similar approach. They trace at great length what he regards as the universal consensus of the Christian church on the meaning of Lev 18, which captures for him the original meaning of the Hebrew. Lev 18:6 established that marriage makes a woman not just near of kin to a man but 'flesh of his flesh', a principle that makes marriage to blood relatives of a spouse unthinkable. For Pusey, attempts to dodge this principle by inferring from Lev 18:18 permission to marry a woman after her sister's death are unworthy rabbinical haggling. His *Letter* sighed over the

> undutiful captious spirit … which pleads for self-indulgence in every thing which the very letter of Scripture does not absolutely in set words prohibit; which will do nothing, give up nothing, unless it 'find it in the bond', though it be ever so plain, that the whole class of actions to which it belongs is included even in the letter'.[53]

Pusey's quotation from William Shakespeare's (1564–1616) *The Merchant of Venice* stigmatises Christian resort to the Talmudic principle that what is not expressly forbidden in the letter of the text is permitted. In his eyes, this was no less a gaming of the law than Shylock's literalism, which demands a pound of flesh from Antonio regardless of extenuating circumstances. He returned to it in the preface to a reprint of his testimony before the Royal Commission. Explaining that it 'would be a very narrow Pharisaic interpretation of Holy

---

52    Timothy Larsen, *A People of One Book: The Bible and the Victorians* (Oxford, 2011), ch. 1.

53    Edward Bouverie Pusey, *Letter on the Proposed Change in the Laws Prohibiting Marriage Between Those Near of Kin* (London, 1842), 7–8.

Scripture which would so insist upon the letter, as to conceive everything, not in so many words forbidden in the letter, to be permitted, although equivalent to that which is forbidden', he explained that such an interpretation was 'professedly borrowed from the Jews, and resting upon their authority, yet more like the argument of a Jew with which most minds are familiar, 'It is not in the bond', than that of teachable minds wishing to know the mind of God'.[54]

Other high church leaders represented the difference between Jewish and Christian readings of Lev 18 as one between letter and spirit, a trope which dated back to Martin Luther (1483–1546). Christopher Wordsworth, the Bishop of Lincoln (1807–1885), led opposition to deceased wife's sister bills in the Lords and published a tract against them which went into multiple editions. For Wordsworth, Christ's disputes with the Sadducees on the resurrection and with the Pharisees on divorce exposed a basic difference between his ability to seize the spirit of Mosaic legislation about marriage and their unworthy appeals to its letter.[55] If Christian students needed further inducement not to 'prefer the broken cisterns of the Talmud to the living waters of Scripture', they should remember that in emulating 'Jewish expositors of Scripture', they would be 'followers of those who said that Jesus of Nazareth was justly crucified'.[56] Conservative Anglicans like Wordsworth criticised not just Jewish hermeneutics but what another tract writer, the Reverend Francis Pott (1832–1909), acidly called 'Jewish morals'. Pott claimed that the ease with which Jewish men could obtain divorces was proof that they were 'horribly lax' in their attitude to marriage and dredged up allegations from the early modern Hebraist John Lightfoot that adultery was rife among the Jews by the time the Talmud was written.[57]

The one exception to the conservative Anglican contempt for Jewish expositions of Leviticus was their interest in the Karaites. A treatise by the early modern antiquarian John Selden (1584–1654) taught them that the Karaites of eighth-century Mesopotamia claimed that marriage with a deceased wife's

---

54 Edward Bouverie Pusey, *Marriage with a Deceased Wife's Sister Prohibited by Holy Scripture as Understood by the Church for 1500 Years: Evidence Given before the Commission Appointed to Inquire into the State and Operation of the Law of Marriage as Relating to the Prohibited Degrees of Affinity* (Oxford, 1849), iv–v.

55 Christopher Wordsworth, *On Marriage with a Deceased Wife's Sister*, new ed. (London, 1883 [1876]), 8.

56 Ibid., 18. His citation of Jeremiah 2:13 ('For my people have committed two evils; they have forsaken me the fountain of living waters, and hewed them out cisterns, broken cisterns, that can hold no water') further spikes the polemic against Jewish faithlessness.

57 Philadelphus (Francis Pott), *Marriage with a Deceased Wife's Sister: A Brief General Review of the Arguments and Pleas on this Subject* … (London, 1885), 18 and Appendix 18.

sister went against Leviticus. Citing Selden's *De Uxor Ebraica*, Pusey argued in his evidence to the Royal Commission that 'the Karaites laid down' the principle that the man and wife being one, 'those of kin to the wife were forbidden to the husband as his own'.[58] Not only was it convenient to find Jewish witnesses against Jewish tradition, but they could cast the dispute between Karaite and Talmudic Jews as a variation on the struggle against Roman Catholicism, with the Karaites being 'Protestant' in their adherence to the scriptures. There had not been many Karaites, Francis Pott allowed, but their superiority 'compared with that of the more or less 'orthodox' Jew, is not to be measured by their numbers, but by the superiority of direct over indirect testimony to the meaning of Holy Scripture'.[59] As late as 1912, Frederick William Puller (1843–1938) claimed in a tract against the implementation of the new marriage law that the Karaites had nobly stood out against the 'Rabbanites' and 'Talmudici'. In this they echoed Jesus, who had reproved 'the ceremonial rigorism and the moral laxity of the Jewish teachers of His day'.[60] The continued prestige accorded the Karaites was in contrast with their rough handling by marriage reformers, who mentioned their interpretation of only to dismiss it as eccentric.[61]

What did it mean to read Lev 18 in a spiritual way, as conservatives urged? When confronting a fragmentary or repetitive list of proscriptions, it was important to draw out the principle uniting them, rather than profiting from their gaps. As Wordsworth put it, 'to ask for an express text for everything we do, or forbear to do, is to tempt God, and to disparage His Word as imperfect, and to despise the gift of Reason which we have from Him'.[62] This meant inferring from the examples given in Lev 18:18 the degrees of affinity which ruled out marriage, even those not itemised in the text – precisely the work done by Parker's Table. In 'all interpretations of the law the general drift of the whole must be considered, and be used as the clue for its exposition', Wordsworth urged:

> and that in right constructions of law, that which is doubtful is to be elucidated by means of what is clear, and not that which is clear be obscured

---

58    Pusey, *Marriage with a Deceased Wife's Sister Prohibited by Holy Scripture*, 9.

59    [Pott], *Marriage with a Deceased Wife's Sister*, 19–20.

60    F.W. Puller, *Marriage with a Deceased Wife's Sister: Forbidden by the Laws of God and of the Church* (London, 1912), 48–49. For the immediate context of Puller's book see Bruce S. Bennett, 'Banister v. Thompson and Afterwards: The Church of England and the Deceased Wife's Sister's Act', *Journal of Ecclesiastical History* 49 (1998), 669–82.

61    See, e.g., McCaul, *Ancient*, 46; Chauvin and Lamy in Allen, ed., *Opinions of the Hebrew and Greek Professors of the European Universities*, 56, 63–64.

62    Wordsworth, *On Marriage with a Deceased Wife's Sister*, 6–7.

by that which is doubtful. And we assert that such a variance as has just been recited is at variance with the whole context of the law, by which a man is expressly forbidden to contract marriage with the *kindred* of his wife, as has already been shown, and in which a *sister* is especially mentioned as near of kin.[63]

The synonyms that conservatives employed for the 'spirit' of the text reveal their faith in its hidden, ultimate rationality: they spoke of its 'tenour' or 'drift'. That was because they regarded its true author not as Moses – the Hebrew Montesquieu – but God, who could lay down the law in mysterious ways. The pamphleteer Henry H. Duke (1816–1888) concluded that while 'ordinary human laws' needed to have 'precision and clearness of expression', 'Divine legislation has another scope and purpose, lying, perhaps, beyond this ... It is to be made of service for the probation of souls, and for the training and disciplining human creatures in the observance of a royal law of obedience for conscience and for love's sake'.[64]

The tenour of Lev 18 was Christocentric. The writings of James Candlish (1835–1897), the Free Church professor of Hebrew who was one of the most persistent opponents of marriage law reform, illustrate this well.[65] Candlish came to the fore in the Scottish opposition to James Stuart-Wortley's (1805–1881) deceased wife's sister bills (1849–50) and remained prominent thereafter, standing out against a softening of his Church's opposition to legalization. Dalhousie's anthology provoked him into writing a tract-length dissection of its errors. Despite his intransigence, Candlish was no troglodytic conservative and had often stood up for the limited practice of higher criticism.[66] In an 1877 tract occasioned by the controversy over William Robertson Smith's investigations into the authorship of the Pentateuch, Candlish argued that in declaring 'all the books of the Old and New Testament' to be the 'Word of God', the Westminster Confession freed rather than constrained biblical critics.[67] Because the Confession did not exactly describe how the Bible was the Word of God, to insist that Robertson Smith's investigations had cast doubts

---

63    Ibid., 13.
64    Henry Duke, *The Question of Incest Relatively to Marriage with Sisters in Succession* (London, 1883), 25.
65    David Barrie, *Sin, Sanctity, and the Sister-in-Law*, 18.
66    See Valerie Wallace and Colin Kidd, 'Biblical Criticism and Scots Presbyterian Dissent in the Age of Robertson Smith', in *Dissent and the Bible in Britain, c.1650–1950,* eds Scott Mandelbrote and Michael Ledger-Lomas (Oxford, 2013), 233–55.
67    James Candlish, *The Authority of Scripture Independent of Criticism* (Edinburgh, 1877).

upon its inspiration was to succumb to just the kind of 'rationalistic principle' they reproved in others.[68] 'Our only safe method is to learn from the study of itself what the Bible actually is, and to judge from the nature of each part and its relation to others what is literal and what is figurative, what historical and what poetical'.[69] This meant that Robertson Smith was within his rights to raise doubts about the Mosaic authorship of Deuteronomy. To claim that much of it had been written down at a later date did not affect its inspiration. Nor, once one recognised that the 'literary habits of Orientals' differed from those of modern Europeans, did this amount to an allegation of 'pious fraud'. Candlish urged 'calm and confident waiting' about the higher criticism of the Pentateuch.[70]

Yet this calmness deserted Candlish when criticism undermined the moral authority of the Mosaic law. Although his rebuttal of Dalhousie's volume did engage in some linguistic fencing, his main tactic was to give theology the whip hand over philology in interpreting the Pentateuch.[71] For Candlish, 'professors of theology, or ethics, or jurisprudence' had more to say than the specialists in 'grammatical exegesis' drafted by Dalhousie, who in their narrow focus on Lev 18:18 had failed to take a 'conjoint view of the teaching of Scripture as a whole'.[72] What could such an approach, which moved away from 'proof texts' to think about scripture as a whole, establish? Candlish insisted that it could locate one consistent teaching across Old and New Testaments. When Paul said in his letter to the Ephesians – a work dear to Candlish – that 'for this cause shall a man leave his father and mother, and shall be joined unto his wife, and they two shall be one flesh' (5:31), he proclaimed that Christians held true to the 'one flesh' teaching expounded by Gen 2:24 and also cited by Jesus (Matt 19:5 and Mark 10:8).[73] Under the one flesh doctrine, marriage made affinity as important as consanguinity. As an outcrop of scripture, therefore, Lev 18 must argue that marriage to a deceased wife's sister violated God's law. For Candlish as for Pusey, scholars who followed 'Jewish Talmudists' in arguing that anything Lev 18:18 did not rule out was permitted missed this insight.[74]

---

68     Ibid., 15–16.
69     Ibid., 18.
70     Ibid., 20–26.
71     Idem, 'The Real Bearings of the Opinions of the Professors of Hebrew and Greek on the Scriptural Law of Prohibited Degrees of Marriage,' in *Tracts Issued by the Marriage Law Defence Union*, vol. 1, *Scriptural* (London, 1889), tract 25, pp. 177–96, at 180.
72     Ibid., 182.
73     Ibid., 192–93.
74     Ibid., 188.

Candlish's insistence that Christ's reading of Genesis supplies the frame for Leviticus was a recurrent theme in Anglican controversial writing until the end of the nineteenth century. The Reverend W.F. Hobson (ca. 1827–1892) started his tract by asking what 'Our Lord' said upon the subject, answering that he had confirmed Genesis when he said that on marriage man and woman 'are no more twain, but one flesh' (Matt. 19:5,6).[75] One could pass 'from our Lord's words straight to the Levitical ordinances' and find there that '*the wife's sister is a man's own sister,* and that *because* he has become 'one flesh' with the two sisters'.[76] Hobson even imagined how Jesus might have padded out the terse injunction of Lev 18:18: '*I* say unto you that *the living sister is the husband's sister,* and that even as through *blood*'. This led Hobson to the conclusion that the debate was 'not simply a point of Jewish polity and social regulations ... No! it is nothing less than the granting of a greater licence where Christ our Lord afresh laid restraint'.[77]

Making Christ the chief commentator on Leviticus dealt with another thorny problem for conservatives: levirate marriage. Because they regarded the ban on men marrying their brother's widow (Lev 18:16) as convertible with a ban on women marrying their sister's widower, conservatives were foxed by Deut 25:5–10. Scholarly reformers such as McCaul had frequently noted that in enjoining men to marry their brother's widow if she was still childless, this text contradicted Lev 18:16 and suggested that kinship barriers to marriage might be flexible, not absolute. The Sadducees had raised this very law in their dispute with Christ on the resurrection, when they asked as whose wife a woman who had successively married seven brothers would be resurrected. Conservatives spilled much ink on arguing that a closer investigation of Deuteronomy's Hebrew showed that the widow was supposed to marry not her husband's brother, but merely another member of his tribe. The Reverend W.B. Galloway cited Rabbis D.A. De Sola (1796–1860) and M.J. Raphall's (1798–1868) 1843 translation of the Mishnah treatise 'Yebamoth' as 'historical evidence' that the Jews had always understood Deut 25:5–10 in this way, twitting McCaul for his ignorance.[78] But an easier approach was to argue that this apparent exception

---

75  W.F. Hobson, 'The Christian Law of Marriage: What does Our Lord Say upon the subject?', in *Tracts Issued by the Marriage Law Defence Union,* vol. 1, *Scriptural* (London, 1889), tract 20, pp. 145–52, at 147.

76  Ibid., 148.

77  Ibid., 151.

78  William Galloway, *The Unlawfulness of the Marriage of Brother and Sister-in-law: In the Light of the Word of God; with Ancient Evidence Hitherto Generally Overlooked* (London, 1870), 22–27, citing D.A. De Sola and M.J. Raphall, trans., *Eighteen Treatises from the Mishna* (London, 1843).

only held under the Jewish dispensation, when it was important that no family should be allowed to disappear from God's contract with the Hebrew nation. Once Christ had made salvation universal, this exception to the moral law was surplus to requirements.

## 4    'Developed Jews': Restating the Authority of Leviticus

This reliance on Christ as the authoritative interpreter of Mosaic law was so emphatic that opponents of sororate marriage were sometimes tempted to decouple their religious case from minute assertions about the text of Leviticus, just as reformers increasingly did. Liberal Anglican bishops had taken that step early on. In 1851, Samuel Hinds (1793–1872), the Bishop of Norwich, confessed that 'I would not trust to a Jew for the meaning of a doctrinal scripture, or for the interpretation of a prophetic scripture; but the presumption in favour of his rightly interpreting a Scripture direction respecting marriage customs is such as would require some very strong internal evidence to overthrow it'.[79] Given the ubiquity of marriage to a deceased wife's sister in Judaism, he preferred to rest his opposition on the surer ground of the integrity of the Christian family. Connop Thirlwall (1797–1875), the Bishop of St David's, made the same move in this debate. A man deeply versed in German philology, Thirlwall argued that even if one accepted that Lev 18 'belonged to the moral law', it 'did not at all follow, that every particular ordinance on this subject should possess the character of an immutable moral law'. Because any given verses might address the social problems of Moses's own time rather than all time, it was 'only a matter of inference and construction' to draw out systematic proscriptions from them.[80]

Hinds and Thirlwall's party was a minority in the church at mid-century, but thirty years later, the Bishop of Peterborough William Connor Magee (1821–1891) startled churchmen and delighted Dalhousie when he gave up the assistance of Lev 18:18 in a speech against the second reading of his 1882 bill.[81] The Duke of Marlborough did much the same when he led the successful efforts to defeat the third reading of Dalhousie's bill on its reintroduction in 1883. The Duke could not find an explicit condemnation of these marriages in Leviticus, but that did not matter because

---

79    *Hansard*, 3rd Series, 114 (1851), col. 958.
80    Ibid., col. 953.
81    *Hansard*, 3rd Series, 270 (1882), col. 795.

they had arrived at a different state of things, and, with the sanction of Christianity and of the New Testament, they might say that old things had passed away, and that they had a higher law, and that a higher morality had been introduced. That was the ground on which they ought to test the religious sanction of these marriages; and that, he thought, was a far higher ground of objection than obscure passages of the Old Testament, or the mere presence or absence of an express prohibition.[82]

The eighth duke of Argyll, George Campbell (1823–1900), took a similar tack in backing up Marlborough. A Presbyterian peer who was also an austere scientific theist, Argyll made a religious but not a scriptural case against sororate marriage. Using language which echoed Pusey's diatribes against Pharisaism, he alleged that

> it was not the opponents of the Bill, but its promoters who might be accused of Judaism. The latter appealed to the absence of direct prohibition. They desired to be guided always by some petty and verbal direction. That was the spirit of the Jewish Dispensation; it was not the spirit of the Christian Dispensation. 'Thou shalt' and 'Thou shalt not' – that was the language of the Jewish Dispensation. On the contrary, Christianity adopted a wholly different system.[83]

Yet Argyll's efforts to find a 'general principle' merely confused the lines of battle. 'The noble Duke had taunted them with entertaining feelings of Judaism', marvelled the Earl of Kimberley, John Wodehouse (1826–1902), a briskly anticlerical Liberal peer. 'He had thrown to the winds the argument from the whole Old Testament, and said he went upon some broader principle of his own'.[84] For Kimberley, the bulwark of Leviticus had crumbled, but the phalanx of bishops who spoke on Argyll's side of the debate quickly closed up this breach between Christianity and the Mosaic law. Harold Browne (1811–1891) the Bishop of Winchester thundered that

> every orthodox Church and every tolerably orthodox sect in Christendom held the doctrine that the moral law of the Old Testament was binding upon Christians, and that the New Testament was the development of the Old. There was a famous saying accepted in the Primitive Church – *Novum*

---

82  *Hansard*, 3rd Series, 280 (1883), cols. 1655–56.
83  Ibid., col. 1665.
84  Ibid., col. 1675.

*Testamentum latet in Veteri Testamento, Vetus Testamentum patet in Novo*
[the New Testament lies hidden in the Old and the Old is made plain in
the New].

The argument from Leviticus stood where it always had done. Leviticus 18:6
proclaimed the 'general principle' for which Argyll argued; Lev 18:7–17 illustrated it with so many examples that efforts to overturn its teaching by quoting Lev 18:18 in isolation failed.[85] Nor did Browne have any truck with the
suggestion that Leviticus did not apply to them because they 'were not Jews'.
Borrowing the recently deceased Benjamin Disraeli's (1804–1881) quip that he
was a 'developed Jew', Browne suggested that 'truly, they were all developed
Jews. All the privileges and all the moral obligations of the Jewish Fathers still
were theirs, only expanded and intensified'.[86]

Four years later, even Argyll, who had once stigmatised marriage reformers
as overly Jewish in their approach, now lamented the 'philosophic contempt'
with which the Jews were regarded as a 'little, insignificant people settled in
the extreme Western part of Asia'. That was misguided, because the Jews had
survived the vicissitudes of the centuries – an argument that should tell with
the House of Lords, 'who were supposed to revere the hereditary principle
... What were their titles of nobility compared with the descent of the Jews?'
Whereas years before, Argyll had played off the Jewish against the Christian
Dispensation, he now invited the Lords to respect the 'Jewish Marriage Laws',
whose exacting restrictions had guaranteed their 'preservation' as a people.[87]

The debate on marriage with a deceased wife's sister had brought scholars from Michaelis to Lagarde into the thick of political conflict. But 'debate'
is a misnomer, because despite skirmishes over details, the opponents of
reform hunkered in their trenches and let their enemy's fire pass overhead.
Though Candlish, for instance, conceded that the reformers had knocked
out the marginal reading of Lev 18:18, Wordsworth and other Anglicans still
paraded the early modern Hebraists and divines who had supported it, urging that Dalhousie had staked everything on an 'ambiguous and obscure'
text.[88] Edward White Benson (1829–1896), the Archbishop of Canterbury, did

---

85    Ibid., col. 1672–3.
86    Ibid., col. 1673.
87    *Hansard*, 3rd Series, 305 (1886), col. 1803.
88    Bishop of Lincoln (Christopher Wordsworth), 'Lev. XVIII. v. 18. "A Wife to her Sister."
      Explained', in *Tracts Issued by the Marriage Law Defence Union*, vol. 1, *Scriptural* (1889),
      tract 22, pp. 155–56, at 155; 'Speech of the late Bishop (Thirlwall) of St David's', in ibid., tract
      24, pp. 173– 76, at 175. See also George Trevor, *The Scriptural Argument against Marriage
      with a Deceased Wife's Sister* (London, 1884), 11–12.

the same, telling the Lords in 1886 that it was 'an error to suppose that the controversy turned upon one verse in Leviticus', not least because there were 'two schools among the Jews' on how it should be construed. Without agreeing with the Karaites exactly, he preferred given this supposed uncertainty to ask whether the 'Levitical Law. ... sanctioned the idea of the family we had in England'.[89] Benson's citation of the history of scholarship may look disingenuous, but it nerved his party to throw out the bills repeatedly sent up to them by the Commons.

This essay has argued that the interpretation of the Mosaic law was never an academic question in Victorian Britain: the very legislative definition of marriage and the family were at stake. Though the testimonials in Dalhousie's book showed how political conflicts could bring scholarship into public life, the insistence of religious conservatives that they, rather than his professors, knew how to read the law of Moses showed that the ecclesiastical contours of Victorian life could be durable earthworks against secularizing approaches to the Pentateuch. 'I thank God that the word "Scriptural" still bears in England, to some extent, the meaning of "moral", and that what is laid down in Scripture does come to us with the force of a moral commandment', Benson had argued in helping to defeat Dalhousie's 1883 bill.[90] In Victorian Britain, news of the death of scripture could be slow to arrive.

---

89    *Hansard*, 3rd Series, 305 (1886), col. 1815.
90    *Hansard*, 3rd Series, 280 (1883), col. 172.

# CHAPTER 7

# Moses and the Left

*Traces of the Torah in Modern Jewish Anarchist Thought*

*Carolin Kosuch*

In his treatise *Jerusalem oder über religiöse Macht und Judenthum* (*Jerusalem, or on Religious Power and Judaism*, 1783), Jewish philosopher Moses Mendelssohn (1729–1786) drew a distinction between historical truths and eternal truths in Judaism. While he stated that the historical truth would be connected to the Jewish nation, which ceased its existence after the Temple was destroyed in the year 70 CE, he emphasized the presence of eternal truths which he believed would be available to all human beings at any time independently of any religious canon. But other than religions based on divine revelation, Judaism, to Mendelssohn, seemed *the* religion of tolerance and humanity because it allowed accessing the eternal truths by means of pure reason. Mosaic law, which, following Mendelssohn, was, is, and stays mandatory for all those who are born as Jews – Jesus included – represented a way of reason to reach out to these eternal truths. Mendelssohn stressed that these laws were intended to be put into practice. They do not require faith but offer an exemplary rational way that would help to realize God's final goal for humankind: happiness. Still, this particular Jewish way, which Mendelssohn wanted to maintain at any cost, does not stand in the way of civil equality and brotherly union. On the contrary, difference furthers tolerance and therefore is desirable for a community, he claimed.[1]

Mendelssohn's thoughts reflect the Jewish Enlightenment (*Haskalah*). In the course of the late eighteenth and early nineteenth centuries, they heralded the idea that in the presence and future the Jewish segregation in civil, economic, and cultural life could be overcome by 'improving' (*'verbessern'*) German Jews and by integrating them in the history of their Christian neighbors.[2] Such

---

1  Moses Mendelssohn, *Jerusalem, oder über religiöse Macht und Judentum. Mit dem Vorwort zu Manasse ben Israels* Rettung der Juden *und dem Entwurf zu* Jerusalem *sowie einer Einleitung, Anmerkungen und Register*, ed. Michael Albrecht, Philosophische Bibliothek 565 (Hamburg, 2005 [1783]). If not indicated otherwise, all translations are my own.

2  Many *Maskilim* after Mendelssohn included religion to this list. Other than him, they felt no longer bound to Mosaic Law.

plans were not uncontested but challenged by German Romanticism, which focused on the different histories and religions of distinct peoples.[3] Jewish emancipation and acculturation became a state project. As Hannah Arendt (1906–1975) has pointed out, Jews were integrated as particular into the predominant and therefore generalized (Christian) culture and history. She quotes Johann Gottfried Herder (1744–1803), who claimed that the Palestine of the Jews should be where they live and where they would act in a noble sense for the better of their societies.[4]

This essay aims to study traces of the Mosaic law and the Jewish textual canon in modern German-Jewish radical leftist thought. Gustav Landauer (1870–1919) and Erich Mühsam (1878–1934), two prominent German-Jewish anarchists, in their political theory, combined anarchist, socialist, and artistic strands. On a broader level, their unique anarchism seems to have mixed enlightened and romantic ideas of Jewish acculturation in a post-emancipatory era of formally accomplished legal equality.[5] Their anarchism, as will be shown, answered in a particular way to the problems modernity had imposed on Jews.

In a first step it will be discussed why acculturated, educated German Jews of bourgeois and affluent origin turned to anarchism – a theory of revolution striving to alter completely the status quo. Second, Mosaic law and figures of the Jewish canon as part of their anarchist theory will be presented and interpreted. These references in Landauer's and Mühsam's anarchism seem to echo the rabbinic and Kabbalist principle of *tikkun olam*, as will be analyzed in a third section. In a multidirectional process of revisiting and mingling Christian, ancient Greek, and secular references, and in light of the growing antisemitism that undermined the promised equality in the German Empire, these anarchists created a new, secular approach to Jewish tradition and law in modernity.[6] This chapter aims at studying such shifting significances and highlights the continuous value of Mosaic law in modern Jewish anarchist thought.

---

3   Hannah Arendt, 'Aufklärung und Judenfrage', repr. in *Wir Juden. Schriften 1932 bis 1966*, eds Marie Luise Knott and Ursula Ludz (Munich, 2019 [1932]), 11–30; David Sorkin, *The Transformation of German Jewry, 1780–1840* (New York, 1987); Andreas Gotzmann, *Eigenheit und Einheit. Modernisierungsdiskurse des deutschen Judentums der Emanzipationszeit* (Leiden, 2002); Christoph Schulte, *Die jüdische Aufklärung. Philosophie, Religion, Geschichte* (Munich, 2002).

4   Arendt, 'Aufklärung und Judenfrage', 29.

5   Steven M. Lowenstein et al., *Deutsch-jüdische Geschichte in der Neuzeit*, vol. 3, *Umstrittene Integration 1871–1918* (Munich, 1997).

6   Daniel Weidner, 'Säkularisierung', in *Enzyklopädie jüdischer Geschichte und Kultur*, vol. 5, ed. Dan Diner (Stuttgart, 2014), 295–301.

1      Anarchism, Marxism, and Jewish Radicals in the Nineteenth
       Century

Anarchism as a philosophy, not a political theory, was already debated both affirmatively and negatively by ancient Greek philosophers such as Plato (429–347 BCE) or Zeno (ca. 495–430 BCE). But it was in Europe during the eighteenth and nineteenth centuries that modern anarchism – under the influence of Enlightenment, Romanticism, and early socialist ideas – developed as a multifaceted political movement comprising such disparate varieties like terrorism and pacifism, syndicalism and anarcho-communism.[7] Modern anarchists envisioned an egalitarian society in which the state and its institutions – notably the police forces and the military machinery – would become superfluous.[8] The history of modern anarchisms is complex and took different courses depending on the region and period. German anarchism was an idea only a small minority sympathized with. It had a particularly hard time under Otto von Bismarck (1815–1898), whose Anti-Socialist Laws (1878–1890) forced socialist organizations into hiding, where they further radicalized.

Even though modern anarchism developed as part of the rich spectrum of European socialisms, it started to differentiate from other socialisms in the midst of the century. Specifically, anarchists opposed some of the basic ideas of Karl Marx (1818–1883) and finally separated from the Marxists in 1872 after long and intense disputes and once having been expelled by Marx in the course of the First International. It was the first self-designated 'anarchist' Pierre-Joseph Proudhon (1809–1865), who rejected Marx's Hegelian-inspired political concepts because of their determinism and strict systematizing reading of the historical process. Proudhon was convinced that freedom and state socialism would be ill-matched principles. Instead, he suggested the immediate introduction of cooperatives and credit unions owned by the people that should replace the state, the monetary system, and capitalism.[9]

Cooperation, conviviality, and mutual aid were also at the core of Pyotr Alexeyevich Kropotkin's (1842–1921) anti-Darwinian and anti-Marxist anarchism. He advocated for a network of autonomously operating socialist communities in which the individual should strive for a maximum of happiness and fulfillment. These self-sufficient unities were destined to replace the

---

7   Carl Levy, 'Social Histories of Anarchism', *Journal for the Study of Radicalism* 4 (2010), 1–44.
8   Peter Seyferth, ed., *Den Staat zerschlagen! Anarchistische Staatsverständnisse* (Baden-Baden, 2015).
9   Walter Theimer, *Geschichte des Sozialismus* (Tübingen, 1988); Anne-Sophie Chambost, *Proudhon. L'Enfant terrible du socialisme* (Paris, 2009).

capitalist industry that Kropotkin considered pathogen.[10] Other than Marxists, Kropotkin together with many other anarchists held that the revolution was in feasible reach and not postponed to an uncertain moment in the distant future.

Anarchism's general focus on the emancipated individual and the emphasize it put on humanist egalitarianism did not contradict hostile feelings anarchists entertained towards specific social groups including, foremost and as part of the anarchist anti-capitalism and anti-etatism, those holding positions of administrative, military, religious, financial, and academic-cultural power. However, leading anarchists also expressed antisemitic attitudes towards the Jewish minority, starting with Proudhon but voiced as well by Mikhail Aleksandrovich Bakunin, (1814–1876), another influential anarchist radical who was in constant conflict with Marx. To him, the 'whole Jewish world' formed an 'exploitative sect, a people of bloodsuckers, a single, devouring parasite, united and connected not only across boarders but also across different political camps, – this Jewish world today serves largely both Marx and Rothschild'.[11] Besides, and together with Sergey Gennadiyevich Nechayev (1847–1882) and others, Bakunin did not shrink form promoting the violent 'propaganda of the deed', terrorism, and assassinations of prominent potentates.[12]

Considering these tendencies, it may come as a surprise that German Jews who grew up in a bourgeois setting in the last decades of the nineteenth century felt drawn to anarchist positions. Gustav Landauer and Erich Mühsam, two of the most known German anarchists, were born in the 1870s into acculturated Jewish families in the German Empire which had formally granted legal equality to the Jewish minority. Their social upbringing was a typical bourgeois one with values like duteousness, the appreciation of higher education,

---

10   Jim Mac Laughlin, *Kropotkin and the Anarchist Intellectual Tradition* (London, 2016).

11   Mikhail Aleksandrovich Bakunin, 'Persönliche Beziehung zu Marx – Auszug', repr. in *'Antisemit, das geht nicht unter Menschen'. Anarchistische Positionen zu Antisemitismus, Zionismus und Israel*, vol. 1, *Von Proudhon bis zur Staatsgründung*, eds Jürgen Mümken and Siegbert Wolf (Lich/Hessen, 2013 [1871]), 80–84, at 83.

12   Wolfgang Bock, 'Terrorismus und politischer Anarchismus im Kaiserreich: Entstehung, Entwicklung, rechtliche und politische Bekämpfung', in *Anarchismus. Zur Geschichte und Idee der herrschaftsfreien Gesellschaft*, ed. Hans Diefenbacher (Darmstadt, 1996), 143–68. Marx, too, perpetuated certain antisemitic clichés which formed a 'cultural code' (S. Volkov) in the nineteenth century: cf. Karl Marx, 'Zur Judenfrage', *Deutschfranzösische Jahrbücher* 1–2 (1844), available at http://www.mlwerke.de/me/me01/me01_347.htm (accessed 6 December 2022); see further Gareth Stredman Jones, *Karl Marx. Die Biographie* (Frankfurt am Main, 2017).

social commitment, and civic engagement.[13] Mühsam's and Landauer's fathers succeeded as merchants: both were esteemed and engaged members of their communities and cultivated a style of living befitting their social status. In the families' every-day life only little reference to Jewish custom or the Jewish canon surfaced. Landauer remembered that his upbringing had been 'as little religious as possible',[14] and Mühsam's cousin Paul, who lived with the Mühsam's for a period of time, recalled his uncle, the head of the family and Erich Mühsam's father, 'by no means' had been

> the type of assimilation Jew who deliberately was trying to hide his Jewishness. [...] He simply did not attach any importance to it towards the general public, but sought to integrate himself into the Christian environment as much as possible. Celebrating of Christmas with Christmas tree and gifts was obligatory, and in the presence of the staff, everything Jewish, even the word Jew, was fearfully avoided. In religious terms, he was liberal and of a free spirit. He let me take part in the Christian religious education in school without hesitation.[15]

The somewhat surprising turn from this well-established and secure bourgeois sphere to anarchism young Landauer and Mühsam took was triggered by three factors. The first concerned their growing opposition to their patriarchal fathers. In line with the gendered social order and attributions of their time and standing, these fathers presided over their families and, sometimes harshly, disciplined their youngest sons because of their deviation from the expectations set in them. Erich Mühsam recollected about his childhood:

---

13      Peter Gay, 'Begegnungen mit der Moderne – Deutsche Juden in der deutschen Kultur', in *Juden im Wilhelminischen Deutschland, 1890–1914*, eds Werner Mosse and Arnold Paucker (Tübigen, 1976), 241–312; Dieter Langewiesche, 'Liberalismus und Judenemanzipation im 19. Jahrhundert', in *Juden in Deutschland. Emanzipation, Integration, Verfolgung und Vernichtung*, eds Peter Freimark, Alice Jankowski, and Ina Lorenz (Hamburg, 1991), 148–63; Till Van Rahden, 'Von der Eintracht zur Vielfalt. Juden in der Geschichte des deutschen Bürgertums', in *Juden, Bürger, Deutsche. Zur Geschichte von Vielfalt und Differenz, 1800–1933*, eds Andreas Gotzmann, Rainer Liedtke, and Till van Rahden (Tübingen, 2001), 9–32; Simone Lässig, *Jüdische Wege ins Bürgertum: Kulturelles Kapital und sozialer Aufstieg im 19. Jahrhundert* (Göttingen, 2004).

14      Landauer to Ida Wolf, 15 June 1891, International Institute of Social History Amsterdam, Landauer Papers, no. 100, Early Writings.

15      Paul Mühsam, *Ich bin ein Mensch gewesen. Lebenserinnerungen*, ed. Ernst Kretzschmar (Gerlingen, 1989), 15–16.

When I think of the unspeakable beatings by which every natural feeling was beaten out of me I feel something like hatred. [...] And always my father who prided himself of his educational method, the pride of this man, who could not accept that his children weren't all the same, that three were the way he wanted them to be, good, hardworking, obedient, and only I was completely different.[16]

Already during their adolescence both Landauer and Mühsam started to rebel against this dominant father-figure and the fatherly bourgeois culture of their youth.

They did so, second, by means of philosophy and literature. Their reading of Friedrich Schiller (1759–1805) and Johann Wolfgang von Goethe (1749–1832) was slowly supplemented by Friedrich Nietzsche's (1844–1900) concept of the superhuman, his harsh criticism of state and religion, as well as his claim for new, self-defined values. They also received Max Stirner's (1806–1856) idea of the total emancipation of the individual from all inner and outer heteronomy. Encouraged by these radical writings, Landauer and Mühsam broke with the expectations of their families: by quitting university without degree and marrying a Protestant worker without the blessing of his father (Landauer); by refusing to work hard in school and to engage in a permanent employment (Mühsam); and by joining radical literary and political circles in Berlin, frequented by their peers all struggling with generational conflicts (both). In these circles, Landauer and Mühsam encountered the writings of Proudhon, Bakunin, and Kropotkin and slowly familiarized with radical socialism and anarchism.[17] In the surroundings of Berlin, alternative ways of living and working together were tried out and attempts were made to mingle art, spirituality, and life by creating new forms of artistic expression (Figure 16).[18] Interestingly, Landauer felt not deterred by the antisemitic comments of philosophers such as Proudhon or Eugen Dühring (1833–1921). He simply adopted selected teachings from their systems of thought he found useful for his own ideas. Landauer developed them further on in his own anarchist approach while paying little attention to their animosities which he treated as 'merely excuses

---

16    Erich Mühsam, *Tagebücher, 1910–1924*, ed. Chris Hirte (Munich, 1994), 17–18.
17    Carolin Kosuch, *Missratene Söhne. Anarchismus und Sprachkritik im Fin de Siècle* (Göttingen, 2015).
18    Gertrude Cepl-Kaufmann and Rolf Kauffeldt, *Berlin-Friedrichshagen, Literaturhauptstadt um die Jahrhundertwende. Der Friedrichshagener Dichterkreis* (Munich, 2015).

and irrelevancies [...] neither reasonable nor nice'.[19] Mühsam, on the other hand, started to enter the Berlin and Munich bohemian circles gathering in cafés. There, he established the persona of the anti-bourgeois: poor, drunken, promiscuous, shabby, and expressionist (Figure 17). His anarchism was much more 'practical' and part of his personal resistance against the bourgeois conventions. However, he continued to be friends with Landauer whose anarchist teachings he received.[20]

The third reason for Landauer's and Mühsam's turn to anarchism concerns some lingering problems the Jewish emancipation and acculturation had left unresolved: Both were well aware of the public and controversial debates about the 'Jewish Question', the growing antisemitism of their time, and they also felt confronted with certain problems of belonging. For one thing, both were addressed and sometimes also mocked as Jews from the outside. For another, they were raised in a German setting with only fragmented knowledge of and little attachment to the Jewish tradition. And for yet a third one, they considered themselves as both Germans *and* Jews.[21] These tensions of belonging became part of their anarchism. Finding a solution was essential to their rebellion against the bourgeois German-Jewish world of their fathers with equality only superficially completed in light of continuous threats and discriminations.

## 2    Modern Anarchists and Jewish Law

German-Jewish revolutionaries based their critique of the bourgeois culture not only on anarchism but, entwined with this political theory, also on the Jewish tradition which they tried to approach in a new way in the era of acculturation. In doing so, they translated the ethics of the Mosaic Law and certain figures of the Jewish canon – read through an anarchic, Nietzschean, and neoromantic lens – into politics and social rebellion to denounce the injustice, inequality, and poverty they sensed in society. Martin Buber (1878–1965) proved a central inspiration for their recourse to the Jewish tradition; he was friends

---

19    Gustav Landauer, 'Referat über Eugen Dührings "Kursus der National-und Sozialökonomie", repr. in *Anarchismus*, ed. Siegbert Wolf, Ausgewählte Schriften 2 (Lich/Hessen, 2009 [1892]), 107–14, at 107.

20    See Kosuch, *Missratene Söhne.*

21    Paul Breines, 'The Jew as Revolutionary: The Case of Gustav Landauer,' *Leo Baeck Institute Year Book* 12 (1967), 75–84; Christ Hirte, 'Erich Mühsam und das Judentum', in *Erich Mühsam und das Judentum*, ed. Jürgen Wolfgang Goette (Lübeck, 2002), 52–70.

FIGURE 16   Landauer (third from the left, sitting), surrounded by members of the
*Friedrichshagener Dichterkreis* (Poet's Circle of Friedrichshagen, 1892). The circle
attracted an illustrious public of writers, bohemians, philosophers, artists, and
political activists, among them popularizer of Darwin, monism, and freethought
Wilhelm Bölsche and the philosopher, monist, freethinker, and pantheist Bruno
Wille. In 1893, Landauer took over the editorship of the magazine *Der Sozialist*
(The Socialist), which had been started by a political branch of the circle in 1891.
Subsequently, *Der Sozialist* became a mouthpiece for Landauer's anarchism.
(INTERNATIONAL INSTITUTE OF SOCIAL HISTORY AMSTERDAM, PHOTO
COLLECTION, B 7/102)

with Landauer and Mühsam since their young adulthood.[22] Buber's writings
on Judaism and Jewish history encouraged especially Landauer to deal with
Judaism more profoundly and to integrate a non-Zionist, but affirmative interpretation of Jewishness into his political philosophy.[23]

---

22   Ruth Link-Salinger, 'Friends in Utopia: Martin Buber and Gustav Landaue', *Midstream* 24
(1978), 67–72.
23   Carolin Kosuch, 'Retrieving Tradition? The Secular-Religious Ambiguity in Nineteenth
Century German Jewish Anarchism', in *Negotiating the Secular and the Religious in the
German Empire: Transnational Approaches*, ed. Rebekka Habermas (New York, 2019),
147–70.

FIGURE 17    Caricature of Erich Mühsam visiting one of his favorite Munich cafés
             (Café Stefanie), drawn by Hanns Bolz, *Der Komet* 1, no. 22 (July 1911).
             The caricature's subtitle refers to the often expressed, but always teasing
             overconfidence of the writer and poet Mühsam: 'If I had known that
             Zeus would make so many mistakes in his creation, I would have offered
             him my assistance'

In their works, German-Jewish anarchists repeatedly resorted to biblical figures.[24] Mühsam's poems, in particular, address Cain, the misjudged fratricide; Moses, the hopeful but deceived lawgiver; and Jesus, the Jew from Nazareth and rebel who intended nothing more than to make the poor happy.[25] In his writings, these characters all appear as underrated, tragic, and disappointed personalities fighting in vain for justice and equality. What is more, they all struggle with God. By this, they question the status quo: Mühsam's Moses, for instance, led the Israelites to the doors of the Promised Land just like Mühsam himself attempted to guide the people into a future land of anarchic promises. God, however, who gave both of them a sense of longing, does not answer their calls and does not meet their hopes. His silence made Mühsam warn God that one day the people could start to take the initiative and realize their dreams and visions in this world: 'God! Beware that men does not rise against you!'[26]

German-Jewish anarchists codified freedom and justice as the cornerstones of the future society. These values not only are inherent to the Torah but reappear as ethical principles also in the secular-political context.[27] Influenced by Jean-Jacques Rousseau's (1712–1778) idea of the original goodness of man, in anarchism, it is his 'natural social conscience' that tells the political activist what is the right thing to do, not the religious (or state) law: 'Any explanation of what justice is, seems superfluous. For the ability to distinguish between right and wrong is a natural gift inherent to man. [...] The knowledge of justice and injustice is the social consciousness in man. Without it, foreign misery could not even touch us as a matter of our concern'.[28]

---

24    Gustav Landauer, 'Kiew', repr. in idem, *Zeit und Geist. Kulturkritische Schriften 1890–1919*, eds Rolf Kauffeldt and Michael Matzigkeit (Munich, 1997 [1913]), 224–28; Ernst Toller, *Die Wandlung* (Potsdam, 1919).

25    Erich Mühsam, *Brennende Erde. Verse eines Kämpfers* (Munich, 1920), 20; Carolin Kosuch, '"Ein Jude zog aus von Nazareth ...": Erich Mühsams Wahlverwandtschaft mit Bruder Jesus', *PaRDeS. Zeitschrift der Vereinigung für Jüdische Studien* 21: 'Jesus in the Jewish Culture of the 19th and 20th Century' (2015), 123–40.

26    Erich Mühsam, *Wüste – Krater – Wolken* (Berlin, 1914).

27    The Torah can be interpreted as an infinite duty. It might be taken as a law in the process of becoming that calls for study but does not set any defined limits. Because of this, the law itself contains moments of anarchy, without beginning and without fixed specifications; it is a task for the individual, less a commandment, and thus provides freedom. The question of justice is left to be answered by man alone: cf. Thomas Dörr, 'An-archie und (talmudisches) Gesetz. Erich Mühsams Judentum', in *Erich Mühsam und das Judentum*, ed. Jürgen Wolfgang Goette (Lübeck, 2002), 71–84.

28    Erich Mühsam, *Die Befreiung der Gesellschaft vom Staat. Was ist kommunistischer Anarchismus?* Fanal, Special Edition (Berlin, 1933), 259.

In *Kain*, the magazine Mühsam authored and edited (1911–1919, published irregularly and interrupted by the First World War), the anarchist took on the persona of the first outlaw – Cain – to relate to the needs of those forgotten and cast out from society. As Cain, Mühsam publicly decried their misery and intended to replace the bourgeois system he considered the cause of their current status with an idealized, just community of true equality. Mühsam's Cain is an angry, rebellious, and pioneering figure, attempting to improve, even revolutionize the world (Figure 18).

The first issue of *Kain* opens with an editorial poem, 'Cain', in which Mühsam gave voice to his personal sense of justice and social duty by condemning the double standards of the bourgeois society he had been raised in. The fatherly God he speaks about is not only the Jewish but also the Christian God. Above all, Cain embodies Mühsam's struggle with the bourgeois patriarchal system under which he suffered since childhood.[29] This system made use of both religion and its emblems to substantiate the given social hierarchies and power structures. In short, Mühsam's accusing lines directed towards the bourgeois world of his origin mirror his Jewish, Christian, and bourgeois cultural imprints as a German Jew in the age of acculturation. His Cain angrily complains:

> [...] Just bring sacrifices to the God of righteousness and goodness,
> who fills your huts with delicious fruit,
> who wraps your body in warming furs!
> Let young lambs bleed for his glory!
> Thank the God of the rich for your wealth!
> And close the barn to the hunger of the poor!
> Who God hates, you may judge as bad!
> What your God allows to grow in the fields is yours!
> Only you are worthy to resemble the image of God!
> But upon me pours the wrath of the righteous!
> Come on! I'm not afraid anymore! Here I stand, ready to
> Fight!
> Your clenched fists don't scare me!
> Fratricides yourselves – and a thousand times worse!
> From your pyre smokes my heart blood's steam.
> Don't I carry a human face just like you?
> I stand upright before you and claim my part! ...

---

29      Kosuch, *Missratene Söhne.*

FIGURE 18    Title of Mühsam's magazine *Kain:*
*Zeitschrift für Menschlichkeit* with the
emblem of a promethean Cain. Mühsam's
Prometheus-Cain is pictured – probably
following the image of the contemporary
Muskeljude (muscular Jew) – as an
unchained, powerful, and strong light
bringer ready to blast his confinement, not
as a weak, feeble petitioner (cf. Monika
Rüthers, 'Von der Ausgrenzung zum
Nationalstolz. "Weibische" Juden und
"Muskeljuden"', in *Der Traum von Israel.*
*Die Ursprünge des modernen Zionismus*,
ed. Heiko Haumann [Weinheim, 1998],
319–29). The cover of the magazine is in
bright yellow: It catches the attention of
the readers (and buyers) and is considered
the color of heresy, a fact that Mühsam
possibly made use of to underline his
rebellious attitude

Give me freedom and land! – and as brother forever
Cain returns to you, for the salvation of mankind![30]

The Cain Mühsam had created in his literary work is a modern adaption of
the figure of Prometheus and echoes the characteristics of the godly titan.
Like Prometheus, Cain is unwilling to sacrifice healthy cattle which he feels
sorry for and wants to keep for the living;[31] like Prometheus opposed Zeus,
Cain resists God in order to educate, enlighten, emancipate, and free the people from tyranny; and like Prometheus, Cain is a critic of the current order
which he strives to alter. However, Mühsam's human Cain seems much more
radical than his ancient Greek mythological blueprint. Different from the story
of the brothers Prometheus and Epimetheus, Mühsam's Cain kills his brother
Abel: the leitmotif of the poem is formed by the story of the Torah, not by
the Greek myth.[32] Equipped with such features, Mühsam's Cain appears as an
anti-normative figure questioning the law as fundament of the social order.[33]
He is an *an-archist* (from the ancient Greek ἀν ἀρχή, without ruler), someone
who opposes authority and foreign domination, claiming instead autonomy
and self-determination.[34]

Later, this Cain turned into another Prometheus on the cover of Mühsam's
journal *Fanal* (*Torch*, issued regularly between 1926–1931, followed by special
editions) (Figure 19). In this journal, Mühsam published political essays with
anarchist-communist leanings that critically commented on both the capitalist and communist regimes of his time while upholding an anarchist agenda
focused on integrity, thinking outside the box, and giving a voice to imprisoned
combatants or other forgotten underprivileged. The design of these emblem-
atic promethean figures Mühsam chose for the covers of his magazines deliberately borrowed from the Greek mythology: not from the Roman law of the father
associated with the bourgeois system of the time that he opposed. His choice
certainly was inspired by the broader German philhellenism of the eighteenth

---

30    Erich Mühsam, 'Kain', *Kain*, April, 1–4 (1911), 4.
31    Mühsam, 'Kain', 2. Other than in the Torah, Mühsam's Cain is a farmer who owns some
      cattle.
32    On the broad reception of Prometheus in the nineteenth century, see Caroline Corbeau-
      Parsons, *Prometheus in the Nineteenth Century: From Myth to Symbol* (London, 2013).
33    The story of Cain features no significant reference to Mosaic Law. In fact, it seems that
      Mühsam, an anarchist, chose lawlessness as point of reference and alter ego. However,
      with this choice, he still refers to the Law of the Thora – by deliberately opting for its
      counterpart.
34    On the philosophy of anarchist anti-authoritarianism, see Paul McLaughlin, *Anarchism
      and Authority: A Philosophical Introduction to Classical Anarchism* (Aldershot, 2007).

and nineteenth centuries. Yet Mühsam, an author of Jewish descent, also re-framed the German obsession with ancient Greece and its antisemitic under-tones. In Mühsam's work, Greek culture adds to a modern reading of one of the most known stories of the Torah (and the Old Testament)[35] – not reverse – to mold the characteristics of the lyrical Cain.[36] This ostentatious recourse to fig-ures of the Jewish tradition was, as a side note, also a means to differ from oth-ers in the rich cultural and political landscape of the German *fin de siècle* – not least a market of competing artistic groups and full of struggles to gain a voice among the many by means of provocation and exceptional ideas.

Interestingly, this recourse to the Jewish tradition found its echo also in the anarchist philosophy itself, worked out in detail by Gustav Landauer. During his youth, Landauer had studied the writings of Proudhon, Kropotkin, Bakunin, Leo Tolstoy (1828–1910), Stirner, but also Christian mystics such as Meister Eckhart (1260–1328).[37] Following the Russian anarchists and their emphasis on the Russian village collective, which they believed was filled with a Christian spirit, Landauer took the Middle Ages as the ideal time for com-munal life he wished to improve and build anew in an anarchic sense.[38] To him, anarchism meant true socialism, freed from the burdens of party policy and parliamentarian compromises. His focus was on the self-reliant individ-ual who had left behind any supposed burdensome tradition and authority. According to Landauer, pioneers should start to exit the capitalist state in a non-violent way by forming exemplary, free, and self-sustained rural commu-nities with exponentiating effects, connected in a loose network of material and spiritual exchange.

---

35    Cain is a figure of both the Jewish and Christian canon. But in his writings, Mühsam seems to have addressed much more the God of the Torah (and the Old Testament) than the Christian God of the New Testament. Furthermore, Jesus, to him, was explicitly a Jewish rebel. Thus, despite manifest overlaps, Mühsam moved in a Jewish religious imagery.

36    As Bernd Witte has shown, the German enthusiasm for Ancient Greek culture and phi-losophy in the eighteenth and nineteenth centuries coincided with the hostile exclusion of symbols, texts, and traditions of the Jewish culture. This struggle was also about mon-otheism and polytheism and their value for modernity: Bernd Witte, *Moses und Homer. Griechen, Juden, Deutsche. Eine andere Geschichte der deutschen Kultur* (Berlin, 2018). By equipping a 'modern', capable-of-action Cain with the features of Prometheus, Mühsam underlined the continuous value of the Jewish tradition for European history and culture he wished to influence.

37    Joachim Willems, *Religiöser Gehalt des Anarchismus und anarchistischer Gehalt der Religion? Die jüdisch-christlich-atheistische Mystik Gustav Landauers zwischen Meister Eckhart und Martin Buber* (Albeck, 2001); Kosuch, *Missratene Söhne*.

38    Cf. Andrzej Walicki, *A History of Russian Thought: From the Enlightenment to Marxism* (Stanford, 1979), 268–79.

**„FANAL"**

**HERAUSGEBER: ERICH MÜHSAM**

FIGURE 19     Flaming and torch-bearing figure on the front page of Mühsam's journal
*Fanal* (Torch) (1926)

Landauer's anarchist society was built on the voluntary coexistence of
independent people engaged in crafts, farm, and manual labor as well as in
intellectual and cultural activities who would be able to manage their own
affairs and decide about their own material and spiritual needs without being
forced or feeling obligated to obey by external laws and conventions.[39] As we
shall see in a moment, elements of the Mosaic Law formed an integral part of
this anarchist vision which resurfaced particularly in Landauer's idea of the
*Sozialistischer Bund* (Socialist League, 1908–1913). To gather people willing
to settle in the countryside and to start socialism, Landauer – supported by
Mühsam and Buber – issued leaflets and (again) the journal *Der Sozialist* (The
Socialist, published as organ of the league from 1909 till 1915). He conceptualized the league and its subunits as tools which should slowly replace the capitalist society by a 'movement of spirit',[40] based on mutual aid, cooperation, the

---

39    Gustav Landauer, 'A Few Words on Anarchism', repr. in *Revolution and Other Writings: A
      Political Reader*, ed. and trans. Gabriel Kuhn (Oakland, 2010 [1897]), 79–83.
40    Gustav Landauer, 'The Twelve Articles of the Socialist Bund: Second Version', repr. in
      *Revolution and Other Writings: A Political Reader*, ed. and trans. Gabriel Kuhn (Oakland,
      2010 [1912]), 215–16, at 216.

exchange of goods and skills instead of money, and by people willing to share, to educate themselves, and to live creative and 'authentic' lives.[41] These clearly neo-romantic and utopian ideas assumed the best in man and were framed by twelve articles Landauer circulated – probably in remembrance of the Twelve Articles formulated during the German Peasant's War in 1525, an attempt from below to set up a federal order based on human and civil rights.[42] About 1,000 settlers followed Landauer's call and established communities in Germany and Switzerland. In 1911, Landauer (himself not a settler) released his *Aufruf zum Sozialismus* (*Call to Socialism*), a programmatic text[43] in which, after criticizing harshly the Marxist system of thought, the anarchist philosophy pivotal for the

41    The search for authenticity or the 'essence' of reality and living was crucial to the nine-
      teenth century with its focus on history and culture of memory. Anarchism, Landauer's
      in particular, marked no exception from this. Because of the multi-optionality modernity
      had generated, constantly urging the individual to select among the offers provided by a
      rich supply of life plans and worldviews, authenticity was something of a paradox: chased
      after and idealized, but never to achieve. In the end, the fragile modern self had no fixed
      identity but was subjected to permanent change. Landauer's modern anarchism tried to
      square the circle: His philosophy was centered on the modern individual with its con-
      fusing contradictory facets and needs, a complexity highly appreciated and welcomed
      by Landauer. He integrated this individual in a social structure that he conceptualized
      as flexible enough to cope with shifts and changes but – with liberty as its base – still
      grounded in shared principles necessary for a functioning society. This community
      ceased its existence once its members chose so. 'Let us imagine a town that experiences
      both sunshine and rain. [...] Neither do we want to force all individuals under a common
      roof nor do we want to end up in fistfights over umbrellas. When it is useful, we can share
      a common roof – as long as it can be removed when it is not useful. At the same time, all
      individuals can have their own umbrellas, as long as they know how to handle them. And
      with regard to those who want to get wet – well, we will not force them to stay dry' (Gustav
      Landauer, 'Anarchism – Socialism', repr. in *Revolution and Other Writings: A Political
      Reader*, ed. and trans. Gabriel Kuhn [Oakland, 2010 (1895)], 70–74, at 71). In the end, this
      was a hyper-modern, ultra-liberal way to think social life based on freedom which, ulti-
      mately, had no final cause or justification but was a desire, choice, and promise upheld
      by its own appeal and the individual's willingness to give meaning to this form of order.
      Outdated ideas or structures no longer supported by the living, in contrast, would make
      no sense and, according to Landauer, had to vanish: cf. Kosuch, *Missratene Söhne*. These
      anarchic principles combined with the will and ability to decide independently about
      directions and changes anew at every crossroad, according to Landauer, would enable the
      politically emancipated individual to live an authentic life.
42    Gustav Landauer, 'The Twelve Articles of the Socialist Bund: Second Version', repr. in
      *Revolution and Other Writings: A Political Reader*, ed. and trans. Gabriel Kuhn (Oakland,
      2010 [1912]), 215–16; cf. Peter Blickle, *Die Revolution von 1525*, 4th ed. (Munich, 2004).
43    Landauer's thoughts on socialism proved highly influential on a younger generation of
      radicals such as Ernst Toller who studied the *Aufruf* and took it literally as a call to action.

league was presented in detail. As a basic rule of the anarchic society, Landauer stipulated that property should be redistributed every fifty years. He quoted and revived this principle of the Torah in a prophet-like language:

Let us [socialists, C.K.] act like a Job among the nations, who in suffering came to action; abandoned by God and the world in order to serve God and the world.

[...] Private property is not the same thing as ownership; and I see in the future private ownership, cooperative ownership, and community ownership in most beautiful flowering; [...] No final security measures for the millennium or for eternity are to be made, but a great, comprehensive equalization and the creation of the will to repeat this equalization periodically.

'Then you are to sound the trumpet throughout your land on the tenth day of the seventh month as the day of equalization ...'

'And you are to sanctify the fiftieth year and proclaim a free year in the land to all that live therein; for it is your year of jubilee; then everyone among you is to come back to his property and to his family'.

'That is the jubilee year, when every man is to regain what belongs to him'.

Let him who has ears, hear.

You shall sound the trumpet through all your land!

The voice of the spirit is the trumpet that will sound again and again and again, as long as men are together. Injustice will always seek to perpetuate itself; and always as long as men are truly alive, revolt against it will break out.

Revolt as constitution; transformation and revolution as a rule established once and for all; order through the spirit as intention; that was the great and sacred heart of the Mosaic social order.

We need that again: a new rule and transformation by the spirit, which will not establish things and institutions in a final form, but will declare itself as permanently at work in them. Revolution must be a part of our social order, must become the basic rule of our constitution.[44]

---

(See letter Toller to Landauer [1917] reprinted in Hansjörg Viesel, *Literaten an der Wand. Die Münchner Räterepublik und die Schriftsteller* [Frankfurt am Main, 1980], 337).

44   Gustav Landauer, *Aufruf zum Sozialismus*, ed. Siegbert Wolf, Ausgewählte Schriften 11 (Lich/Hessen, 2015 [1911]), 142–43.

Mosaic Law, following Landauer, never had lost its significance, even if it was no longer taken as a divine revelation.[45]

This intertwinement of political anarchism and Landauer's reading of Judaism became even more explicit when he stated that the Jews would be the true revolutionaries, and that the 'Jewish question' would only be resolved once humanity as a whole would be redeemed:

> In the new nation which is in the process of becoming, there are a large number of Jews; but these Jews feel as one, as a covenant which has to fulfil its vocation to humanity; and the more they feel this in themselves, the more Zion is already alive for them. For what is the nation other than a covenant of those who, united in themselves by a unifying spirit, feel a special task for mankind? To be a nation means to have a duty. What is described here is a new entity, something like a nation in the making, which, as a new community for building the beginnings of a just and free society unleashing creative forces, outrageously opposes all old nation-states, dynastic states, states of injustice and violence. Like a wild cry over the world and like a whisper from the inside, a strong voice tells us that the Jew can only be redeemed together with humanity and that it is one and the same thing: to wait for the Messiah in exile and diaspora and to be the Messiah of the peoples.[46]

This view, neither Orthodox nor Zionist, was also shared by Mühsam; both anarchists opted for a third way to solve the open questions of their time.[47] The Jew, in Landauer's concept, appeared as a promethean figure like Mühsam's Cain precisely because of his particularity and marginality. In anarchism, this not-belonging to the majority of society – which also had been stressed by Herder – actually turned into a strength vital for overcoming a flawed, oppressive, and excluding system. As acculturated and emancipated German citizens of Jewish descent Mühsam and Landauer reinterpreted (and also rediscovered) ideas of the Torah for their political philosophy of revolution, initiated

---

45   Gustav Landauer, 'Etwas über Moral', in *Anarchismus*, ed. Siegbert Wolf, Ausgewählte Schriften 2 (Lich/Hessen, 2009 [1893]), 37–41.

46   Gustav Landauer, 'Sind das Ketzergedanken?', repr. in idem, *Zeit und Geist. Kulturkritische Schriften 1890–1919*, eds Rolf Kauffeldt and Michael Matzigkeit (Munich, 1997 [1913]), 216–23, at 220.

47   Erich Mühsam, 'Zur Judenfrage', *Die Weltbühne* 16/49 (2 December 1920), 643–47; Breines, 'The Jew as Revolutionary'.

and spearheaded by marginal but empowered figures. By this interpretation, both placed Mosaic Law – the law of a persecuted minority – at the center of renewal that would bring about a better future for the whole of society.

## 3    Tikkun Olam

German-Jewish anarchists, Mühsam and Landauer were activists in revolt, searching for an idealized new order based on humanity. In their everyday lives, they faced the disparity in the German Empire, which had promised civil equality but, in reality, continuously excluded and discriminated against its Jewish citizens. Notwithstanding their advancing acculturation, German-Jews, at certain points, still painfully hit the glass ceiling in society and work life and were confronted with the growth of antisemitic tendencies in politics and culture.[48] Together with a generational sense of revolt in the *fin de siècle* this experience of not-belonging in private and public added to the anarchists' questioning of the status quo.[49]

The future society envisioned by modern German-Jewish leftist radicals was neither congruent with the coming of the Messiah in Jewish eschatological thought nor with the return of Jesus and the Christian concept of the new world in the post-apocalyptic era.[50] But despite the negligible role religious traditions played in their day-to-day lives, religious ideas and figures, including the Mosaic Law, seemed to have reappeared – fragmented and newly interpreted – in their politicized notion of that earthly, free human kingdom in the making which they tried to further with their commitment.[51] This sort of revitalization points to a 'messianic paradox',[52] because the anarchists'

---

48    Peter Pulzer, *Jews and the German State: The Political History of a Minority 1848–1933* (Oxford, 1992).

49    Albert Camus, *Der Mensch in der Revolte. Essays* (Reinbek bei Hamburg, 1997), 25; Michael Löwy, *Erlösung und Utopie. Jüdischer Messianismus und libertäres Denken. Eine Wahlverwandtschaft* (Berlin, 1997), 57; see further Kosuch, *Missratene Söhne*.

50    On the shift from religious to political worldviews and their continuous entanglement under new auspices, see Eric Voegelin, *Die politischen Religionen*, 3rd ed., ed. Peter J. Opitz (Munich, 2007 [1938]).

51    Kosuch, 'Retrieving Tradition?'

52    Elke Dubbels, *Figuren des Messianischen in Schriften deutsch-jüdischer Intellektueller, 1900–1933* (Berlin, 2011), 275.

political messianism relied on a *particular* (Jewish) tradition as the base of the imagined future *universal* (trans/post-national) humanity.[53]

The specifics of this political-religious entanglement resound in Mühsam's anarchist notion of freedom: 'Freedom is a religious term. Whoever is a revolutionary because he wants to reach freedom is a religious man. To be revolutionary without being religious means to strive for goals other than freedom by revolutionary means'.[54] Michael Löwy has compared such ideas to the Rabbinic and Kabbalist principle of תיקון עולם (*tikkun olam*), that is, the improvement or healing of the world – originally by God, but in the Kabbalist reading also by human activities – to overcome the separation from God in anticipation of the messianic age.[55] *Tikkun olam* further includes the engagement for social justice to bring the world closer to perfection: a central claim also in the anarchisms of Landauer and Mühsam.[56] With their focus on the Middle Ages, biblical as well as mythological figures and narratives, and future-oriented projects like the *Sozialistischer Bund*, the anarchisms of those modern German Jews comprised restorative-romantic and utopian-revolutionary elements.[57] This polarity seems to put into a new frame the Jewish eschatology and the idea of *tikkun olam*, which both call for the restoration of an earlier ideal state and imagine a future utopia. The Jewish messianic period, as Gershom Scholem (1897–1982) has noted, is conceived as a visible, public event *in* history.[58] It is 'an irruption in history', not a transcendent mystery.[59] Taken as a worldly event of total change it seems quite similar to the concept of revolution in anarchism, which is also a tangible historical incident triggering the genesis of a new society. As Landauer put it:

> The joy of revolution is not only a reaction against former oppression. It lies in the euphoria that comes with a rich, intense, eventful life. What is

53      Paul Mendes-Flohr, 'Messianic Radicals: Gustav Landauer and Other German-Jewish Revolutionaries', in *Gustav Landauer: Anarchist and Jew*, eds Paul Mendes-Flohr and Anya Mali (Berlin, 2014), 14–42.

54      Erich Mühsam, 'Bismarxismus', *Fanal*, February (1927), 65–71, at 65.

55      Michael Löwy, 'Jewish Messianism and Libertarian Utopia in Central Europe (1900–1933)', *New German Critique* 8, Special Issue 2: 'Germans and Jews' (1980), 105–15, at 105–06; Gerhard Wehr, *Europäische Mystik zur Einführung* (Hamburg, 1995), 125–34.

56      Michael L. Morgan, 'Tikkun olam', in *Enzyklopädie jüdischer Geschichte und Kultur*, vol. 6, ed. Dan Diner (Stuttgart, 2015), 102–06.

57      Gershom Scholem, *The Messianic Idea in Judaism and Other Essays on Jewish Spirituality* (New York, 1971), 23–45; Kosuch, *Missratene Söhne*.

58      Scholem, *The Messianic Idea in Judaism and Other Essays on Jewish Spirituality*, 21–58.

59      Löwy, 'Jewish Messianism and Libertarian Utopia in Central Europe', 107.

essential for this joy is that humans no longer feel lonely, that they experience unity, connectedness, and collective strength. [...] We say that
everything must be turned upside down! We refuse to wait for the revolution in order to begin the realization of socialism; we begin the realization of socialism to bring about the revolution! [...] We want to directly
link the production of consumer goods to the needs of the people. We
want to create the basic form of a new, real, socialist, free, and stateless
society, in other words, a community.[60]

With the dawn of the Messianic period, some Jewish religious texts speak of
the annulment of the Torah laws, because the 'Torah of the Messiah' would
spread a general, deep, and spiritual understanding of the divine order, which
is why the instructive, mandatory, and bonding character of these laws no
longer would be required.[61] In German-Jewish anarchism, quite the same, the
revolution destroys the previous structure of the world, and sweeps away its
institutions, the established power relations, statutes, and laws, along with
social hierarchies to initiate the creation of a new, free community based on
the self-determination of the empowered individual. This revolution is mingled with the romantic idea of an all-encompassing 'spirit', a Hegelian notion
complemented by Russian radical philosophy and Henri Bergson's (1859–1941)
idea of *joie de vivre*, frequently referred to by Landauer and his fellows.[62]
    Religious law and modern anarchist theory, thus, seem to be entangled and
have affected each other resulting in a new approach to politics in the modern
age. This 'elective affinity' between political anarchism and Jewish messianism
further intensified through the Neo-Romanticism of the late nineteenth century.[63] The Mosaic Law reappeared as part of this affinity: it was reinterpreted
and placed in a modern political context. In modern anarchism, though, the
selectively cited parts of the Mosaic Law did not serve the purpose to erect a
new (political or historical) law codifying the steps towards a potential revolution. Rather, the Jewish religious law, set into a new frame, helped to validate

60    Gustav Landauer, 'Revolution', 'What does the Socialist Bund Want', and 'The Settlement',
      repr. in *Revolution and Other Writings: A Political Reader*, ed. and trans. Gabriel Kuhn
      (Oakland, 2010 [1907]), 110–85, 188–90, 196–200, at 151, 188, 197.
61    Scholem, *The Messianic Idea in Judaism and Other Essays on Jewish Spirituality*, 41–42;
      Peter Schäfer, 'Die Torah der messianischen Zeit', *Zeitschrift für die neutestamentliche
      Wissenschaft* 65 (1974), 27–42.
62    Eugene Lunn, *Prophet of Community: The Romantic Socialism of Gustav Landauer*
      (Berkeley, 1973), 179–81; Kosuch, 'Retrieving Tradition?'
63    Löwy, *Erlösung und Utopie*, 21; idem, 'Messianismus', in *Enzyklopädie jüdischer Geschichte
      und Kultur*, vol. 4, ed. Dan Diner (Stuttgart, 2013), 147–51.

the plausibility of the spontaneous and creative aspects inherent to anarchism and its revolution. At this point, Walter Benjamin's (1892–1940) 'splinters of messianic time'[64] come to mind, which seem to have been preconceived in the anarchic revolution: conceptualized not as a one-time event but as a possibility always present in the course of history, ready to be realized whenever there were people willing to bring it to life.[65] This particular concept gave agency to those attempting to change the world for the better – *tikkun olam* rephrased in anarchism.

## 4    Conclusion

Nineteenth-century modernity opened up a rich plurality of new positions and identities in the dense social and cultural fabric of Jewish and German coexistence. Anarchism – by nature a libertarian and anti-normative political idea – in this multi-optionality proved one among many choices, albeit with one reservation. It was the road taken by a small minority only, on both the Jewish and the German side. Anarchists of any background did struggle with the neo-Roman laws and values of their time that allowed the state and its institutions a strong influence over its subjects.[66] Particularly, they criticized the predominant patriarchal family laws with their restrictive sexual norm, but also private property and inheritance laws, which were protected by state laws. These laws ensured the continuity of social hierarchies as well as the persistence of the political status quo. As Johann Most (1846–1906), one of the most radical anarchists of the German Empire, underlined, it was the *Eigentumsbestie* ('Beast of Property') that hindered those without property to claim their fair share. Private property secured by state law, according to Most and his fellow anarchists, obstructed the redistribution of the means of production and, in consequence, impeded the whole of society to move towards equality, autonomy, and a sufficient amount of leisure time filled with cultural and intellectual activities for everyone in a social order of self-ruled communities.[67]

As pointed out in this essay, German-Jewish anarchists built upon such basic anarchist ideas. But in the era of the seemingly concluded Jewish emancipation

---

64    Walter Benjamin, *Über den Begriff der Geschichte*, repr. in idem, *Werke und Nachlass – Kritische Gesamtausgabe*, vol. 19, ed. Gérard Raulet (Berlin, 2010 [1942]), 28.

65    Landauer, 'Revolution', in *Revolution and Other Writing*, 116.

66    Ruth Kinna and Alex Prichard, 'Anarchism and Non-Domination', *Journal of Political Ideologies* 24/3 (2019), 221–40.

67    Johann Most, *Die Eigenthumsbestie* (New York, 1887).

and acculturation, the anarchism they advocated differed from other anarchisms of the time, particularly concerning the sources of this political theory. Whereas German anarchists such as Johann Most attempted to push through a radical atheist stance because they were convinced the alliance of throne and altar would support the system of exploitation they decried which, in consequence, led them to polemicize against Jewish and Christian religious experts and teachings in a sweeping blow,[68] Landauer and Mühsam put forth more differentiated arguments, even though they shared the criticism of religious institutions and clerics. In their writings, central figures of the Jewish canon and elements of the Jewish law reappeared in a new frame and reading. This seems an act of revolution in itself because both anarchists were raised in families acculturated to the German Protestant bourgeoisie. With their explicit referencing to Jewish sources they prized a heritage debated highly controversial in the German cultural landscape of the nineteenth century on which their fathers had put silence in public.

However, this choice and the sophisticated interpretation of sources mirrored their belonging not only in a cultural-religious but also in a class-sense. Johann Most was a powerful voice propagating anarchism; however, he grew up in humble conditions. Before radicalizing and turning to anarchism, he had joined the workers' movement and entertained socialist convictions. Most of Germany's anarchists came from comparable proletarian or petty bourgeois backgrounds; many of them took their route to anarchism through the worker movement. In this respect, Landauer and Mühsam stood out. As educated members of the bourgeoisie they became radicalized in Berlin's alternative literary circles in which a cultural approach to anarchism prevailed. Despite young Landauer's brief membership in a splinter group of social democracy (*Verein Unabhängiger Sozialisten*, Association of Independent Socialists, 1891–1894), their anarchism was not workerist but intellectual in tendency, mixed with philosophical, avant-gardist, and artistic elements.[69] Major intersections with proletarian anarchism existed regarding economics, state, and law. But Landauer's and Mühsam's poetical, fictional, satirical, cultural, literary, and theatre critical work that – together with their recourse to the Jewish tradition – blended in with their anarchisms made of their political theory an idea of its own. In this particular interpretative framework, they echoed and answered to arguments previously expressed by Mendelssohn and Herder and, by this, furthered currents originating in the *Haskalah* and Romanticism. Just

---

68    Johann Most, *Die Gottespest* (New York, 1883).

69    Carolin Kosuch, ed., *Anarchism and the Avant-Garde: Radical Arts and Politics in Perspective* (Leiden, 2019).

like in the former, in anarchism, too, the Jewish law and particularity appeared both as a means and duty of practice and action to work for the better of mankind and to increase its happiness. Jewish law, the concept of *tikkun olam*, and political anarchism mingled to form a new, secular-messianic promise of true equality.

As anarchists of bourgeois and Jewish descent, Landauer and Mühsam irritated their bourgeois families and their mostly non-Jewish, proletarian political environment. They took the particularity assigned to them literally and formed a lifestyle and political attitude from it. By this, they became a target for hostilities, notably during their active political engagement in the Bavarian Council Republic of 1919.[70] Their writings and actions were highly appreciated by their followers but were pursued by their opponents with vehemence. This included personal (and in both cases ultimately deadly) threats because of their political conviction and Jewish origin.

---

70    Carolin Kosuch, 'Räterepublik', in *Enzyklopädie jüdischer Geschichte und Kultur*, ed. Dan
      Diner, vol. 5 (Stuttgart, 2014), 96–101.

# Afterword

## Moses and the Modern Germans: the Lawgiver in a Philhellenic Age

Suzanne Marchand

Moses was a hard man. He learned the wisdom of the Egyptians (Acts 7:22), but he was never tempted to assimilate. Exemplary of moral justice, he was horrified by any form of licentiousness or deviation from God's laws. He was a fully autonomous, fully self-confident individual, capable of acts requiring great willpower and bravery. His own self-possession, self-righteousness, and asceticism as well as his anti-rhetorical, spare eloquence made him a charismatic leader whose commitment to his cause and belief in his God persuaded the Israelites of Egypt to follow him, despite the improbability of his promises. And he did, in fact, free the Israelites from slavery and find for them a place to call home. Things that might have been gray areas to others he saw strictly in black and white. He was capable of smiting his enemies, and even his friends (Exod 32), in pursuit of divinely sanctioned justice. What European or American could not admire that? Minus the whiskey and the weapons, he was the Clint Eastwood of biblical times.

Early modern Christians, like the ancient Israelites, discovered that having such a father figure was useful, especially in times of turmoil. Secular rulers were sure that their people needed commandments; the ruled believed that their leaders at least ought to live by God's laws as well. As the Old Testament was read as irrefutable history, and the history of the Israelites thought to be exemplary as well as preparatory for Christian civilizations, no one could forget Moses – or fail to see him as a heroic, if superseded, ancestor, leader, prophet, and lawgiver. But as the Israelites also knew, it was no easy matter to please such a leader and often arduous to live under his law. Perhaps modern moviegoers feel the same way about Eastwood. In times of great peril or moral turpitude, he is a vital leader and scourge; when clear and present dangers to body and soul have passed, it is more pleasant to return to our old, less taxing pursuits and to exile him, once again, to the wilderness.

The aim of this 'afterword' is to not to extend spurious cinematic comparisons with biblical accounts of the most famous of Hebrew prophets. Nor do I wish to summarize the contributions to this fine volume, already so beautifully done in Paul Michael Kurtz's introduction. Instead, I here feel liberated to take quite a different, and rather playful, approach, tackling the question of the diminution of Moses' power and centrality in German-speaking culture by a different means – and that is by assessing some of the *external* forces that diluted his role and his teachings, including the rise of a rival cultural complex offering very different models for the good life, both for polities and for individuals. These models were almost equally as venerable as biblical models, but they could perhaps only really come to the fore with the end of violent confessional strife and the advent of enlightened absolutism. The models I refer to are, of course, those drawn from classical antiquity, especially from ancient Greece.

I hasten to say that what eighteenth-or nineteenth-century Germans thought about ancient Greece, and the ways in which the Greeks were deployed as cultural heroes, has as little to do with the ancient Greeks themselves, as did stereotypes of Moses and the ancient Israelites at the time. But they did stand for something different – indeed, for an alternative conception of Europe's ancestry and the purposes and destinies of its peoples and states. What brought the refurbished model to the fore (one cannot call this a 'new' model, of course, since western Europeans, especially since the Renaissance, had always understood themselves as heirs, too, to Rome – and through Rome to Greek culture), I will suggest, was not merely the advent of highly uneven and imperfect forms of toleration but the expansion of the arts and sciences and the increasing consumption of luxury goods that transformed the German lands after about 1740, all of which began to shift the balance between discussions of law toward discussions of various forms of freedom.

I will make my case, in part, by invoking some of those who developed the new philhellenism – by design opposed to older, stricter, theological models – and by surveying some of the values and images that demonstrate shifts in cultural meanings. I realize this is a rather idiosyncratic way to illuminate what are essentially theological and political questions, some of them of great and disastrous consequence. By leaving aside the advent of modern antisemitism in this essay, I do not mean to occlude the critical importance of the racist thinking and radicalized persecutions of the later nineteenth century. But those have been discussed by so many others, including myself, that in this essay I want to take a different tack and engage, instead, in some wider cultural comparisons and evolving oppositions. I am convinced that we cannot understand Moses among the Moderns – and even the modern Germans – if we do not take stock of the alternative ancestries and models available to writers, scholars, and artists at a particular time and why, at some moments in history, some fade while others move to center stage.

1    Moses, Early Modern Man of the Hour

Early modern German intellectuals knew their Bibles intimately and were well aware of the ambiguities of St. Paul (ca. 5–64 CE) on the proper Christian attitude toward Jewish law. Jesus came to free humankind from the law, but the law does seem to bind the living (Rom 7), and we are not totally free to break it at will, as that would entail sin. Protestant thinkers would have also known the tormented discussions of Martin Luther (1483–1546) on the question, in which he concluded (to simplify grossly) that while neither obedience to the law nor obedience to church teachings were sufficient for grace ('grace alone availeth'), the true Christian would nonetheless embrace the law as a result of his or her free choice and love of God. While Paul was his major touchpoint for such discussions, Luther, naturally, could not avoid discussing the Hebrew Bible's great lawgiver himself.[1]

Luther's view of Moses, spelled out in a 1525 lecture, is that his laws – even the commandments – are not binding for Christians: the important laws are already given in nature and inscribed on human hearts. Moses, for Luther, was a mere teacher and one exclusively of the Jewish people. Christians are not to regard him as their lawgiver 'unless he agrees with both the New Testament and the natural law'.[2] After the proclamation of the gospel, the law of Moses is 'dead', yet Luther insists on not 'sweeping him under the rug', as he did formulate some excellent rules, which would also be exemplary for rulers themselves: 'I would even be glad if [today's] lords ruled according to the example of Moses. If I were emperor, I would take from Moses a model for [my] statutes'.[3] Even more importantly, Moses, for Luther, is indispensable in providing something *not* available in nature:

> the promises and pledges of God about Christ. This is the best thing. [...] And it is the most important thing in Moses which pertains to us. The first thing, namely, the commandments, does not pertain to us. I read Moses because such excellent and comforting promises are there recorded, by which I can find strength for my weak faith. For things take place in the

---

1   Luther's writings on Moses were extensive. For the newest insights into his 1522 lectures, see Miles Hopgood, *How Luther Regards Moses: The Lectures on Deuteronomy*, Refo500 Academic Studies 98 (Göttinen, 2023).

2   Martin Luther, 'How Christians Should Regard Moses,' sermon from 27 August 1525, translated by Martin H. Bertram in *Luther's Works (American Edition)*, vol. 35, *Word and Sacrament 1*, ed. Theodore Bachmann (Philadelphia, 1960), 161–174, at 165.

3   Ibid., 166.

kingdom of Christ just as I read in Moses that they will; therein I find also my sure foundation.[4]

Thus Luther, notorious antisemite that he was, even while saying Christians could and should act as if his laws were not binding, could not do without Moses. His laws were exemplary, and his prophecies were essential for faith.

This version of having one's cake and eating it too runs through subsequent Christian and especially German Protestant culture, which is featured so centrally in this book. It contains a historical conviction: Moses is dead, but the historical event of his prophecies and their content must be considered true for the faithful, and the text he transmitted is sacred and true for Christians as for Jews – even though later theologians would endlessly dispute *which* biblical truths were necessary for faith. Luther's line of thought also contains a critical attitude toward religious law: it must be imposed on oneself, not dictated from above or outside. Luther's tortured discussions of the freedom of the Christian, or the freedom of the will in bondage, are in a sense all about dealing with the legacies of Moses, as he saw them: how to create obedience and encourage good works without external forms of compulsion that make them hypocritical or purely means-oriented. Moses could not be 'swept under the rug', Luther reiterated. He had to continue to be a model for the Christian prince and for Christian subjects who imposed his laws on themselves.

In Luther's world of violent religious polemics and absolutist princes, and in the even more war-torn century to follow, Europeans were not terribly squeamish about Moses' (or God's) smiting of enemies for the sake of righteousness. Nor was anyone surprised by dictatorial leadership. Just about everyone had a king, or in Italy a tyrannical prince. It was that prince's main job to defend the realm and maintain moral order; even if he or she enjoyed great wealth and privilege, few expected such things to be redistributed or enjoyed by all. Humanists might pine for Roman republican virtues, but this remained a world in which lawgivers were more revered than liberators, even if Moses might deserve both titles. Humanists could also uphold some aspects of other ancient pagan cultures, or even risk discussing Moses' rootedness in Egyptian culture, the beginnings of a deep historicization of Israelite culture that would flower in the eighteenth century.[5] However, few were willing to doubt either

---

4   Ibid., 168–69.
5   See here Jan Assmann, *Moses the Egyptian: The Memory of Egypt in Western Monotheism* (Cambridge, MA, 1998), and more recently, Mordechai Feingold, "'The Wisdom of the Egyptians': Revisiting Jan Assmann's Reading of the Early Modern Reception of Moses', *Aegyptiaca: Journal of the History of Reception of Ancient Egypt* 4 (2019), 99–124.

his priority or Europeans' need for such an ancestor. In a world where toleration of heretics, infidels, witches, and political radicals was regarded as a danger to the realm and where the cultivation of arts and sciences as well as the
consumption of luxuries remained a matter for a tiny elite, adherence to laws
seemed to be a vital aspect of moral personhood and national cohesion. Moses
was considered the most trustworthy historian as well as God's chosen mouthpiece and legislator. He had no rivals and few critics and featured centrally in
works of art by major sculptors and painters, both Catholic and Protestant.

What happened to Moses in the course of the eighteenth, nineteenth, and
early twentieth centuries in the German-speaking lands is, of course, the subject of this volume, in which scholars from a wide variety of disciplinary backgrounds reflect on crucial theological, historical, sociological, and philological
transformations. Moses was not exactly 'swept under the rug', they argue, but
certainly became a more controversial figure, and his laws conceived as less
and less binding (or even as fully obsolete and indicative of Judaism's 'desiccation' and 'hide-bound' legalism). Historicized by biblical scholars, criticized by
enlightened reformers, and relativized by orientalists, Moses was stripped of a
reverence even Luther could not deny him. I endorse this line of thinking and
applaud the authors for their deep explorations of this important subject. But
in what follows, I want to explore the ways in which, especially in the northern German states, another cultural model displaced that of the Israelites: the
ancient Greeks.

## 2      Making a New Model: the Greeks

In recent years, I have begun to appreciate just how strange late enlightened
and nineteenth-century German philhellenism looks in contrast to the world
of the sixteenth and seventeenth centuries. In that period, one of Moses' ubiquity, many absolutist rulers still traced their ancestry to Aeneas, and Homer's
poetry was revered, especially after 1700. Yet most Greek poetry, history, and
philosophy was filtered through the dominant Christian and Latin culture
of the day. If most educated persons had good Latin, few had good Greek, so
they read Greek works in Latin or vernacular translations. Greek mythology
was largely known through Ovid, and then often in vernacular and illustrated
editions. So too Greek philosophy was generally mediated by Christian or neo-
Platonic readings, while Greek history remained exclusively part of universal
history. The first stand-alone Greek history in Europe was the 1707 *Graecian
History* by the Whig Temple Stanyan (1675–1752), and its first volume found few

readers until its second volume appeared, in 1739.[6] Almost no one (including Stanyan) approved of Athenian democracy, which led to mob rule and decadence. Some approved of the Spartans, because they had a monarchy and eschewed luxuries, Alexander the Great (336–323 BCE), who conquered much of Asia, or Solon (ca. 630–560 BCE), who famously ridiculed the Lydian king Croesus (died ca. 546 BCE) for his attachment to earthly luxuries. To discuss the artistic accomplishments of the pagan Greeks would have seemed odd, even a bit unsettling, to seventeenth-century ears. To plan a journey to Greece (still under Ottoman control) would have sounded like a reckless venture into the barbaric unknown, rather than a visit to the lands of Europe's cultural heritage.

Renaissance Italians, of course, had been investigating and admiring things Greek since at least the fourteenth century and developing a taste for classical antiquity, which they spread over Europe in the form of not only translations and scholarly treatises but also the major and minor arts. The classics – and especially classical mythology – offered insights into the operations of the secular world and alternative techniques for representing humans, divine beings, and the natural world as well as opportunities to explore and depict secular stories suitable to courtly patrons. Artists and luxury craftsmen reveled in their newfound freedom to invoke new scenes, emotions, and body parts. Indeed, as Malcolm Bull has argued, familiarity with classical themes diffused throughout Europe largely 'through an accumulation of expensive yet seemingly trivial exchanges; the distribution of pornography and wedding presents, and the acquisition of things such as picnic dishes and jewelry, and garden ornaments for people's holiday homes.'[7] 'It may not sound like a cultural revolution', he concludes, 'but that's what it turned out to be'.[8] The Italians, of course, also led the way in collecting actual Greek antiquities, and by the end of the seventeenth century, British and French agents were avidly seeking manuscripts, sculptures, coins, and other antiquities throughout the Mediterranean. But until

---

6  Temple Stanyan, *The Grecian History*, vol. 1, *Containing the Space of about 1684 Years* (London, 1707), vol. 2, *From the End of the Peloponnesian War to the Death of Philip of Macedon, Containing the Space of Sixty-eight Years* (London, 1739); see also Giovanna Ceserani, 'Narrative, Interpretation, and Plagiarism in Mr. Robertson's 1778 *History of Ancient Greece*', *Journal of the History of Ideas* 66/3 (2005), 415–19. Even the major seventeenth-century modern Greek histories do not discuss Greek pagan events until the Hellenistic era, or beyond the moments at which they intersect with Hebraic history (Pascalis M. Kitromilides, *Enlightenment and the Revolution: The Making of Modern Greece* [Cambridge, MA, 2013], 67–68).

7  Malcolm Bull, *The Mirror of the Gods: Classical Mythology in Renaissance Art* (London, 2005), 84–85.

8  Ibid.

after Jacob Spon (1647–1685) and George Wheeler (1651–1724) documented the survival of Athenian antiquities in their publications of the 1670s and 1680s, no one was sure what had survived of the Greek classical past or whether Greek antiquity could, or should, be divided from its Roman reimaginings.

Poorer and less commercially active in southern regions, central Europeans were slower to obtain material inspiration from the Mediterranean world. Even into the mid-eighteenth century, German courts were exceedingly poor in ancient art. Although there had been quite a number of learned German scholars of Greek in the sixteenth century, the teaching of Greek seems to have fallen off in the course of the depredations of the seventeenth century. By the early eighteenth century, experts were few and far between. Johann Winckelmann (1717–1768) in the 1730s and 1740s had a hard job even finding any books in classical Greek or any teachers who could do an adequate job of teaching it to him (what they did know, of course, was New Testament Greek). A few Germans went on the Grand Tour, and some students to Bologna or Padua to study medicine. But on the whole, German princes did not travel southward much, even to Rome. Able to read – and mostly write – in Latin, the educated elite were not entirely cut off from the Republic of Letters. Still, their domiciles were very sparely decorated; this was even more the case for the middling classes, right down into the post-1848 era.

The rather sudden advent of enlightened absolutism, however, around 1740, made a real difference in beginning to free subjects from clerical overlordship (chiefly in the interests of the monarchy), and so too did the trickling into the Germanies of Italian-, French-, English-, and Dutch-style consumers luxuries, such as sugar, coffee, and household decorations. Very swiftly in the middle of the eighteenth century, German princes turned in the direction of this-worldly consumption, building vast new residence palaces full of mirrors, paintings, and porcelain figurines, at which German artisans excelled in making. German artists, and aristocrats, now went to Italy in ever larger numbers, to enjoy themselves and see the beauties of the ancient world, so as to create or appreciate art in their own day. Neoclassicism did not properly come in until these artists began to trickle back in the 1770s or 1780s, as court artists began to turn away from biblical and toward mythological themes.

Of course, German philhellenism was the product of many factors, including the imitation of Roman, British, and French philhellenisms and the proto-nationalist emphasis on the cultural originality of the fragmented Teutons – as compared to the overbearing, decadent French, depicted as the late Romans of the day.[9] But it must tell us something that the newly fashionable pagans

9   See here Suzanne Marchand, *Down from Olympus: Archaeology and Philhellenism in Germany, 1750–1970* (Princeton, 1996).

in general and Greeks in particular were connected with a certain kind of freedom of the arts and sciences and of this-worldly pleasures, while Jews remained culturally linked to the law, to that great (but tedious) German virtue of *Sparsamkeit*, or thrift, and to God's frightful majesty and justice (again, the Clint Eastwood virtues). It is hard to imagine that anyone before even the 1780s would have dared to say, as did Friedrich Schiller (1759–1805), that 'it is through Beauty that we arrive at Freedom'.[10] For one thing, in the *Ständestaaten* of the Old Regime what one owned and valued were privileges, not rights or freedoms, and rulers were not particularly keen on the trumpeting of freedoms, which might disrupt stable hierarchies, or this-worldly beauty, which might smack of licentiousness and waste. By the century's end, however, Schiller's sentiments had become perfectly comprehensible – at least for a playwright and friend of Immanuel Kant (1724–1804). That is to say, the demise of Moses and the law may be deeply linked not only to changes in biblical criticism and the abstraction of moral law in Kant's Second Critique but also to increasing cultural acceptance of the virtues of pursuing aesthetic pleasures, which made all the older strictures about modesty, frugality, and humility seem out of date and unappealing.

Sixteenth-and seventeenth-century thinkers, of course, had already discussed the constitution of ideal polities and compared ancient lawgivers, including Moses, Solon, Lycurgus, Zoroaster, and Muhammed. This discussion continued into the eighteenth century, culminating in *The Spirit of the Laws* by Montesquieu (1689–1755), in which many local factors, including climate and customs, lay the foundation for individual polities' best practices.[11] But it is striking how much the German discussion by the 1770s and 1780s is about the artists, rather than the lawmakers, as society's molding force – a turn that not only favored the Greeks but even a certain version of Greece, that is, not the Greece of Solon (ca. 630–560 BCE) but of Phidias (ca. 490–430 BCE). This is clear in Winckelmann's 1764 *History of Ancient Art*, where he links Greek freedom not to political changes but to the exaltation of art and the artist as well as the unencumbered movement of the (male) body. Unquestionably, Winckelmann did think that political freedom provided a context where beauty could be expressed, enjoyed, and admired. In a key passage, he writes,

---

10    'weil es die Schönheit ist, durch welche man zu der Freiheit wandert': Friedrich Schiller, 'Ueber die ästhetische Erziehung des Menschen, in einer Reyhe von Briefen', *Die Horen. Eine Monatsschrift, von einer Gesellschaft verfaßt und herausgegeben von Schiller* 1/1 (1795), Letter 2, available online through Project Gutenberg at https://www.projekt-gutenberg.org/schiller/aesterz/aesterz.html (accessed 17 July 2023).

11    [Montesquieu], *De l'esprit des loix* ..., new ed., 3 vols (Geneva, 1749).

'With regard to the constitution and government of Greece, freedom was the chief reason for their art's superiority'.[12] But he then suggests that freedom can be had even in times of monarchy or even tyranny, as long as no single emperor, or God, monopolized national greatness:

> Freedom always had its seat in Greece, even beside the thrones of the kings, who ruled paternally before the enlightenment of reason allowed the people to taste the sweetness of full freedom; and Homer called Agamemnon a 'shepherd of the people' to indicate the latter's love for them and concern for their welfare. Though tyrants installed themselves soon after, they succeeded only in their native lands, and the entire nation never recognized a sole ruler. Thus, the right to be great among his people never rested on one person alone, nor could one person immortalize himself to the exclusion of others.[13]

Is this a coded critique of Judeo-Christian monotheism? It is hard to say, though elsewhere Winckelmann does denounce Near Eastern and Etruscan art for having been restricted to serving dynastic or religious ends, which meant 'their artistic spirit was bound to accepted forms by superstition'.[14] In any event, in the next passages he leaves no uncertainty about the benefits of Greek freedom, which inhere chiefly in that people can exercise their own minds and that artists, who can now work for eternity (rather than for the church or the king), replace the rich as society's leaders. Why, they can even be lawmakers or command armies! This is a world of eternal youth, with Nature alone as its teacher. And in this idyllic world:

> The artist's honor and good fortune did not depend on the stubbornness of an ignorant pride, and their works were not conceived according to wretched taste or the malformed eyes of a judge puffed up by flattery and fawning. Rather, the wisest among the people judged and rewarded the artist and his works in the assembly of all Greeks. [...] Thus, the artists worked for eternity, and the rewards of their work put them in a position to elevate their art beyond all regard for profit and recompense.[15]

---

12    Johann Joachim Winckelmann, *History of the Art of Antiquity*, trans. Harry Francis Mallgrave (Los Angeles, 2006 [1764]), 187.
13    Ibid.
14    Ibid., 150, also 130.
15    Ibid., 189.

These are the dreams, one might say, of a man weary of a world in which artists (and their scholarly champions) had to depend on fat-headed aristocrats
with bad taste – a world, like that of his childhood, where 'Sparsamkeit' and
acceptance of status norms was expected. Freedom here meant the ability to
do what we call 'thinking outside the box', that box being orthodox piety and
Baroque court society. When Winckelmann settled in Rome, to live and work
among artists, he felt he had been made a free man. It did not much matter
to him that Rome was still the fiefdom of the Pope or that he had to convert
to Catholicism; at least he was now free to cultivate his own tastes and write
for scholarly eternity, two forms of freedom his German successors would certainly cherish, too.

Reading carefully, we might say that what Winckelmann offers is a new kind
of national eternity, one available to societies that allow at least mental freedom
to their artists – and then embrace them as heroes. The freedom does precede
the beauty, but beauty is definitely the end goal and might even be possible
under conditions of particularist tyranny. It is not possible, however, under conditions of superstition or of Nero-like (or Mosaic?) theocracy. It is also of eminent importance that the art made is representational, especially of the nude
body: this, to Winckelmann and so many of his admirers, was the incarnation of
Greek greatness, especially since the nudes depicted could be ordinary citizens,
not exclusively kings or gods. One of the failings of Egyptian art, in his view,
was its poor grasp of anatomy; a ceaseless objection to the exemplariness of the
Jews, in the next century, would be that they made no graven images and hence
really had no art to speak of. It is another hallmark of the later eighteenth century to have made a culture's desire to produce anatomically correct sculpture
a major factor in its evaluation. But after Winckelmann, this surely becomes a
familiar refrain, indeed.

Passing on to Schiller, I think we can say that political and spiritual freedom
were, for him, the goal not the means, as implied in the phrase I quoted earlier: 'it is through Beauty that we arrive at Freedom'. It might be worth mentioning, too, that Schiller was raised in a very pious Protestant household and
expected to become a cleric. Instead, he became a military doctor, a post he so
despised that he went AWOL, after which he was forbidden from publishing in
Württemberg. He then left that polity and ended up in Weimar, where he was
an early fan of the French Revolution. He pulled back sufficiently from radical
politics to receive an appointment as professor of history at Jena and, finally, to
be ennobled in 1802, three years before his death. For him, the Greeks clearly
pushed out the Bible, so much so that his poem 'Die Götter Griechenlandes'
('The Gods of Greece') quite obviously suggests the death of paganism was a

bad thing for humanity.[16] For Schiller – friend of Johann Wolfgang von Goethe (1749–1832), Friedrich August Wolf (1759–1824), Wilhelm von Humboldt (1767–1835), and other notable philhellenes of the era – the purpose of Greek beauty was to offer an immortal, absolute model of Truth that draws the artist, or the individual more generally, beyond his corrupt and particular era and his everyday needs and concerns and thereby makes possible the pursuit of the Ideal and the free plasticity of youth, which no longer needs the severe discipline of tradition, age, and law.

There is a strong Kantian element here, of course. For Schiller, as for Kant in his Third Critique, Beauty is the propaedeutic that trains us to move from necessity to freedom, from the physical world to the moral realm. But it is Schiller who articulates most clearly, in his *Aesthetic Letters*, the role of the imagination and art in this crucial process, and he makes the Greeks the quintessential example of the noble culture whose aesthetic products (tragedy, epic, architecture, art) have the power to rescue human dignity from its current state of degradation. That was a big ask for the Greeks, but Schiller thought they were up to it. Notice that here, as in Winckelmann, Greek religion is hardly mentioned – presumed, in fact, to be wholly subordinated to aesthetic concerns. Nor does either Winckelmann or Schiller mention more than in passing Greek freedom as having been born from the defeat of the 'Orientals' in the Persian Wars or the lawmaking power of the Greek citizen as something desirable for a culture to achieve. The exemplary quality of the Greeks lies in their art, not in their politics, though the Athenian principle of the equality of citizens in public life underwrote the promise of meritocratic treatment for those with education and talent. For these cultural revolutionaries, a new 'chosen' people has been selected as a model, one that Wilhelm von Humboldt would soon make the core of his new university and his reformed, deliberately secularized, *Gymnasien*.

## 3      Moses and the Culture of Philhellenism: a Poor Fit

I would submit, without further elaborating the point, that it was this sort of Greek freedom which informed the cultural ideals of at least a significant portion of the *Goethezeit*'s elite. This model figured freedom as more individual than collective, more aesthetic than political. Lurking behind aesthetic liberation was an escape from the dour Lutheranism of the past and the impoverished sensual world of the penurious German educated classes. Readers of this volume will know, too, that such a model took shape alongside

---

16      Friedrich Schiller, 'Die Götter Griechenlandes', *Der Teutsche Merkur* 61 (1788), 250–260.

theological innovations which historicized and, in some cases, demoted the ancient Israelites. The Enlightenment in general disparaged particularist laws in favor of universal ones and ridiculed rituals that had no function save to undergird superstition, sometimes also known as religious conformity. Even supporters of the *Haskalah* shared many of these rationalizing inclinations.[17] None of this was good news for Moses, who increasingly dropped out of the visual repertoire of the leading artists and makers of the decorative arts.

Nor was the celebration of nudity, which one found not only in the libertine world of Madame de Pompadour (1721–1764) but also, in a different way, in Winckelmann and later eighteenth-century German aesthetics more generally. Moses, in his heavily draped modesty, could not be shown in the nude: even Michelangelo (1475–1564) dared only to bare an arm and a knee (Figure 20). This befit the prophet's wise old age as well as his ascetic lifestyle. Not so, of course, the eternally young Apollo and Mercury, much less with Venus, whose chief charm was to be beautifully naked, or Diana, constantly discovered by Actaeon in her bath. Renaissance artists, as noted above, had reveled in depicting at least partial nudity, but until the later eighteenth century, very few Germans – most of whom were rural town-dwellers – had seen even sketches of such risqué monuments. After this time, however, the very courtly residences described above had begun to teem with nudes, especially in courtly gardens, such as those of Frederick the Great (1712–1786) at Sanssouci, where the bronze nude dubbed the 'Praying Boy' (Figure 21) was stationed just outside the king's study window. Thanks largely to plaster casts, a vision of the white, pure statuary of the ancient world spread through northern climes, exemplified here by the 800+ replicas that the Saxon Elector purchased from the estate of Anton Raphael Mengs (1728–1779) in 1780 and by the many other cast collections in the holdings of princes and elite societies and increasingly open to the general public.[18] Porcelain figurines, particularly the unpainted 'biscuit' porcelains of the post-1780 period, were cheaper and much more portable versions of these classicizing idols. As I have argued elsewhere, they helped popularize a vision of the pagan world still scandalous to the pious defenders of underclass asceticism and obedience.[19] If a few figurines of Moses, the crucifixion, or the saints

---

17    On Haskalah's relationship to other contemporary forms of German biblical criticism, see David Sorkin, *The Berlin Haskalah and German Religious Thought: Orphans of Knowledge* (London, 2000).

18    On plaster casts, see the essays in Rune Frederiksen and Eckart Marchand, eds., *Plaster Casts: Making, Collecting, and Displaying from Classical Antiquity to the Present*, Transformationen der Antike 18 (Berlin, 2010).

19    See Suzanne Marchand, 'Porcelain: Another Window on the Neoclassical Visual World', *Classical Receptions Journal* 12/2 (2020), 200–230.

were on offer, their numbers were dwarfed by the vast number of Venuses, Apollos, and lightly erotic shepherd and nymph pairs. Thus did Renaissance classicism reach the northern middling classes, with all of its secular delights and subtle messages.

But if ideals of Greek beauty and national freedom were being carried forward in casts, paintings, architecture, and porcelain, what about images of the Mosaic Law? This is all very anecdotal, but in the many books I have read and exhibits I have seen of later eighteenth-and nineteenth-century paintings, casts, and porcelains, precious few examples deal with Moses, especially Moses as a lawgiver, after the 1750s. He does appear in the fresco cycle 'History of Mankind' by Wilhelm von Kaulbach (1804–1874) in the Neues Museum of Berlin (1860), in a side panel, and as a spectral presence observing 'The Destruction of Jerusalem'. But it is Homer as well as the Huns and the heroes of the Reformation who take up most of the space. The one major painting treating the subject I know of was, tellingly, by the Jewish artist Daniel Moritz Oppenheim (1800–1882), and it mimics seventeenth-century models.

Perhaps Moses was not sufficiently historical to appeal to history painters or adequately literary to offer the plasticity of an Oedipus or a Cassandra. After Jean-Auguste-Dominique Ingres (1780–1867), mythological painting was, in any case, on the decline. Moses was not a particularly sentimental figure, making him not terribly appealing even to bourgeois Christians: it is telling that the most popular nineteenth-century depictions of him seem to have been as a baby, found among the bullrushes by kindly women. Johann Friedrich Overbeck (1789–1869) portrayed a handsome young Moses saving Zipporah, one of the daughters of Reuel/Jethro, in 1850. In his 1871 'Victory O Lord!', John Everett Millais (1829–1896) depicted a severely weakened Moses, his arms held aloft by Aaron and Hur to ensure the defeat of the Amalekites (Exod 17), but like the Overbeck, this was exceptional in the painter's corpus – and quickly forgotten (Figure 22). There are some later twentieth-century images of Moses and the tablets, but these seem to belong to a different world, clearly meant to be didactic material for children rather than artistic representations of a major scene from Europe's religious and mythographic past.

Similarly, as Theodore Ziolkowski has shown, Moses plays a very small role in nineteenth-century literature. When invoked, he is often off stage, and the drama shifts to other characters and to the Egyptian landscape, with fictional females often added to bring romance to the biblical stories. Those most earnest in conjuring Moses as a hero tended

FIGURE 20    Statue of Moses by Michelangelo, for the tomb of Pope Julius II in the
church of San Pietro in Vincoli, Rome
PHOTOGRAPH BY JÖRG BITTNER (UNNA); IMAGE COURTESY OF
WIKIMEDIACOMMONS, CC BY-SA 3.0. HTTPS://COMMONS
.WIKIMEDIA.ORG/WIKI/FILE:%27MOSES%27_BY_MICHELANGELO
_JBU140.JPG

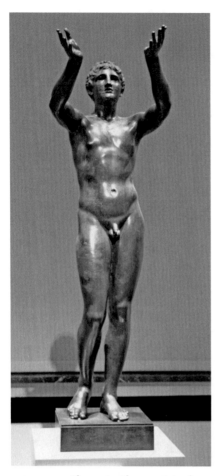

FIGURE 21    The ancient bronze statue known as
'Praying Boy' or 'Berlin Adorante'; now
housed in the Altes Museum, Berlin
PHOTOGRAPH BY ISMOON;
IMAGE COURTESY OF
WIKIMEDIACOMMONS, CC BY-SA
4.0. HTTPS://COMMONS.WIKIMEDIA
.ORG/WIKI/FILE:ALTES_MUSEUM
_-_BETENDER_KNABE-.JPG

to be the oppressed: revolutionaries, African Americans, Ukrainian nationalists.[20] Perhaps the most prominent of nineteenth-century German depictions appears in *Joshua,* a novel by Georg Ebers (1837–1898). Again, Moses

---

20    Theodore Ziolkowski, *Uses and Abuses of Moses: Literary Representations since the Enlightenment* (Princeton, 2016), 38–54, 105–08.

FIGURE 22    The oil painting 'Victory O Lord', by John Everett
             Millais in 1871
             WORK IN THE PUBLIC DOMAIN;
             PHOTOGRAPH BY PAUL BARLOW; COURTESY
             OF WIKIMEDIACOMMONS

plays a bit part, but Ebers' heroic attempts to sentimentalize stories from the Old Testament are worth mentioning. Ebers is now completely forgotten, but in the later nineteenth century, he was a highly popular writer of historical romances as well as a leading Egyptologist.[21] In the novel, Ebers – whose high-society Jewish parents had converted to Christianity in the 1820s – makes every effort to humanize the Israelites as well as their Egyptian captors and to explain the sufferings and hardships of both peoples. He actually attributes more

---

21    See Suzanne Marchand, 'Georg Ebers, Sympathetic Egyptologist', in *For the Sake of Learning: Essays in Honor of Anthony Grafton*, eds Ann Blair and Anja-Silva Goering, 2 vols, Scientific and Learned Cultures and Their Institutions 18 (Leiden, 2016), 917–932.

harshness of character to Moses' sister, Miriam, who, after all, the Egyptians had abused by sending her beloved husband to the mines. Perhaps it is worth describing the novel's conclusion, in which Joshua – he of the biblical hardened heart – decides that the law cannot be the final goal: 'No, again no! The Law could not afford the people who had grown so dear to him all that he desired for them. Something else was needful to make their future lot as noble and fair as he had dreamed it might be, on his way to the mines'.[22]

That something else is hinted to him through a vision of Miriam with the tablets, and it is juxtaposed with one of a lovely child resembling an Egyptian woman, whose love had saved him from the mines earlier in the novel. In his dream, this girl is accompanied by a lamb and speaks three unintelligible words, before offering him a palm as a sign of peace. When he wakes, Joshua cannot remember the words, but they turn out to be a prophesy of a new Jehoshua, born many centuries later in Bethlehem, and Ebers closes the novel by telling us that the three words were: 'Love, Mercy, Redemption!'[23] This is obviously Christian supercessionism, but I think it points to something, again, about nineteenth-century German society's discomfort with the law and its need for beauty: in this case, in the form of a lovely child and a sheep. The times no longer demanded unyielding severity. Even Joshua could dream of beauty and peace.

## 4    Conclusion

Were we to spend more time among nineteenth-century biblical scholars and antisemites, we would find that Ebers' sentimental Moses was exceedingly rare: the conventional attributes assigned to Moses and ancient (and often modern) Jews more generally were much less ambivalent and humane. Instead, for many Christians Moses now stood for an outdated, intolerant, particularistic, sclerotic, violent, and fanatical form of belief, one perhaps remarkable in its ethical rigor – even Ernest Renan (1823–1892) and Matthew Arnold (1822–1888) thought so – but absolutely unfit for Europeans in the present. The Greeks, on the other hand, had come to embody the virtues that the nineteenth century now admired: beauty, youth, and personal if not political freedom. Already in the 1820s, too, a new discourse on Greek military valor, focusing on the battles against the Persians, whether won (Marathon) or lost

---

22    Georg Ebers, *Joshua: A Story of Biblical Life*, trans. Clara and Margaret Bell, 2 vols, Collection of German Authors 48 (Leipzig, 1890), 2:260.

23    Ibid., 2:262–63.

(Thermopylae), gave another gloss to discussions of Greek freedom. Moses, by contrast, especially as lawgiver, was not suited to any of these associations. His modern cultural functions had peaked in the seventeenth century, and he could, indeed, be 'swept under the carpet', with his descendants persecuted in ways even Luther would have found appalling.

Looking back on his youthful disdain for the biblical hero, Heinrich Heine (1797–1856) attributed his contempt to 'the Hellenic spirit [that] predominated within me, and I could not forgive the lawgiver of the Jews for his hatred against all figurative representations, against plastic art'.[24] Even for the youthful Heine, Moses had been too narrowminded a nationalist and too severe in his condemnation of artistic beauty. He could be invoked to remind citizens of their moral duties, but it was much more desirable for this society to perceive God's power in his goodness and mercy than in his wrath and his commandments. Like Heine, post-Enlightenment Germans (or at least the educated elite) liked to see themselves as descendent from the artistic ancient Greeks – who allowed for this-worldly indulgences and foibles – more than the sober, severe Hebrews of Moses. Their visual universe was populated with Venuses, cherubs, and modern Europeans, not ancient Jews. These were subtle, if powerful, provocations to seek personal happiness rather than righteousness, to value the creation of beauty more than fulfillment of the law.

In an impressionistic piece such as this one, the author ought not to draw grand general conclusions. But as the reader closes this book of reflections on Moses among the modern Germans, it is worth noting that representations of even this extremely important cultural figure always exist in worlds populated by other heroes and villains, lawgivers and lawbreakers, father-figures and mother goddesses. We choose, we admire, we change. And as we change, so too do our idols, sacred and profane. It is one of the ambitions of this book, as its editor states in the introduction, to survey the harmony and tensions of the cultural roles of Moses as both 'a father of Judaism and a framer of European civilization'.[25] It is one of my hopes that these parting thoughts offer some food for thought about why the greatest of the Hebrew prophet's latter role is nowadays so regularly forgotten. But Moses remains in the West's cultural memory bank, for Jews most particularly but also for Christians. Who knows when we might need his severity, his righteousness, his charismatic leadership again.

---

24    Quoted in Ziolkowski, *Uses and Abuses of Moses*, 55.
25    Paul Michael Kurtz, 'Introduction: Moses in Modernity', above.

# Bibliography

Abel, Caspar. *Hebräische Alterthümer, Worinnen* ... (Leipzig, 1736).

Abel, Casper. *Teutsche und Sächsische Alterthümer* ... (Braunschweig, 1729).

Allen, T. Paynter, ed. *Opinions of the Hebrew and Greek Professors of the European Universities, of Bible Revisers, and of Other Eminent Scholars and Commentators on the Scriptural Aspect of the Question Regarding the Legalisation of Marriage with a Deceased Wife's Sister*, 2nd ed. (London, 1884).

Almog, Yael. *Secularism and Hermeneutics*, Intellectual History of the Modern Age (Philadelphia, 2019).

Altmann, Alexander. *Die trostvolle Aufklärung. Studien zur Metaphysik und politischen Theorie Moses Mendelssohns* (Stuttgart-Bad Cannstatt, 1982).

Amengual, Gabriel. 'Der Begriff der Sittlichkeit. Überlegungen zu seiner differenzierten Bedeutung', *Hegel-Jahrbuch* 1 (2001), 197–203.

Amram, David Werner. 'A Lawyer's Studies in Biblical Law', *Green Bag* 14/2 (1902), 83–84; 14/5 (1902), 231–33; 14/7 (1902), 343–46; 14/10 (1902), 490–93; 15/1 (1903), 41–44; 15/6 (1903), 291–94.

Amram, David Werner. 'Ancient Conveyance of Land', *Green Bag* 10/2 (1898), 77–78.

Amram, David Werner. 'Chapters from the Ancient Jewish Law', *Green Bag* 4/1 (1892), 36–38; 4/10 (1892), 493–95; 6/9 (1894), 407–08.

Amram, David Werner. 'Chapters from the Biblical Law', *Green Bag* 12/2 (1900), 89–92; 12/4 (1900), 196–99; 12/8 (1900), 384–87; 12/9 (1900), 483–85; 12/10 (1900), 504–06; 12/11 (1900), 585–89; 12/12 (1900), 659–61; 13/1 (1901), 37–40; 13/2 (1901), 70–74; 13/4 (1901), 198–202; 13/6 (1901), 313–16; 13/8 (1901), 406–08; 13/10 (1901), 493–96; 13/12 (1901), 592–94.

Amram, David Werner. *Leading Cases in the Bible* (Philadelphia, 1905).

Amram, David Werner. 'Some Aspects of the Growth of Jewish Law', *Green Bag* 8/6 (1896), 253–56; 8/7 (1896), 298–302.

Amram, David Werner. *The Jewish Law of Divorce According to Bible and Talmud, with Some References to its Development in Post-Talmudic Times* (Philadelphia, 1896).

Anderson, Nancy. 'The "Marriage with a Deceased Wife's Sister Bill" Controversy: Incest Anxiety and the Defense of Family Purity in Victorian England', *Journal of British Studies* 21 (1982), 67–86.

Arendt, Hannah. 'Aufklärung und Judenfrage [1932]', repr. in *Wir Juden. Schriften 1932 bis 1966*, eds Marie Luise Knott and Ursula Ludz (Munich, 2019), 11–30.

Arkush, Allen. *Moses Mendelssohn and the Enlightenment* (Albany, 1994).

Arnold, Bill T., and David B. Weisberg, 'A Centennial Review of Friedrich Delitzsch's "Babel und Bibel" Lectures', *Journal of Biblical Literature* 121 (2002), 441–57.

Asad, Talal. *Formations of the Secular: Christianity, Islam, Modernity* (Stanford, 2003).

Assmann, Jan. *Moses the Egyptian: The Memory of Egypt in Western Monotheism* (Cambridge, MA, 1998).

Auerochs, Bernd. 'Poesie als Urkunde. Zu Herders Poesiebegriff', in *Johann Gottfried Herder. Aspekte seines Lebenswerks*, eds Marin Keßler and Volker Leppin (Berlin 2005), 93–114.

Avineri, Shlomo. *Hegel's Theory of the Modern State*, Cambridge Studies in the History and Theory of Politics (Cambridge, 1972).

Bakunin, Mikhail Aleksandrovich. 'Persönliche Beziehung zu Marx – Auszug', repr. in *'Antisemit, das geht nicht unter Menschen'. Anarchistische Positionen zu Antisemitismus, Zionismus und Israel*, vol. 1, *Von Proudhon bis zur Staatsgründung*, eds Jürgen Mümken and Siegbert Wolf (Lich/Hessen, 2013 [1871]), 80–84.

Banier, Antoine. *Explication historique des fables, où l'on découvre leur origine & leur conformité avec l'histoire ancienne, & où l'on rapporte les époques des héros & des principaux événemens dont il est fait mention*, 2 vols (Paris, 1711).

Banier, Antoine. *La mythologie et les fables expliquées par l'histoire*, 8 vols (Paris, 1738).

Barrie, David. *Sin, Sanctity, and the Sister-in-Law: Marriage with a Deceased Wife's Sister in the Nineteenth Century* (London, 2018).

Bastian, Adolf. 'Ethnische Elementargedanken in der Lehre vom Menschen [1895]', repr. in idem, *Ausgewählte Werke*, vol. 7, eds Peter Bolz and Manuela Fischer, Historia Scientiarum (Hildesheim, 2007).

Batnitzky, Leora. *How Judaism Became a Religion. An Introduction to Modern Jewish Thought* (Princeton, 2011).

Bauer, Georg Lorenz. *Hebräische Mythologie des alten und neuen Testaments, mit Parallelen aus der Mythologie anderer Völker, vornemlich der Griechen und Römer*, 2 vols (Leipzig, 1802).

Baumgardt, David. 'The Ethics of Lazarus and Steinthal', *Leo Baeck Institute Year Book* 2 (1957), 205–17.

Baumgart, Peter. 'Absolutismus ein Mythos? Aufgeklärter Absolutismus ein Widerspruch? Reflexionen zu einem kontroversen Thema gegenwärtiger Frühneuzeitforschung', *Zeitschrift für historische Forschung* 27 (2000), 573–89.

Beal, Jane, ed. *Illuminating Moses: A History of Reception from Exodus to the Renaissance*, Commentaria 4 (Leiden, 2014).

Behrman, C.F. 'The Annual Blister: A Sidelight on Victorian Social and Parliamentary History', *Victorian Studies* 11 (1968), 483–50.

Belke, Ingrid. 'Steinthals *Allgemeine Ethik*', in *Chajim H. Steinthal. Sprachwissenschaftler und Philosoph im 19. Jahrhundert*, eds Hartwig Wiedebach and Annette Winkelmann, Studies in European Judaism 4 (Leiden, 2002), 189–236.

Belton, Roshunda Lashae. 'A Non-Traditional Traditionalist: Rev. A. H. Sayce and His Intellectual Approach to Biblical Authenticity and Biblical History in Late-Victorian Britain' (PhD thesis, Louisiana State University, Baton Rouge, 2007).

Benes, Tuska. *In Babel's Shadow: Language, Philology, and the Nation in Nineteenth-Century Germany* (Detroit, 2008).

Benjamin, Walter. *Über den Begriff der Geschichte*, repr. in idem, *Werke und Nachlass – Kritische Gesamtausgabe*, vol. 19, ed. Gérard Raulet (Berlin, 2010 [1942]).

Bennett, Bruce S. 'Banister v. Thompson and Afterwards: The Church of England and the Deceased Wife's Sister's Act', *Journal of Ecclesiastical History* 49 (1998), 669–82.

Ben-Pazi, Hanoch. 'Moritz Lazarus and the Ethics of Judaism', *Daat: A Journal of Jewish Philosophy and Kabbalah* 88, Special Issue: 'Wissenschaft des Judentums: Judaism and the Science of Judaism. 200 Years of Academic Thought on Religion' (2019), 91–104.

Bergel, Joseph. *Die Eheverhältnisse der alten Juden im Vergleiche mit den Griechischen und Römischen* (Leipzig, 1881).

Berkovits, Eliezer. *God, Man and History: A Jewish Interpretation* (New York, 1959).

Berkovits, Eliezer. *Not in Heaven: The Nature and Function of Jewish Law* (New York, 1983).

Berkovits, Eliezer. *Women in Time and Torah* (Hoboken, 1990).

Bernays, Jacob. *Jugenderinnerungen und Bekenntnisse* (Berlin, 1900).

Bernstein, Richard J. *Freud and the Legacy of Moses*, Cambridge Studies in Religion and Critical Thought 4 (Cambridge, 1998).

Berry, Christopher J. *Social Theory of the Scottish Enlightenment* (Edinburgh, 1997).

Bin-Gorion (Berdyczewski), Micha Josef, ed. *Joseph und seine Brüder. Ein altjüdischer Roman*, repr. (Berlin, 1933 [1917]).

Binstock, Louis. 'Mosaic Legislation and Rabbinic Law', *Loyola Law Journal* 10 (1929), 13–19.

Blair, Hugh. *A Critical Dissertation on the Poems of Ossian, the Son of Fingal* (London, 1763).

Blair, Hugh. *The Works of Ossian, the Son of Fingal* (London, 1765).

Blänkner, Reinhard. 'Absolutismus'. *Eine begriffsgeschichtliche Studie zur politischen Theorie und zur Geschichtswissenschaft in Deutschland (1830–1870)* (Frankfurt, 2011).

Blickle, Peter. *Die Revolution von 1525*, 4th ed. (Munich, 2004).

Blitz, Hans-Martin. *Aus Liebe zum Vaterland. Die deutsche Nation im 18. Jahrhundert* (Hamburg, 2000).

Blitz, Hans-Martin. '"Gieb, Vater, mir ein Schwert!" Identitätskonzepte und Feindbilder in der "patriotischen" Lyrik Klopstocks und des Göttinger "Hain"', in *Machtphantasie Deutschland. Nationalismus, Männlichkeit und Fremdenhaß im Vaterlandsdiskurs deutscher Schriftsteller des 18. Jahrhunderts*, eds Hans Peter Herrmann et al. (Frankfurt, 1996), 80–122.

Bochart, Samuel. *Geographia Sacra, seu Phaleg et Canaan* (Caen, 1646).

Bock, Wolfgang. 'Terrorismus und politischer Anarchismus im Kaiserreich: Entstehung, Entwicklung, rechtliche und politische Bekämpfung', in *Anarchismus. Zur*

*Geschichte und Idee der herrschaftsfreien Gesellschaft*, ed. Hans Diefenbacher (Darmstadt, 1996), 143–68.

Boyarin, Daniel. *A Radical Jew: Paul and the Politics of Identity*, Contraversions: Critical Studies in Jewish Literature, Culture, and Society (Berkeley, 1994).

Braemer, Andreas. 'The Dilemmas of Moderate Reform: Some Reflections on the Development of Conservative Judaism in Germany 1840–1880', *Jewish Studies Quarterly* 10 (2003), 73–87.

Brämer, Andreas. *Judentum und religiöse Reform. Der Hamburger Israelitische Tempel 1817–1938*, Studien zur jüdischen Geschichte 8 (Hamburg, 2000).

Brämer, Andreas. 'Jüdische "Glaubenswissenschaft" – Zacharias Frankels rechtshistorische Forschung als Herausforderung der Orthodoxie', in *Die Wissenschaft des Judentums. Eine Bestandsaufnahme*, eds Thomas Meyer and Andreas Kilcher (Paderborn, 2015), 79–94.

Brämer, Andreas. *Rabbiner Zacharias Frankel. Wissenschaft des Judentums und konservative Reform im 19. Jahrhundert*, Netiva 3 (Hildesheim, 2000).

Brämer, Andreas. '"Wissenschaft des Judentums" und "Historische Rechtsschule": Zwei Briefe Zacharias Frankels an Carl Josef Anton Mittermaier', *Aschkenas. Zeitschrift für Geschichte und Kultur der Juden* 7 (1997), 173–79.

Breines, Paul. 'The Jew as Revolutionary: The Case of Gustav Landauer', *Leo Baeck Institute Year Book* 12 (1967), 75–84.

Breslauer, S. Daniel. *Creating a Judaism Without Religion: A Postmodern Jewish Possibility*, Studies in Judaism (Lanham, 2001).

Breuer, Edward. *The Limits of Enlightenment: Jews, Germans, and the Eighteenth-Century Study of Scripture* (Cambridge, MA, 1996).

Breuer, Mordechai. *Modernity within Tradition: The Social History of Orthodox Jewry in Imperial Germany* (New York, 1992).

Breuer, Mordechai. *The 'Torah-Im-Derekh-Eretz' of Samson Raphael Hirsch* (Jerusalem, 1970).

Britt, Brian. *Rewriting Moses: The Narrative Eclipse of the Text*, Journal for the Study of the Old Testament Supplement Series 402 (London, 2004).

Brooks, Thom. 'Is Hegel a Retributivist?', *Hegel Bulletin* 25 (2004), 113–26.

Buchholz, P. *Die Familie in rechtlicher und moralischer Beziehung nach mosaisch-talmudischer Lehre, allgemein faßlich dargestellt* (Breslau, 1867).

Buchwalter, Andrew. 'Hegel, Modernity, and Civic Republicanism', *Public Affairs Quarterly* 7 (1993), 1–12.

Bull, Malcolm. *The Mirror of the Gods: Classical Mythology in Renaissance Art* (London, 2005).

Burckhardt, Jacob. *Force and Freedom: Reflections on History*, trans. James Hastings Nichols (New York 1943). First published as *Weltgeschichtliche Betrachtungen*, ed. Jakob Oeri (Berlin, 1905).

Camus, Albert. *Der Mensch in der Revolte. Essays* (Reinbek bei Hamburg, 1997).

Cancik-Kirschbaum, Eva, and Thomas L. Gertzen, eds, *Der Babel-Bibel-Streit und die Wissenschaft des Judentums*. Investigatio Orientis 6 (Münster, 2021).

Candlish, James. *The Authority of Scripture Independent of Criticism* (Edinburgh, 1877).

Candlish, James. 'The Real Bearings of the Opinions of the Professors of Hebrew and Greek on the Scriptural Law of Prohibited Degrees of Marriage', in *Tracts Issued by the Marriage Law Defence Union*, vol. 1, *Scriptural* (London, 1889), tract 25, pp. 177–96.

Cannadine, David. *Ornamentalism: How the British Saw Their Empire* (London, 2001).

Carhart, Michael M. *The Science of Culture in Enlightenment Germany*, Harvard Historical Studies 159 (Cambridge, MA, 2007).

Carus, Friedrich August. *Psychologie der Hebräer* (Leipzig, 1809).

Cepl-Kaufmann, Gertrude, and Rolf Kauffeldt, *Berlin-Friedrichshagen, Literaturhauptstadt um die Jahrhundertwende. Der Friedrichshagener Dichterkreis* (Munich, 2015).

Ceserani, Giovanna. 'Narrative, Interpretation, and Plagiarism in Mr. Robertson's 1778 *History of Ancient Greece*', *Journal of the History of Ideas* 66/3 (2005), 413–36.

Chambost, Anne-Sophie. *Proudhon. L'enfant terrible du socialisme* (Paris, 2009).

Chanes, Jerome A., and Mark Silk, eds. *The Future of Judaism in America*, Studies of Jews in Society 5 (Cham, Switzerland, 2023).

Charpin, Dominique. *Writing, Law, and Kingship in Old Babylonian Mesopotamia* (Chicago, 2010).

Chase, Karen, and Michael Levenson. *The Spectacle of Intimacy: A Public Life for the Middle-Class Family* (Princeton, 2001).

Clark, Victoria. *Allies for Armageddon: The Rise of Christian Zionism* (New Haven, 2007).

Clifford, Hywel. 'Moses as Philosopher-Sage in Philo', in *Moses in Biblical and Extra-Biblical Traditions*, eds Axel Graupner and Michael Wolter, Beihefte zur Zeitschrift für die alttestamentliche Wissenschaft 372 (Berlin, 2007), 151–67.

Cline, Eric H. *Biblical Archaeology: A Very Short Introduction* (Oxford, 2009).

Cohen, Hermann. 'Religion und Sittlichkeit. Eine Betrachtung zur Grundlegung der Religionsphilosophie [1907]', repr. in idem, *Jüdische Schriften*, vol. 3, *Zur jüdischen Religionsphilosophie und ihrer Geschichte*, ed. Akademie für die Wissenschaft des Judentums (Berlin, 1924), 98–168.

Cohen, Jack J. *Judaism in a Post-Halakhic Age*, Reference Library of Jewish Intellectual History (Boston, 2010).

Cohn, Georg. *Die Gesetze Hammurabis. Rektoratsrede gehalten am Stiftungsfeste der Hochschule Zürich den 29. April 1903* (Zurich, 1903).

Cooper-White, Pamela. 'Freud's Moses, Schoenberg's Moses, and the Tragic Quest for Purity,' *American Imago* 79/1 (2022), 89–122.

Corbeau-Parsons, Caroline. *Prometheus in the Nineteenth Century: From Myth to Symbol* (London, 2013).

Corrodi, Heinrich. 'Ob in der Bibel Mythe zu finden sind?', *Beiträge zur Beförderung des vernünftigen Denkens der Religion* 18 (1794), 1–73.

Cregan-Reid, Vybarr. 'Discovering Gilgamesh: George Smith and the Victorian Horizon of History', in *The Victorians and the Ancient World: Archaeology and Classicism in Nineteenth-Century Culture*, ed. Richard Pearson (Cambridge, 2006), 109–23.

Cullhed, Anna. 'Original Poetry: Robert Lowth and Eighteenth-Century Poetics', in *Sacred Conjectures: The Context and Legacy of Robert Lowth and Jean Astruc*, ed. John Jarick (New York, 2007), 25–47.

Curtis, Michael. *Orientalism and Islam: European Thinkers on Oriental Despotism in the Middle East and India* (Cambridge, 2009).

Davis, Thomas W. *Shifting Sands: The Rise and Fall of Biblical Archaeology* (Oxford, 2004).

De Sola, D.A., and M.J. Raphall, *Eighteen Treatises from the Mishna* (London, 1843).

de Wette, Wilhelm Martin Lebrecht, *Aufforderung zum Studium der hebräischen Sprache und Literatur* (Jena, 1805).

de Wette, Wilhelm Martin Lebrecht. *Beiträge zur Einleitung in das Alte Testament*, 2 vols (Halle, 1806–07).

de Wette, Wilhelm Martin Lebrecht. *Ueber Religion und Theologie. Erläuterungen zu seinem Lehrbuche der Dogmatik* (Berlin, 1815).

Delitzsch, Friedrich. *Babel and Bible: Two Lectures on the Significance of Assyriological Research for Religion, Embodying the Most Important Criticisms and the Author's Replies, Profusely Illustrated*, trans. Thomas J. McCormack and W.H. Carruth (Chicago, 1903).

Delitzsch, Friedrich. *Babel und Bibel. Dritter (Schluss-)Vortrag* (Leipzig, 1905).

Delitzsch, Friedrich. *Babel und Bibel. Ein Rückblick und Ausblick* (Stuttgart, 1904).

Delitzsch, Friedrich. *Babel und Bibel. Ein Vortrag* (Leipzig, 1902).

Delitzsch, Friedrich. *Zweiter Vortrag über Babel und Bibel* (Stuttgart, 1903).

Dörr, Thomas. 'An-archie und (talmudisches) Gesetz. Erich Mühsams Judentum', in *Erich Mühsam und das Judentum*, ed. Jürgen Wolfgang Goette (Lübeck, 2002), 71–84.

Droge, Arthur J. *Homer or Moses? Early Christian Interpretations of the History of Culture*, Hermeneutische Untersuchungen zur Theologie 26 (Tübingen, 1989).

Dubbels, Elke. *Figuren des Messianischen in Schriften deutsch-jüdischer Intellektueller, 1900–1933* (Berlin 2011).

Duke, Henry Hinxman. *The Question of Incest Relatively to Marriage with Sisters in Succession* (London, 1883).

Dunkelgrün, Theodor. 'The Philology of Judaism: Zacharias Frankel, the Septuagint, and the Jewish Study of Ancient Greek in the Nineteenth Century', in *Classical Philology and Theology: Entanglement, Disavowal, and the Godlike Scholar*, eds Catherine Conybeare and Simon Goldhill (Cambridge, 2020), 63–85.

Duschak, Moritz. *Das mosaisch-talmudische Eherecht, mit besonderer Rücksicht auf die bürgerlichen Gesetze* (Vienna, 1864).

Duschak, Moritz. *Das mosaisch-talmudische Strafrecht. Ein Beitrag zur historischen Rechtswissenschaft* (Vienna, 1869).

Duvernoy, Claude. *Le prince et le prophète* (Jerusalem, 1966).

Düwel, Klaus, and Harro Zimmermann, 'Germanenbild und Patriotismus in der deutschen Literatur des 18. Jahrhunderts', in *Germanenprobleme in heutiger Sicht*, ed. Heinrich Beck (Berlin, 1986), 358–95.

Dyck, Joachim. *Athen und Jerusalem. Die Tradition der argumentativen Verknüpfung von Bibel und Poesie im 17. und 18. Jahrhundert* (Munich, 1979).

Ebers, Georg. *Joshua: A Story of Biblical Life*, trans. Clara and Margaret Bell, 2 vols, Collection of German Authors 48 (Leipzig, 1890).

Eichhorn, Johann Gottfried. *Urgeschichte*, ed. Johann Philipp Gabler, 3 vols (Altdorf-Nürnberg, 1790–95). First published anonymously as 'Urgeschichte. Ein Versuch', in 2 Parts, *Repertorium für Biblische und Morgenländische Litteratur* 4 (1779), 129–256.

Eisen, Arnold. *Rethinking Modern Judaism* (Chicago, 1998).

Ellenson, David. 'Antinomianism and its Responses in the Nineteenth Century', in *The Cambridge Companion to Judaism and Law*, ed. Christine Hayes (Cambridge, 2017), 260–86.

Ellern, Hermann, and Bessi Ellern. *Herzl, Hechler, the Grand Duke of Baden and the German Emperor, 1896–1904, documents found ... reproduced in facsimile* (Tel Aviv, 1961).

Erlewine, Robert. *Monotheism and Tolerance: Recovering a Religion of Reason* (Bloomington, IN, 2010).

Feingold, Mordechai. '"The Wisdom of the Egyptians": Revisiting Jan Assmann's Reading of the Early Modern Reception of Moses', *Aegyptiaca: Journal of the History of Reception of Ancient Egypt* 4 (2019), 99–124.

Feldman, Burton, and Robert D. Richardson, *The Rise of Modern Mythology, 1680–1860* (Bloomington, IN, 1972).

Feldman, David. *Englishmen and Jews: Social Relations and Political Culture, 1840–1914* (New Haven, 1994).

Feldman, Louis H. *Philo's Portrayal of Moses in the Context of Ancient Judaism*, Christianity and Judaism in Antiquity 15 (Notre Dame, 2007).

Feuchtwang, David. 'Moses und Hammurabi', *Monatsschrift für Geschichte und Wissenschaft des Judentums* 48 (1904), 385–99.

Figueira, Dorothy M. 'Oriental Despotism and Despotic Orientalisms', in *Bucknell Review* 38/2: 'Anthropology and the German Enlightenment: Perspectives on Humanity', ed. Katherine M. Faull (1995), 182–99.

Flynne, Elisabeth L. 'Moses in the Visual Arts', *Interpretation* 44/3 (1990), 265–76.

Förster, Gerhard. *Das mosaische Strafrecht in seiner geschichtlichen Entwickelung*, Ausgewählte Doktordissertationen der Leipziger Juristenfakultät (Leipzig, 1900).

Frakes, Robert M. *Compiling the* Collatio Legum Mosaicarum et Romanarum *in Late Antiquity*, Oxford Studies in Roman Society & Law (Oxford, 2011).

Fränkel, Emil. *Das jüdische Eherecht nach dem Reichscivilehegesetz vom 6. Februar 1875* (Munich, 1891).

Frankel, Zacharias. 'Das Talmudstudium', *Monatsschrift für Geschichte und Wissenschaft des Judentums* 18 (1869), 347–57.

Frankel, Zacharias. *Die Eidesleistung der Juden in theologischer und historischer Beziehung* (Dresden, 1840).

Frankel, Zacharias. 'Grundlinien des mosaisch-talmudische Eherechts', in *Jahresbericht des jüdisch-theologischen Seminars 'Fraenckelscher Stiftung'. Breslau, am Gedächtnisstage des Stifters, den 27. Januar 1860* (Breslau, 1860).

Frederiksen, Rune, and Eckart Marchand, eds. *Plaster Casts: Making, Collecting, and Displaying from Classical Antiquity to the Present*, Transformationen der Antike 18 (Berlin, 2010).

Fredriksen, Paula. *Paul: The Pagans' Apostle* (New Haven, 2017).

Frei, Hans W. *The Eclipse of the Biblical Narrative: A Study of Eighteenth and Nineteenth Century Hermeneutics* (New Haven, 1974).

Freudentheil, Wilhelm Nicolaus. 'Ueber die Siegslieder der Hebräer', in idem, *Nachträge zu Sulzers allgemeiner Theorie der schönen Künste. Charaktere der vornehmsten Dichter aller Nationen; nebst kritischen und historischen Abhandlungen über Gegenstände der schönen Künste und Wissenschaften von einer Gesellschaft von Gelehrten*, vol. 4/2 (Leipzig, 1795), 253–70.

Frew, Charlotte. 'Sister-in-Law Marriage in the Empire: Religious Politics and Legislative Reform in the Australian Colonies 1850–1900', *Journal of Imperial and Commonwealth History* 41 (2013), 194–210.

Friedman, Isaiah. *Germany, Turkey, and Zionism, 1897–1918* (Oxford, 1977).

Frishman, Judith. 'True Mosaic Religion: Samuel Hirsch, Samuel Holdheim and the Reform of Judaism', in *Religious Identity and the Problem of Historical Foundation*, eds Judith Frishman, Willemien Otten, and Gerard Rouwhorst (Leiden, 2004), 195–222.

Frishman, Judith. *Wat heeft het christendom van het jodendom overgenomen? Abraham Geiger en de geschiedschrijving van het rabbijns jodendom*, Inaugural lecture, Katholieke Theologische Universiteit (Utrecht, 1999).

Fuchs, Walther Peter. *Großherzog Friedrich I. von Baden und die Reichspolitik, 1871–1907*, vol. 4, *1898–1907*, Veröffentlichungen der Kommission für geschichtliche Landeskunde in Baden-Württemberg, Series A/32 (Stuttgart, 1980).

Fuchs, Walther Peter. *Studien zu Großherzog Friedrich I. von Baden*, Veröffentlichungen der Kommission für geschichtliche Landeskunde in Baden-Württemberg, Series B/ 100 (Stuttgart, 1995).

G., I. Review of *Studies in Biblical Law* by Harold M. Wiener, *Harvard Law Review* 18/5 (1905), 408–09.

Gale, Susan Gaylord. 'A Very German Legal Science: Savigny and the Historical School', *Stanford Journal of International Law* 18/1 (1982), 123–46.

Galloway, William. *The Unlawfulness of the Marriage of Brother and Sister-in-law: In the Light of the Word of God; with Ancient Evidence Hitherto Generally Overlooked* (London, 1870).

Gardt, Andreas, ed. *Nation und Sprache. Die Diskussion ihres Verhältnisses in Geschichte und Gegenwart* (Berlin, 2000).

Gaskill, Howard. 'German Ossianism: A Reappraisal?', *German Life and Letters* 42 (1989), 329–41.

Gastambide, A. 'Législateur', *Dictionnaire de la conversation et de la lecture*, 1st ed., vol. 34 (Paris, 1837), 486; 2nd ed., vol. 12 (Paris, 1864), 212.

Gay, Peter. 'Begegnungen mit der Moderne – Deutsche Juden in der deutschen Kultur', in *Juden im Wilhelminischen Deutschland, 1890–1914*, eds Werner Mosse and Arnold Paucker (Tübigen, 1976), 241–312.

Geiger, Abraham. 'Das Judenthum unsrer Zeit und die Bestrebungen in ihm', *Wissenschaftliche Zeitschrift für jüdische Theologie* 1 (1835), 1–12.

Geiger, Abraham. *Der Hamburger Tempelstreit, eine Zeitfrage* (Breslau, 1842).

Geiger, Abraham. 'Neunzehn Briefe über Judenthum, von Ben Uziel (Recension)', *Wissenschaftliche Zeitschrift für jüdische Theologie* 2 (1836), 351–359, 518–548; 3 (1837), 74–91.

Gerhardt, Kristiane. 'Frühneuzeitliches Judentum und "Rabbinismus". Zur Wahrnehmung des jüdischen Rechts in den Zivilisierungsdebatten der Aufklärung', *Trajectoires* 4, 'Postkolonial' (2010), https://journals.openedition.org/trajectoires/473 (accessed 5 December 2022).

Gierke, Otto (von). 'Recht und Sittlichkeit', *Logos. Internationale Zeitschrift für Philosophie der Kultur* 6 (1916/17), 211–64.

Ginsburg, Ruth, and Ilana Pardes, eds. *New Perspectives on Freud's Moses and Monotheism*, Conditio Judaica 60 (Tübingen, 2006).

Goethe, Johann Wolfgang. *Aus meinem Leben. Dichtung und Wahrheit*, 4 vols (Leipzig, 1811–33).

Goethe, Johann Wolfgang. *Leiden des jungen Werthers* (Leipzig, 1774).

Goetschel, Roland. 'Aux origines de la modernité juive: Zacharias Frankel (1801–1875) et l'école historico-critique', *Pardès* 19–20 (1994), 107–132.

Goldhill, Simon. 'What Has Alexandria to Do with Jerusalem? Writing the History of the Jews in the Nineteenth Century', *Historical Journal* 59 (2016), 125–51.

Goldman, Shalom. *Zeal for Zion: Christians, Jews, & the Idea of the Promised Land* (Chapel Hill, 2009).

Goldstein, Bluma. *Reinscribing Moses: Heine, Kafka, Freud, and Schoenberg in a European Wilderness* (Cambridge, MA, 1992).

Gordon, Bruce. 'Creating a Reformed Book of Knowledge: Immanuel Tremellius, Franciscus Junius, and their Latin Bible, 1580–1590', in *Calvin and the Book: The Evolution of the Printed Word in Reformed Protestantism*, ed. Karen E. Spierling, Refo500 Academic Studies 25 (Göttingen, 2015), 95–122.

Gosewinkel, Dieter. 'Citizenship in Germany and France at the Turn of the Twentieth Century: Some New Observations on an Old Comparison', in *Citizenship and National Identity in Twentieth-Century Germany*, eds Geoff Eley and Jan Palmowski (Stanford, 2008), 27–39.

Gottlieb, Michah. *Faith and Freedom: Moses Mendelssohn's Theological-Political Thought* (Oxford, 2011).

Gottlieb, Michah. 'Oral Letter and Written Trace: Samson Raphael Hirsch's Defense of the Bible and Talmud', *Jewish Quarterly Review* 106/3 (2016), 316–351.

Gottlieb, Michah. 'Scripture and Separatism: Politics and the Bible Translations of Ludwig Philippson and Samson Raphael Hirsch', in *Deutsch-jüdische Bibelwissenschaft*, eds Daniel Vorpahl, Sophia Kähler, and Shani Tzoref (Berlin, 2019), 57–73.

Gotzmann, Andreas. *Eigenheit und Einheit. Modernisierungsdiskurse des deutschen Judentums der Emanzipationszeit* (Leiden, 2002).

Gotzmann, Andreas. *Jüdisches Recht im kulturellen Prozeß. Die Wahrnehmung der Halachah im Deutschland des 19. Jahrhunderts*, Schriftenreihe wissenschaftlicher Abhandlungen des Leo Baeck Instituts 55 (Tübingen, 1997).

Gotzmann, Andreas. 'The Dissociation of Religion and Law in Nineteenth-Century German-Jewish Education', *Leo Baeck Institute Year Book* 43 (1998), 103–26.

Graetz, Michael. *The Jews in Nineteenth-Century France: From the French Revolution to the Alliance Israélite Universelle*, trans. Jane Marie Todd, Stanford Studies in Jewish History (Stanford, 1996).

Graf, Wilhelm. *Moses Vermächtnis. Über göttliche und menschliche Gesetze*, 3rd ed. (Munich, 2006).

Grafton, Anthony, and Joanna Weinberg. *'I Have Always Loved the Holy Tongue': Isaac Casaubon, the Jews, and a Forgotten Chapter in Renaissance Scholarship* (Cambridge, MA, 2011).

Graßl, Ignaz. *Das besondere Eherecht der Juden in Oesterreich nach den §§. 123–136 des allgemeinen bürgerlichen Gesetzbuches*, 2nd ed. (Vienna, 1849).

Graves, Robert, and Raphael Patai, *Hebrew Myths: The Book of Genesis* (New York, 1964).

Green, Arthur. *Judaism for the Post-Modern Era*, Samuel H. Goldenson Lecture Delivered December 12, 1994, at the Hebrew Union College–Jewish Institute of Religion, Cincinnati, Ohio (Cincinnati, 1995).

Grégoire, Abbé (Henri). *Essai sur la régénération physique, morale et politique des juifs. Ouvrage couronne par la Société royale des Sciences et des Arts de Metz, le 23 Août 1788* (Metz, 1789).

Grey Griffith, B. Review of *Studies in Biblical Law* by David Daube, *Modern Law Review* 11/2 (1948), 239–40.

Grimme, Hubert. *Das Gesetz Chammurabis und Moses. Eine Skizze* (Cologne, 1903).

Grimme, Hubert. *Mohammed,* Weltgeschichte in Karakterbildern (Munich, 1904).

Großmann, Andreas. 'Recht verkehrt. Hegels Rechtsphilosophie im Neuhegelianismus', in *Recht ohne Gerechtigkeit? Hegel und die Grundlagen des Rechtsstaates*, eds Mirko Wischke and Andrzej Przyłębski (Würzburg, 2010), 191–208.

Grupen, Christian Ulrich. *Teutsche Alterthümer, zur Erleuterung des Sächsischen auch Schwäbischen Land-und Lehn-Rechts …* (Hannover, 1746).

Gullette, Margaret Morganroth. 'The Puzzling Case of the Deceased Wife's Sister: Nineteenth-century England Deals with a Second-Chance Plot', *Representations* 31 (1990), 142–66.

Habermas, Rebekka, ed., *Negotiating the Secular and the Religious in the German Empire: Transnational Approaches,* New German Historical Perspectives 10 (New York, 2019).

HaCohen, Ran. *Reclaiming the Hebrew Bible: German-Jewish Reception of Biblical Criticism*, trans. Michelle Engel, Studia Judaica 56 (Berlin, 2010).

Hammill, Graham. *The Mosaic Constitution: Political Theology and Imagination from Machiavelli to Milton* (Chicago, 2012).

Harke, Jan Dirk. *Das Sanktionssystem des Codex Hammurapi,* Würzburger rechtswissenschaftliche Studien 70 (Würzburg, 2007).

Harper, Robert Francis. *The Code of Hammurabi, King of Babylon About 2250 B.C.: Autographed Text, Transliteration, Translation, Glossary, Index of Subjects, Lists of Proper Names, Signs, Numerals, Corrections, and Erasures with Map Frontispiece and Photograph of Text* (Chicago, 1904).

Harris, Henry Silton. *Hegel's Ladder: A Commentary on Hegel's* Phenomenology of Spirit, 2 vols (Indianapolis, 1997).

Harris, Jay M. *How Do We Know This? Midrash and the Fragmentation of Modern Judaism,* SUNY Series in Judaica: Hermeneutics, Mysticism, and Religion (Albany, 1995).

Hartlich, Christian, and Walter Sachs, *Der Ursprung des Mythosbegriffes in der modernen Bibelwissenschaft* (Tübingen, 1952).

Hartman, David. *A Heart of Many Rooms: Celebrating the Many Voices Within Judaism* (Woodstock, 1999).

Hartman, David. *A Living Covenant: The Innovative Spirit in Traditional Judaism* (New York, 1985).

Hartman, David. *The God Who Hates Lies: Confronting and Rethinking Jewish Tradition* (Woodstock, 2011).

Hartwich, Wolf-Daniel. *Die Sendung Moses. Von der Aufklärung bis Thomas Mann* (Munich, 1997).

Haugen, Kristine Louise. 'Ossian and the Invention of Textual History', *Journal of the History of Ideas* 59 (1998), 309–27.

Hayes, Christine. *What's Divine about Divine Law? Early Perspectives* (Princeton, 2015).

Haym, Rudolf. *Herder nach seinem Leben und seinen Werken*, 2 vols (Berlin, 1880–85).

Hechler, William Henry. 'Christen über die Judenfrage', *Die Welt* 1/2 (1897), 7–9.

Hegel, G. W. F. *Elements of the Philosophy of Right*, ed. Allen W. Wood, trans. H.B. Nisbet (Cambridge, 1991).

Hegel, G. W. F. *Grundlinien der Philosophie des Rechts* (Berlin, 1821).

Hegel, G. W.F. 'The Positivity of Christian Religion', in idem, *Early Theological Writings*, trans. T. M. Knox (Chicago, 1948 [1795/96]), 67–181.

Heidenreich, Marianne. *Christian Gottlob Heyne und die Alte Geschichte*, Beiträge zur Altertumskunde 229 (Leipzig, 2006).

Helfer, Martha B. *The Word Unheard: Legacies of Anti-Semitism in German Literature and Culture* (Chicago, 2011).

Herbermann, Charles G., et al., eds. *The Catholic Encyclopedia: An International Work of Reference on the Constitution, Doctrine, Discipline, and History of the Catholic Church*, 16 vols (New York, 1907–14).

Herder, Johann Gottfried. *Aelteste Urkunde des Menschengeschlechts* (Riga, 1774).

Herder, Johann Gottfried, ed. *Briefe zur Beförderung der Humanität*, 10 vols (Riga, 1793–97).

Herder, Johann Gottfried. *Kritische Wälder, oder Betrachtungen, die Wissenschaft und Kunst des Schönen betreffend, nach Maasgabe neuerer Schriften*, 3 vols (Riga, 1769).

Herder, Johann Gottfried. *Lieder der Liebe* (Leipzig, 1778).

Herder, Johann Gottfried. 'Ueber die Wirkung der Dichtkunst auf die Sitten der Völker in alten und neuen Zeiten. Eine Preißschrift. (1778)', repr. in *Johann Gottfried von Herder's sämmtliche Werke*, Section 2, *Zur schönen Literatur und Kunst*, vol. 9, ed. Johann von Müller (Tübingen, 1807), 353–450.

Herder, Johann Gottfried. 'Ueber Ossian und die Lieder der alten Völker; Auszug einiger Briefe 1773. Aus der Sammlung von deutscher Art und Kunst', repr. in *Johann Gottfried von Herder's sämmtliche Werke*, Section 2, *Zur schönen Literatur und Kunst*, vol. 8, ed. Johann von Müller (Tübingen, 1807), 1–44.

Herder, Johann Gottfried. *Vom Geist der Ebräischen Poesie. Eine Anleitung für die Liebhaber derselben, und der ältesten Geschichte des menschlichen Geistes*, 2 vols (Dessau, 1782–83).

Hermann, Martin Gottfried. *Handbuch der Mythologie aus Homer und Hesiod als Grundlage zu einer richtigen Fabellehre des Alterthums* (Berlin-Stettin, 1787).

Hertz, Joseph Herman. 'Ancient Semitic Codes and the Mosaic Legislation', *Journal of Comparative Legislation and International Law* 10/4 (1928), 207–21.

Heschel, Susannah. *Abraham Geiger and the Jewish Jesus*, Chicago Studies in the History of Judaism (Chicago, 1998).

Heyne, Christian Gottlob. 'Vorrede', in Martin Gottfried Hermann, *Handbuch der Mythologie aus Homer und Hesiod als Grundlage zu einer richtigen Fabellehre des Alterthums* (Berlin-Stettin, 1787).

Hirsch, Samson Raphael. *Der Pentateuch, übersetzt und erläutert*, 5 vols (Frankfurt am Main, 1867–78).

Hirsch, Samson Raphael. *Horeb. Versuche über Jissroels Pflichten in der Zerstreuung* (Altona, 1837).

Hirsch, Samson Raphael (Ben Uziel). *Igrot Tsafon. Neunzehn Briefe über Judenthum* (Altona, 1836). Translated by Bernard Drachman and edited by Jacob Breuer as *The Nineteen Letters on Judaism* (Jerusalem, 1969).

Hirsch, Samson Raphael. *Naftuli Naftali. Erste Mittheilungen aus Naphtali's Briefwechsel* (Altona, 1838).

Hirsch, Samuel. *Die Reform im Judenthum und dessen Beruf in der gegenwärtigen Welt* (Leipzig, 1844).

Hirschfeld, Marc. 'Rabbinic Universalism in the Second and Third Centuries', *Harvard Theological Review* 93 (2000), 101–15.

Hirte, Christ. 'Erich Mühsam und das Judentum', in *Erich Mühsam und das Judentum*, ed. Jürgen Wolfgang Goette (Lübeck, 2002), 52–70.

Hobson, W.F. 'The Christian Law of Marriage: What does Our Lord Say upon the subject?', in *Tracts Issued by the Marriage Law Defence Union*, vol. 1, *Scriptural* (London, 1889), tract 20, pp. 145–52.

Hoffmann, Christhard. *Juden und Judentum im Werk deutscher Althistoriker des 19. und 20. Jahrhunderts*, Studies in Judaism in Modern Times 9 (Leiden, 1988).

Holdheim, Samuel. *Ueber die Autonomie der Rabbinen und das Princip der jüdischen Ehe. Ein Beitrag zur Verständigung über einige das Judenthum betreffende Zeitfragen* (Schwerin, 1843).

Holloway, Steven W. 'Biblical Assyria and Other Anxieties in the British Empire', *Journal of Religion & Society* 3 (2001), 1–19.

Hölscher, Lucian. 'Die Religion des Bürgers. Bürgerliche Frömmigkeit und protestantische Kirche im 19. Jahrhundert', *Historische Zeitschrift* 250 (1990), 595–630.

Hommel, Fritz. *Die Altisraelitische Überlieferung in inschriftlicher Beleuchtung. Ein Einspruch gegen die Aufstellungen der modernen Pentateuchkritik* (Munich, 1897). Translated by Edmund McClure and Leonard Crosslé as *The Ancient Hebrew Tradition as Illustrated by the Monuments: A Protest Against the Modern School of Old Testament Criticism* (London, 1897).

Hommel, Fritz. *Die altorientalischen Denkmäler und das alte Testament. Eine Erwiderung auf Prof. Fr. Delitzsch's "Babel und Bibel"*, 2nd ed. (Berlin, 1903).

Hommel, Fritz. *Grundriss der Geographie und Geschichte des Alten Orients*, part 1, *Ethnologie des Alten Orients. Babylonien und Chaldäa*, Handbuch der klassischen Altertums-Wissenschaft in systematischer Darstellung 3/1.1 (Munich, 1904).

Hopgood, Miles. *How Luther Regards Moses: The Lectures on Deuteronomy*, Refo500 Academic Studies 98 (Göttingen, 2023).

Howard, Thomas Albert. *Religion and the Rise of Historicism: W. M. L. de Wette, Jacob Burckhardt, and the Theological Origins of Nineteenth-Century Historical Consciousness* (Cambridge, 2006).

Huet, Pierre-Daniel. *Demonstratio Evangelica* (Paris, 1679).

Hug, Walther. 'The History of Comparative Law', *Harvard Law Review* 45/6 (1932), 1027–1070.

Hull, Isabel V. '"Persönliches Regiment",' in *Der Ort Kaiser Wilhelms II. in der deutschen Geschichte*, ed. John C. G. Röhl, Schriften des Historischen Kollegs Kolloquien 17 (Munich, 1991), 3–23.

Hume, David. 'On National Characters', in idem, *Essays and Treatises on Several Subjects*, 4 vols (London, 1753), 1:277–300.

Hyamson, Albert Montefiore. *Mosaicarum et romanarum legum collatio: With introduction, facsimile and transcription of the Berlin codex, translation, notes and appendices* (London, 1913).

Igrao, Charles. 'The Problem of "Enlightened Absolutism" and the German States', *Journal of Modern History* 58 (1986), 161–80.

Ilany, Ofri. '"Alle unsere Wanderungen im Orient". Die deutsche Sehnsucht nach dem Orient – Theologie, Wissenschaft und Rasse', in *Tel Aviver Jahrbuch für deutsche Geschichte* (2017), 41–68.

Ilany, Ofri. 'Christian Images of the Jewish State: The Hebrew Republic as a Political Model in the German Protestant Enlightenment', in *Jews and Protestants: From the Reformation to the Present*, eds Irene Aue-Ben-David et al. (Berlin, 2021), 119–35.

Ilany, Ofri. '*Herr Zebaoth* and the German Nation: Bible and Nationalism in the anti-Napoleonic Wars', *Global Intellectual History* 5/1, Special Issue: 'Theology & Politics in the German Imagination, 1789–1848', ed. Ruth Jackson Ravenscroft (2019), 104–24.

Ilany, Ofri. *In Search of the Hebrew People: Bible and Nation in the German Enlightenment*, trans. Ishai Mishroy, German Jewish Cultures (Bloomington, 2018).

Ilting, Karl-Heinz. *Naturrecht und Sittlichkei. Begriffsgeschichtliche Studien*, Sprache und Geschichte 7 (Stuttgart, 1983).

Isaacs, Nathan. Review of *The Origin and History of Hebrew Law* by J.M. Powis Smith, *Harvard Law Review* 45/5 (1932), 949–52.

Jacob, Benno. *Die Thora Moses*, Volksschriften über die jüdische Religion 1/3–4 (Frankfurt, 1912/13).

Jacobs, Sandra. *The Body as Property: Physical Disfigurement in Biblical Law*, Library of Hebrew Bible/Old Testament Studies 582 (London, 2014).

Jacolliot, Louis. *Les législateurs religieux: Manou – Moïse – Mahomet. Traditions religieuses comparés des lois de Manou, de la Bible, du Coran, du ritual égyptien, du Zend-Avesta des Parses et des traditions finnoises* (Paris, 1876).

Jeremias, Johannes. *Moses und Hammurabi*, 2nd ed. (Leipzig, 1903).

Johanning, Klaus. *Der Bibel-Babel-Streit. Eine forschungsgeschichtliche Studie* (Frankfurt, 1988).

John, Michael. *Politics and the Law in Late Nineteenth-Century Germany: The Origins of the Civil Code* (Oxford, 1989).

Johnson, Barbara. *Moses and Multiculturalism* (Berkeley, 2010).

Jones, Gareth Stredman. *Karl Marx. Die Biographie* (Frankfurt am Main, 2017).

Jones, William J. 'Early Dialectology, Etymology and Language History in German-Speaking Countries', in *History of the Language Sciences / Geschichte der Sprachwissenschaften / Histoire des sciences du langage*, eds Sylvain Auroux et al., 3 vols, Handbücher zur Sprach-und Kommunikationswissenschaft 18 (Berlin, 2000–06), 2:1105–1115.

Jowett, Benjamin. 'On the Interpretation of Scripture', in *Essays and Reviews* (London, 1860), 330–433.

Justi, Karl Wilhelm. *National-Gesänge der Hebräer* (Marburg, 1803).

Kain, Philip J. *Hegel and Right: A Study of the 'Philosophy of Right'* (New York, 2018).

Kaiser, Gerhard. *Klopstock. Religion und Dichtung* (Mainz, 1975).

Kant, Immanuel. *Der Streit der Fakultäten* (Königsberg, 1798).

Kant, Immanuel. 'Die Metaphysik der Sitten [1797]', repr. in *Kant's gesammelte Schriften*, Section 1, *Werke*, vol. 6, ed. Königlich Preußische Akademie der Wissenschaften (Berlin, 1907).

Kant, Immanuel. *Die Religion innerhalb der Grenzen der bloßen Vernunft* (Königsberg, 1793).

Kant, Immanuel. *Groundwork of the Metaphysics of Morals: A German-English Edition* [German text from the second original edition (1786)], eds and trans Mary Gregor and Jan Timmermann, Cambridge Texts in the History of Philosophy (Cambridge, 2011).

Kepnes, Steven, ed. *Interpreting Judaism in a Postmodern Age*, New Perspectives on Jewish Studies (New York, 1995).

Kersting, Wolfgang. 'Sittlichkeit, Sittenlehre', in *Historisches Wörterbuch der Philosophie*, vol. 9, eds Joachim Ritter and Karlfried Gründer (Basel, 1995), 907–23.

Kinna, Ruth, and Alex Prichard. 'Anarchism and Non-Domination', *Journal of Political Ideologies* 24/3 (2019), 221–40.

Kitromilides, Pascalis M. *Enlightenment and the Revolution: The Making of Modern Greece* (Cambridge, MA, 2013).

Klautke, Egbert. '*Völkerpsychologie* in 19th-Century Germany: Lazarus, Steinthal, Wundt', in *Doing Humanities in Nineteenth-Century Germany*, ed. Efraim Podoksik, Scientific and Learned Cultures and Their Institutions 28 (Leiden, 2020), 243–63.

Klopstock, Friedrich Gottlieb. *Klopstocks sämmtliche Werke*, 12 vols (Leipzig, 1823).

Koebner, Richard. 'Despot and Despotism: Vicissitudes of a Political Term', *Journal of the Warburg and Courtauld Institutes* 14 (1951), 275–302.

Kogge, Werner, and Lisa Wilhelmi. 'Despot und (orientalische) Despotie – Brüche im Konzept von Aristoteles vis Montesquieu', *Saeculum* 69 (2020), 305–42.

Kohler, George Y. 'Finding God's Purpose: Hermann Cohen's Use of Maimonides to Establish the Authority of Mosaic Law', *Journal of Jewish Thought and Philosophy* 18/1 (2010), 75–105.

Kohler, George Y. *Reading Maimonides' Philosophy in 19th Century Germany: The Guide to Religious Reform*, Amsterdam Studies in Jewish Philosophy 15 (Dordrecht, 2012).

Kohler, Josef. *Aus vier Weltteilen. Reisebilder* (Berlin, 1908).

Kohler, Josef. 'Begriff und Aufgabe der Weltgeschichte [1899]', repr. in idem, *Aus Kultur und Leben. Gesammelte Essays* (Berlin, 1904), 15–22.

Kohler, Josef. 'Das Recht der orientalischen Völker', in *Allgemeine Rechtsgeschichte*, Pt 1, *Orientalisches Recht und Recht der Griechen und Römer*, eds Josef Kohler and Leopold Wenger, Die Kultur der Gegenwart. Ihre Entwicklung und ihre Ziele 2/7.1 (Leipzig, 1914), 49–153.

Kohler, Josef. 'Fragebogen zur Erforschung der Rechtsverhältnisse der sogenannten Naturvölker, namentlich in den deutschen Kolonialländern', *Zeitschrift für vergleichende Rechtswissenschaft* 12 (1897), 427–40.

Kohler, Josef. 'Hammurabis Gesetz [1903]', repr in idem, *Aus Kultur und Leben. Gesammelte Essays* (Berlin, 1904), 58–64.

Kohler, Josef, *Lehrbuch des bürgerlichen Rechts*, vol. 1, *Allgemeiner Teil* (Berlin, 1906).

Kohler, Josef. 'Rechtsgeschichte und Kulturgeschichte', *Zeitschrift für das Privat-und öffentliche Recht der Gegenwart* 12 (1885), 583–93.

Kohler, Josef. Review of *Das Gesetz Hammurabis und die Thora Israels* by Samuel Oettli, *Deutsche Literaturzeitung* 24 (1903), 1543–49.

Kohler, Josef, and Arthur Ungnad. *Hammurabi's Gesetz*, vol. 2, *Syllabische und zusammenhangende Umschrift nebst vollständigem Glossar* (Leipzig, 1909).

Kohler, Josef, and Arthur Ungnad. *Hammurabi's Gesetz*, vol. 3, *Übersetzte Urkunden, Erläuterungen* (Leipzig, 1909).

Kohler, Josef, and Arthur Ungnad. *Hammurabi's Gesetz*, vol. 4, *Übersetzte Urkunden, Erläuterungen (Fortsetzung)* (Leipzig, 1910).

Kohler, Josef, and Arthur Ungnad. *Hammurabi's Gesetz*, vol. 5, *Übersetzte Urkunden, Verwaltungsregister, Inventare, Erläuterungen* (Leipzig, 1911).

Kohler, Josef, and Felix Peiser. *Hammurabi's Gesetz*, vol. 1, *Übersetzung, juristische Wiedergabe, Erläuterung* (Leipzig, 1904).

Kohn, Hans. *Die Idee des Nationalismus. Ursprung und Geschichte bis zur französischen Revolution* (Frankfurt, 1950).

König, Eduard. 'Hammurabis Gesetzgebung und ihre religionsgeschichtliche Tragweite', *Der Beweis des Glaubens. Monatsschrift zur Begründung und Verteidigung der christlichen Wahrheit für Gebildete* 39 (1903), 169–80.

Koschaker, Paul, and Arthur Ungnad, *Hammurabi's Gesetz*, vol. 6, *Übersetzte Urkunden. Mit Rechtserläuterungen* (Leipzig, 1923).

Koselleck, Reinhart. 'Introduction and Prefaces to the *Geschichtliche Grundbegriffe*,' trans. Michaela Richter, *Contributions to the History of Concepts* 6/1 (2011 [1972]), 1–37.

Kosuch, Carolin, ed. *Anarchism and the Avant-Garde: Radical Arts and Politics in Perspective* (Leiden, 2019).

Kosuch, Carolin. '"Ein Jude zog aus von Nazareth …": Erich Mühsams Wahlverwandtschaft mit Bruder Jesus', *PaRDeS. Zeitschrift der Vereinigung für Jüdische Studien* 21: 'Jesus in the Jewish Culture of the 19th and 20th Century' (2015), 123–40.

Kosuch, Carolin. *Missratene Söhne. Anarchismus und Sprachkritik im Fin de Siècle* (Göttingen, 2015).

Kosuch, Carolin. 'Räterepublik', in *Enzyklopädie jüdischer Geschichte und Kultur*, ed. Dan Diner, vol. 5 (Stuttgart, 2014), 96–101.

Kosuch, Carolin. 'Retrieving Tradition? The Secular-Religious Ambiguity in Nineteenth Century German Jewish Anarchism', in *Negotiating the Secular and the Religious in the German Empire: Transnational Approaches*, ed. Rebekka Habermas (New York, 2019), 147–70.

Krapf, Ludwig. *Germanenmythos und Reichsideologie. Frühhumanistische Rezeptionsweisen der taciteischen 'Germania'* (Tübingen, 1979).

Krochmalnik, Daniel. 'Mendelssohn's Begriff "Zeremonialgesetz" und der europäische Antizeremonialismus. Eine begriffsgeschichtliche Untersuchung', in *Recht und Sprache in der deutschen Aufklärung*, eds Ulrich Kronauer and Jörn Garber (Tübingen, 2001), 129–60.

Kuhlmann, Wolfgang. ed. *Moralität und Sittlichkeit. Das Problem Hegels und die Diskursethik* (Frankfurt, 1986).

Kuper, Adam. 'Incest, Cousin Marriage, and the Origin of the Human Sciences in Nineteenth-Century England', *Past and Present* 174 (2002), 158–83.

Kurtz, Paul Michael. 'A Historical, Critical Retrospective on Historical Criticism', in *The New Cambridge Companion to Biblical Interpretation*, eds Ian Boxall and Bradley C. Gregory (Cambridge, 2022), 15–36.

Kurtz, Paul Michael. 'Defining Hellenistic Jews in Nineteenth-Century Germany: The Case of Jacob Bernays and Jacob Freudenthal', *Erudition & the Republic of Letters* 5 (2020), 308–42.

Kurtz, Paul Michael. 'Is Kant Among the Prophets? Hebrew Prophecy and German Historical Thought, 1880–1920', *Central European History* 54 (2021), 34–60.

Kurtz, Paul Michael. *Kaiser, Christ, and Canaan: The Religion of Israel in Protestant Germany, 1871–1918*, Forschungen zum Alten Testament 1/122 (Tübingen, 2018).

Kurtz, Paul Michael. 'Of Lions, Arabs & Israelites: Some Lessons from the Samson Story for Writing the History of Biblical Scholarship', *Journal of the Bible and its Reception* 5/1 (2018), 31–48.

La Vopa, Anthony J. 'Herder's Publikum: Language, Print, and Sociability in Eighteenth-Century Germany', *Eighteenth-Century Studies* 29 (1995), 5–24.

Lahusen, Benjamin. *Alles Recht geht vom Volksgeist aus. Friedrich Carl von Savigny und die moderne Rechtswissenschaft* (Berlin, 2012).

Landauer, Gustav. 'A Few Words on Anarchism', repr. in *Revolution and Other Writings: A Political Reader*, ed. and trans. Gabriel Kuhn (Oakland, 2010 [1897]), 79–83.

Landauer, Gustav. 'Anarchism – Socialism', repr. in *Revolution and Other Writings: A Political Reader*, ed. and trans. Gabriel Kuhn (Oakland, 2010 [1895]), 70–74.

Landauer, Gustav. *Aufruf zum Sozialismus*, ed. Siegbert Wolf, Ausgewählte Schriften 11 (Lich/Hessen, 2015 [1911]).

Landauer, Gustav. 'Etwas über Moral', repr. in *Anarchismus*, ed. Siegbert Wolf, Ausgewählte Schriften 2 (Lich/Hessen, 2009 [1893]), 37–41.

Landauer, Gustav. 'Kiew', repr. in idem, *Zeit und Geist. Kulturkritische Schriften 1890–1919*, eds Rolf Kauffeldt and Michael Matzigkeit (Munich, 1997 [1913]), 224–28.

Landauer, Gustav. 'Referat über Eugen Dührings "Kursus der National- und Sozialökonomie"', repr. in *Anarchismus*, ed. Siegbert Wolf, Ausgewählte Schriften 2 (Lich/Hessen, 2009 [1892]), 107–14.

Landauer, Gustav. 'Revolution', repr. in *Revolution and Other Writings: A Political Reader*, ed. and trans. Gabriel Kuhn (Oakland, 2010 [1907]), 110–87.

Landauer, Gustav. 'Sind das Ketzergedanken?', repr. in idem, *Zeit und Geist. Kulturkritische Schriften 1890–1919*, eds Rolf Kauffeldt and Michael Matzigkeit (Munich 1997 [1913]), 216–23.

Landauer, Gustav. 'The Twelve Articles of the Socialist Bund: Second Version', repr. in *Revolution and Other Writings: A Political Reader*, ed. and trans. Gabriel Kuhn (Oakland, 2010 [1912]), 215–16.

Langewiesche, Dieter. 'Liberalismus und Judenemanzipation im 19. Jahrhundert', in *Juden in Deutschland. Emanzipation, Integration, Verfolgung und Vernichtung*, eds Peter Freimark, Alice Jankowski, and Ina Lorenz (Hamburg, 1991), 148–63.

Langton, Daniel. *Reform Judaism and Darwin: How Engaging with Evolutionary Theory Shaped American Jewish Religion*, Studia Judaica 111 (Berlin, 2019).

Larsen, Timothy. *A People of One Book: The Bible and the Victorians* (Oxford, 2011).

Lässig, Simone. *Jüdische Wege ins Bürgertum. Kulturelles Kapital und sozialer Aufstieg im 19. Jahrhundert* (Göttingen, 2004).

Lazarus, Moritz. *Die Ethik des Judentums*, 2 vols (Frankfurt am Main, 1898–1911).

Lazarus, Moritz. *Treu und Frei. Gesammelte Reden und Vorträge über Juden und Judentum* (Leipzig, 1887).

Lazarus, Moritz. *Was heißt und zu welchem Ende studirt man jüdische Geschichte und Litteratur*, Populär-wissenschaftliche Vorträge über Juden und Judentum 1 (Leipzig, 1900).

Leerssen, Joep. 'Ossian and the Rise of Literary Historicism', in *The Reception of Ossian in Europe*, ed. Howard Gaskill (London, 2004), 109–25.

Legaspi, Michael. *The Death of Scripture and the Rise of Biblical Studies*, Oxford Studies in Historical Theology (Oxford, 2010).

Lehmann, Reinhard G. *Friedrich Delitzsch und der Babel-Bibel-Streit*, Orbis Biblicus et Orientalis 133 (Freiburg, Switzerland, 1994).

Lehmann-Haupt, Carl Friedrich. *Babyloniens Kulturmission einst und jetzt. Ein Wort zur Ablenkung und Aufklärung zum Babel-Bibel-Streit*, 2nd ed. (Leipzig, 1905).

Leo, Heinrich. *Vorlesungen über die Geschichte des Jüdischen Staates; gehalten an der Universität zu Berlin* (Berlin, 1828).

Levenson, Alan T. 'An Adventure in Otherness: Nahida Remy-Ruth Lazarus (1849–1928)', in *Gender and Judaism: The Transformation of Tradition*, ed. Tamar M. Rudavsky (New York, 1995), 99–111.

Levitin, Dmitri. *Ancient Wisdom in the Age of the New Science: Histories of Philosophy in England, c. 1640–1700*, Ideas in Context (Cambridge, 2015).

Levy, Carl. 'Social Histories of Anarchism', *Journal for the Study of Radicalism* 4 (2010), 1–44.

Lévy, Jean-Philippe. Review of *Mosaïc Law in Practice and Study throughout the Ages* by Pieter Jacobus Verdam, *Revue internationale de droit comparé* 12/4 (1960), 891–93.

Liberles, Robert. *Religious Conflict in Social Context: The Resurgence of Orthodox Judaism in Frankfurt am Main, 1838–1877* (Westport, 1985).

Lichtschein, Ludwig. *Die Ehe nach mosaisch-talmudischer Auffassung und das mosaisch-talmudische Eherecht* (Leipzig, 1879).

Lifschitz, Avi. *Language & Enlightenment: The Berlin Debates of the Eighteenth Century*, Oxford Historical Monographs (Oxford, 2012).

Lightner, Clarence A. 'The Mosaic Law', *Michigan Law Review* 10/2 (1911), 108–119.

Lincoln, Bruce. *Theorizing Myth: Narrative, Ideology, Scholarship* (Chicago, 1999).

Link-Salinger, Ruth. 'Friends in Utopia: Martin Buber and Gustav Landauer', *Midstream* 24 (1978), 67–72.

Lohmann, Uta. 'Das bürgerliche Leben als humanistisches Kunstwerk. Reflexionen zum universal-ästhetischen Selbst-und Gesellschaftsbild des jüdischen Kaufmanns David Friedländer und zur Ikonographie der Haskala', *Trumah* 22 (2014), 39–68.

Lohmann, Uta. 'Wissensspeicher, Lehrbuch, Erkenntnisquelle. Zur Rolle der Hebräischen Bibel im Bildungskonzept der Berliner Haskala', in *Deutsch-jüdische Bibelwissenschaft. Historische, exegetische und theologische Perspektiven*, eds Daniel Vorpahl, Sophia Kähler, and Shani Tzoref, Europäisch-jüdische Studien 40 (Berlin, 2019), 77–92.

Lovejoy, Arthur O. '"Nature" as Aesthetic Norm', repr. in idem, *Essays in History of Ideas* (Baltimore, 1948 [1927]), 69–77.

Lowenstein, Steven M. *The German-Jewish Community of Washington Heights, 1933–1983, Its Structure and Culture* (Detroit, 1989).

Lowenstein, Steven M., et al. *Deutsch-jüdische Geschichte in der Neuzeit*, vol. 3, *Umstrittene Integration 1871–1918* (Munich, 1997).

Lowth, Robert. *De sacra poesi Hebræorum, prælectiones academicæ Oxonii habitæ* (Oxford, 1753), ed. Johann David Michaelis (Göttingen, 1758), trans. George Gregory as *Lectures on the Sacred Poetry of the Hebrews*, 2 vol. (London, 1787).

Löwy, Michael. *Erlösung und Utopie. Jüdischer Messianismus und libertäres Denken. Eine Wahlverwandtschaft* (Berlin, 1997).

Löwy, Michael. 'Jewish Messianism and Libertarian Utopia in Central Europe (1900–1933)', *New German Critique* 8, Special Issue 2: Germans and Jews (1980), 105–15.

Löwy, Michael. 'Messianismus', in *Enzyklopädie jüdischer Geschichte und Kultur*, vol. 4, ed. Dan Diner (Stuttgart, 2013), 147–51.

Lunn, Eugene. *Prophet of Community: The Romantic Socialism of Gustav Landauer* (Berkeley, 1973).

Luther, Martin. 'How Christians Should Regard Moses,' Sermon from 27 August 1525, trans. Martin H. Bertram, in *Luther's Works (American Edition)*, vol. 35, *Word and Sacrament 1*, ed. Theodore Bachmann (Philadelphia, 1960), 161–174.

Mac Laughlin, Jim. *Kropotkin and the Anarchist Intellectual Tradition* (London, 2016).

Machaffie, Barbara Zink. '"Monument Facts and Higher Critical Fantasies": Archaeology and the Popularization of Old Testament Criticism in Nineteenth-Century Britain', *Church History* 50 (1981), 316–28.

Mährlein, Christoph. *Volksgeist und Recht. Hegels Philosophie der Einheit und ihre Bedeutung in der Rechtswissenschaft*, Epistemata 286 (Würzburg, 2000).

Mali, Joseph. *The Rehabilitation of Myth: Vico's 'New Science'* (Cambridge, 2002).

Malley, Shawn. *From Archaeology to Spectacle in Victorian Britain: The Case of Assyria 1845–1854* (Farnham, 2012).

Mandl, Max. *Das Sklavenrecht des alten Testaments. Eine rechtsgeschichtliche Studie* (Hamburg, 1886).

Mann, Thomas. *Joseph und seine Brüder*, 4 vols (Berlin, 1933–43).

Manson, T.W. Review of *Studies in Biblical Law* by David Daube, *Cambridge Law Journal* 10/1 (1947), 135–36.

Marchand, Suzanne L. *Down from Olympus: Archaeology and Philhellenism in Germany, 1750–1970* (Princeton, 1996).

Marchand, Suzanne L. 'Georg Ebers, Sympathetic Egyptologist', in *For the Sake of Learning: Essays in Honor of Anthony Grafton*, eds Ann Blair and Anja-Silva Goering, 2 vols, Scientific and Learned Cultures and Their Institutions 18 (Leiden, 2016), 917–932.

Marchand, Suzanne L. *German Orientalism in the Age of Empire: Religion, Race, and Scholarship*, Publications of the German Historical Institute (Washington, D.C., 2009).

Marchand, Suzanne L. 'Porcelain: Another Window on the Neoclassical Visual World', *Classical Receptions Journal* 12/2 (2020), 200–230.

Marks, David Woolf. *'The Law is Light': A Course of Four Lectures on the Sufficiency of the Law of Moses as the Guide of Israel* (London, 1854).

Marx, Karl. 'Zur Judenfrage', *Deutsch-französische Jahrbücher* 1–2 (1844), available online at http://www.mlwerke.de/me/me01/me01_347.htm (accessed 6 December 2022).

McCaul, Alexander. *The Ancient Interpretation of Leviticus XVIII.18, as Received in the Church for more than 1500 Years, A Sufficient Apology for holding that According to the Word of God, Marriage with a Deceased Wife's Sister is Lawful: A Letter to the Rev. W.H. Lyall* (London, 1859).

McGeough, Kevin M. *The Ancient Near East in the Nineteenth Century: Appreciations and Appropriations*, vol. 1, *Claiming and Conquering*, Hebrew Bible Monographs 67 (Sheffield, 2015).

McLaughlin, Paul. *Anarchism and Authority: A Philosophical Introduction to Classical Anarchism* (Aldershot, 2007).

Meek, Ronald L. *Social Science and the Ignoble Savage* (Cambridge, 1976).

Meissner, Bruno. *Könige Babyloniens und Assyriens. Charakterbilder aus der altorientalischen Geschichte* (Leipzig, 1926).

Melander, Henning. 'Könnte man die Bundeslade wiederfinden?', *Die Zeit* 2/16 (22 April 1898), 3–4; 2/17 (29 April 1898), 2–4; 2/18 (6 May 1898), 7–8; 2/19 (13 May 1898), 5–6; 2/20 (20 May 1898), 9.

Mendelssohn, Moses. *Jerusalem, oder über religiöse Macht und Judentum. Mit dem Vorwort zu Manasse ben Israels* Rettung der Juden *und dem Entwurf zu* Jerusalem *sowie einer Einleitung, Anmerkungen und Register*, ed. Michael Albrecht, Philosophische Bibliothek 565 (Hamburg, 2005 [1783]).

Mendelssohn, Moses, and Hirschel Lewin. *Ritualgesetze der Juden, betreffend Erbschaften, Vormundschaften, Testamente, und Ehesachen, in so weit sie das Mein und Dein angehen* (Berlin, 1778).

Mendes-Flohr, Paul. 'Messianic Radicals: Gustav Landauer and Other German-Jewish Revolutionaries', in *Gustav Landauer: Anarchist and Jew*, eds Paul Mendes-Flohr and Anya Mali (Berlin, 2014), 14–42.

Menges, Karl. 'Particular Universals: Herder on National Literature, Popular Literature, and World Literature', in *A Companion to the Works of Johann Gottfried Herder*, eds Hans Adler and Wolf Koepke (Suffolk, 2009), 189–214.

Merkley, Paul Charles. *The Politics of Christian Zionism, 1891–1948* (New York, 1998).

Mertens, Dieter. 'Die Instrumentalisierung der "Germania" des Tacitus durch die deutschen Humanisten', in *Zur Geschichte der Gleichung 'germanisch-deutsch'*.

*Sprache und Namen, Geschichte und Institutionen,* eds Heinrich Beck and Dieter Geuenich (Berlin, 2004), 37–102.

Merx, Adalbert. *Eine Rede vom Auslegen ins besondere des Alten Testaments. Vortrag gehalten zu Heidelberg im wissenschaftlichen Predigerverein Badens und der Pfalz am 3 Juli 1878* (Halle, 1879).

Meyer, Michael A. *Response to Modernity: A History of the Reform Movement in Judaism,* Studies in Jewish History (Oxford, 1988).

Meyer, Seligmann. *Contra Delitzsch! Die Babel-Hypothesen widerlegt* (Frankfurt, 1903).

Michaelis, Johann David. *Mosaisches Recht,* 6 vols (Frankfurt am Main, 1770–75), trans. Alexander Smith as *Commentaries on the Laws of Moses,* 4 vols (London, 1814).

Micraelius, Johannes. *Antiquitates Pomeraniae, Oder Sechs Bücher vom Alten Pommerlande* ..., 6 vols (Leipzig, 1723).

Mittermaier, Carl Joseph Anton. *Grundsätze des gemeinen deutschen Privatrechts, mit Einschluß des Handels-, Wechsel- und Seerechts,* 3rd ed., 2 vols (Landshut, 1827).

Mittleman, Alan L. *A Short History of Jewish Ethics: Conduct and Character in the Context of Covenant* (Chichester, 2012).

Mittleman, Alan L. 'Introduction: Holiness and Jewish Thought', in *Holiness in Jewish Thought,* ed. idem (Oxford, 2018), 1–11.

Momigliano, Arnaldo. 'Biblical Studies and Classical Studies: Simple Reflections upon Historical Method', *Annali della Scuola Normale Superiore di Pisa,* 3rd Series, 11/1 (1981), 25–32.

Momigliano, Arnaldo. *Pagine Ebraiche* (Rome, 1987).

Momigliano, Arnaldo. 'Religious History Without Frontiers: J. Wellhausen, U. Wilamowitz, and E. Schwartz', *History and Theory,* 21/4, Beiheft 21: 'New Paths of Classicism in the Nineteenth Century' (1982), 49–64.

Mommsen, Theodor. *Römisches Strafrecht,* Systematisches Handbuch der Deutschen Rechtswissenschaft (Leipzig, 1899).

Mommsen, Theodor, ed. *Zum ältesten Strafrecht der Kulturvölker. Fragen zur Rechtsvergleichung gestellt von Theodor Mommsen, beantwortet von H. Brunner, B. Freudenthal, J. Goldziher, H.F. Hitzig, Th. Noeldeke, H. Oldenberg, G. Roethe, J. Wellhausen, U. von Wilamowitz-Moellendorff* (Leipzig, 1905).

Morgan, Michael L. 'Tikkun olam', in *Enzyklopädie jüdischer Geschichte und Kultur,* vol. 6, ed. Dan Diner (Stuttgart, 2015), 102–06.

Morgenstern, Matthias. 'Rabbi S. R. Hirsch and his Perception of Germany and German Jewry', in *The German-Jewish Experience Revisited,* eds Steven E. Aschheim and Vivian Liska (Berlin, 2015), 207–230.

Most, Johann. *Die Eigenthumsbestie* (New York, 1887).

Most, Johann. *Die Gottespest* (New York, 1883).

Mühsam, Erich. 'Bismarxismus', *Fanal,* February (1927), 65–71.

Mühsam, Erich. *Brennende Erde. Verse eines Kämpfers* (Munich, 1920).

Mühsam, Erich. *Die Befreiung der Gesellschaft vom Staat. Was ist kommunistischer Anarchismus?* Fanal: Special Edition (Berlin, 1933).

Mühsam, Erich. *Tagebücher, 1910–1924*, ed. Chris Hirte (Munich, 1994).

Mühsam, Erich. *Wüste – Krater – Wolken* (Berlin, 1914).

Mühsam, Erich. 'Zur Judenfrage', *Die Weltbühne* 16/49 (2 December 1920), 643–47.

Mühsam, Paul. *Ich bin ein Mensch gewesen. Lebenserinnerungen*, ed. Ernst Kretzschmar (Gerlingen, 1989).

Müller, David Heinrich (von). *Die Gesetze Hammurabis und ihr Verhältnis zur mosaischen Gesetzgebung sowie zu den XII Tafeln. Der Text in Umschrift, deutsche und hebräische Übersetzung, Erläuterung und vergleichende Analyse* (Vienna, 1903).

Müller, Karl Otfried. *Prolegomena zu einer wissenschaftlichen Mythologie* (Göttingen, 1825).

Myers, David N. *Resisting History: Historicism and Its Discontents in German-Jewish Thought*, Jews, Christians, and Muslims from the Ancient to the Modern World (Princeton, 2003).

Nelson, Eric. *The Hebrew Republic: Jewish Sources and the Transformation of European Political Thought* (Cambridge, MA, 2010).

Neville, Robert. 'Philosophy of Religion and the Big Questions', *Palgrave Communications* 4/126 (2018), https://doi.org/10.1057/s41599-018-0182-9 (accessed 6 December 2022).

Ní Mhunghaile, Lesa. 'James Macpherson und Ossian. Eine literarische Kontroverse zwischen National- und Universalkultur', in *Aufklärung zwischen Nationalkultur und Universalismus*, ed. B. Wehinger (Hannover-Laatzen, 2008), 155–66.

Niehoff, Maren R. *Philo of Alexandria: An Intellectual Biography*, Anchor Yale Bible Reference Library (New Haven, 2018).

Nietzsche, Friedrich. *Nietzsche Werke. Kritische Gesamtausgabe*, section 4, vol. 1, *Richard Wagner in Bayreuth (Unzeitgemäße Betrachtungen IV), Nachgelassene Fragmente, Anfang 1875 bis Frühling 1876*, eds Giorgio Colli and Mazzino Montinari (Berlin, 1967).

Nirenberg, David. *Anti-Judaism: The Western Tradition* (New York, 2013).

Oergel, Maike. *The Return of King Arthur and the Nibelungen*: National Myth in Nineteenth-Century English and German Literature (Berlin, 1998).

Oettli, Samuel. *Das Gesetz Hammurabis und die Thora Israels. Eine religions- und rechtsgeschichtliche Parallele* (Leipzig, 1903).

Olender, Maurice. *The Languages of Paradise: Race, Religion, and Philology in the Nineteenth Century*, trans. Arthur Goldhammer (Cambridge, MA, 1992).

Parry, Graham. *The Trophies of Time: English Antiquarians of the Seventeenth Century* (Oxford, 1995).

Patai, Raphael, ed. *The Complete Diaries of Theodor Herzl*, trans. Harry Zohn, 5 vols (New York, 1960).

Petuchowski, Jacob. *Prayerbook Reform in Europe* (New York, 1968).

Philadelphus (Francis Pott). *Marriage with a Deceased Wife's Sister: A Brief General Review of the Arguments and Pleas on this Subject* ... (London, 1885).

Philipson, David. 'The Rabbinical Conferences, 1844–6', *Jewish Quarterly Review* 17 (1905), 656–89.

Pluche, Noël-Antoine. *Histoire du ciel, considéré selon les idées des poëtes, des philosophes, et de Moïse*, 2 vols (Paris, 1739).

Polaschegg, Andrea, and Michael Weichenhan, eds. *Berlin – Babylon. Eine deutsche Faszination, 1890–1930* (Berlin, 2017).

Post, Albert Hermann. *Grundriss der ethnologischen Jurisprudenz*, vol. 1, *Allgemeiner Teil* (Oldenburg, 1894).

Puller, F.W. *Marriage with a Deceased Wife's Sister: Forbidden by the Laws of God and of the Church* (London, 1912).

Pulzer, Peter. *Jews and the German State: The Political History of a Minority 1848–1933* (Oxford, 1992).

Pusey, Edward Bouverie. *Letter on the Proposed Change in the Laws Prohibiting Marriage Between Those Near of Kin* (London, 1842).

Pusey, Edward Bouverie. *Marriage with a Deceased Wife's Sister Prohibited by Holy Scripture as Understood by the Church for 1500 Years: Evidence Given before the Commission Appointed to Inquire into the State and Operation of the Law of Marriage as Relating to the Prohibited Degrees of Affinity* (Oxford, 1849).

Rabault-Feuerhahn, Pascale. *Archives of Origins: Sanskrit, Philology, Anthropology in 19th Century Germany*, trans. Dominique Bach and Richard Willet (Wiesbaden, 2013).

Rackman, Emanuel. *Modern Halakhah for Our Time* (Hoboken, 1995).

Rackman, Emanuel. *One Man's Judaism: Renewing the Old and Sanctifying the New* (Jerusalem, 2000).

Redekop, Benjamin W. *Enlightenment and Community: Lessing, Abbt, Herder, and the Quest for a German Public*, McGill-Queen's Studies in the History of Ideas 28 (Montreal, 2000).

Reinalter, Helmut. 'Der aufgeklärte Absolutismus – Geschichte und Perspektiven der Forschung', in *Der aufgeklärte Absolutismus im europäischen Vergleich*, eds Helmut Reinalter and Harm Klueting (Vienna, 2002), 11–19.

Renger, Johannes. 'Noch einmal: Was was der "Kodex" Ḫammurapi – ein erlassenes Gesetz oder ein Rechtsbuch?', in *Rechtskodifizierung und soziale Normen im interkulturellen Vergleich*, ed. Hans-Joachim Gehrke, Script-Oralia 15 (Tübingen, 1994), 27–59.

Reusch, Johann J.K. 'Germans as Noble Savages and Castaways: Alter Egos and Alterity in German Collective Consciousness During the Long Eighteenth Century', *Eighteenth-Century Studies* 42 (2008), 91–129.

Reynolds, Henry Revell. *Considerations on the State of the Law Regarding Marriages with a Deceased Wife's Sister* (London, 1840).

Ritter, Joachim. 'Moralität und Sittlichkeit. Zu Hegels Auseinandersetzung mit der kantischen Ethik [1966]', repr. in idem, *Metaphysik und Politik. Studien zu Aristoteles und Hegel* (Frankfurt, 1977), 281–309.

Roemer, Nils H. *Jewish Scholarship and Culture in Nineteenth-Century Germany: Between History and Faith*, Studies in German Jewish Cultural History and Literature (Madison, 2005).

Rogerson, John William. *W.M.L. de Wette, Founder of Modern Biblical Criticism: An Intellectual Biography* (Sheffield, 1992).

Röhl, John C.G. 'Herzl and Kaiser Wilhelm II: A German Protectorate in Palestine?', in *Theodor Herzl and the Origins of Zionism*, eds Ritchie Robertson and Edward Timms, Austrian Studies 8 (Edinburgh, 1997), 27–38.

Röhl, John C.G. *Wilhelm II: Into the Abyss of War and Exile, 1900–1941*, trans. Sheila de Bellaigue and Roy Bridge (Cambridge, 2014).

Röhl, John C.G. *Wilhelm II: The Kaiser's Personal Monarchy, 1888–1900*, trans. Sheila de Bellaigue (Cambridge, 2004).

Rohrbach, Paul. *Die Geschichte der Menschheit* (Königsstein, 1914).

Roscher, Wilhelm. 'Umrisse zur Naturlehre der drei Staatsformen III', *Allgemeine Zeitschrift für Geschichte* 7 (1847), 436–473.

Rose, Henry John. *The Law of Moses Viewed in Connexion with the History and Character of the Jews, with a Defence of the Book of Joshua against Professor Leo of Berlin: Being the Hulsean Lectures for 1833. To Which is Added An Appendix Containing Remarks on the Arrangement of the Historical Scriptures Adopted by Gesenius, de Wette, and Others* (Cambridge 1834).

Rose, Hugh James. *Notices of the Mosaic Law: With Some Account of the Opinions of Recent French Writers Concerning It* (London, 1831).

Rose, Sven Erik. *Jewish Philosophical Politics in Germany, 1789–1848* (Waltham, MA, 2014).

Rosenbloom, Noah H. *Tradition in an Age of Reform: The Religious Philosophy of Samson Raphael Hirsch* (Philadelphia, 1976).

Rubiés, Joan-Pau. 'Oriental Despotism and European Orientalism: Botero to Montesquieu', *Journal of Early Modern History* 9 (2005), 109–80.

Rückert, Joachim. 'Kodifikationsstreit', in *Handwörterbuch zur deutschen Rechtsgeschichte*, 2nd ed., 16th instalment (2012), available online at https://www.hrgdigital.de/HRG.kodifikationsstreit (accessed 6 December 2022).

Ruderman, David. *Missionaries, Converts, and Rabbis: The Evangelical Alexander McCaul and Jewish-Christian Debate in the Nineteenth Century* (Philadelphia, 2020).

Ruderman, David. 'Towards a Preliminary Portrait of an Evangelical Missionary to the Jews: The Many Faces of Alexander McCaul (1799–1863)', *Jewish Historical Studies* 47 (2017), 48–69.

Rüthers, Monika. 'Von der Ausgrenzung zum Nationalstolz: "Weibische" Juden und "Muskeljuden"', in *Der Traum von Israel/ Die Ursprünge des modernen Zionismus*, ed. Heiko Haumann (Weinheim, 1998), 319–29.

Sacks, Elias. *Mendelssohn's Living Script: Philosophy, Practice, History, Judaism* (Bloomington, IN, 2016).

Salzer, Dorothea. *Mit der Bibel in die Moderne. Entstehung und Entwicklung der Gattung jüdische Kinderbibel*, Studia Judaica 122 (Berlin, 2023).

Savigny, Friedrich Carl (von). *System des heutigen Römischen Rechts* (Berlin, 1840).

Savigny, Friedrich Carl (von). *Vom Beruf unserer Zeit für Gesetzgebund und Rechtswissenschaft* (Heidelberg, 1814).

Sayce, Archibald Henry. *Alte Denkmäler im Lichte neuer Forschungen. Ein Überblick über die durch die jüngsten Entdeckungen in Egypten, Assyrien, Babylonien, Palästina und Kleinasien erhältlichen Bestätigungen biblischer Tatsachen* (Leipzig, 1886).

Sayce, Archibald Henry. *Fresh Light from the Ancient Monuments: A Sketch from the Most Striking Confirmations of the Bible from Recent Discoveries in Egypt, Palestine, Assyria, Babylonia*, 2nd ed., By-Paths of Bible Knowledge 11 (London, 1884).

Sayce, Archibald Henry, *The 'Higher Criticism' and the Verdict of the Monuments* (London, 1894).

Sayce, Archibald Henry. *The Witness of Ancient Monuments to the Old Testament Scriptures*, Proceedings of the Conference of the German Association of University Teachers of English 32 (London, 1884).

Schäfer, Peter. 'Die Torah der messianischen Zeit', *Zeitschrift für die neutestamentliche Wissenschaft* 65 (1974), 27–42.

Schechter, Ronald. *Obstinate Hebrews: Representations of Jews in France, 1715–1815*, Studies on the History of Society and Culture (Berkeley, 2003).

Scheil, V. *Mémoires publiés sous la direction de M. J. de Morgan*, 2nd series, vol. 4, *Textes élamites-sémitiques*, Ministère de l'instruction publique et des beaux-arts. Délégation en Perse (Paris, 1902).

Scheil, Vincent. *La loi de Hammourabi (vers 2000 av. J.-C.)* (Paris, 1904).

Schelling, Friedrich. 'Philosophie der Mythologie', in *Friedrich Wilhelm Joseph von Schellings sämmtliche Werke*, Section 2, vol. 2, ed. Carl Friedrich August Schelling (Stuttgart, 1857).

Schelling, Friedrich. 'Philosophie der Offenbarung', in *Friedrich Wilhelm Joseph von Schellings sämmtliche Werke*, Section 2, vols 3–4, ed. Carl Friedrich August Schelling (Stuttgart, 1858).

Schiff, Danny. *Judaism in a Digital Age: An Ancient Tradition Confronts a Transformative Era* (Basingstoke, 2023).

Schiller, Friedrich. 'Die Götter Griechenlandes', *Der Teutsche Merkur* 61 (1788), 250–260.

Schiller, Friedrich. 'Ueber die ästhetische Erziehung des Menschen, in einer Reyhe von Briefen', *Die Horen. Eine Monatsschrift, von einer Gesellschaft verfaßt und herausgegeben von Schiller* 1/1 (1795), 7–48; 1/2 (1795), 55–94; 2/6 (1795), 45–124.

Schmidt, Wolf Gerhard. *'Homer des Nordens' und 'Mutter der Romantik'. James Macphersons Ossian und seine Rezeption in der deutschsprachigen Literatur* (Berlin, 2003).

Schneewind, Jerome B. *The Invention of Autonomy: A History of Modern Moral Philosophy* (Cambridge, 1997).

Scholem, Gershom. *The Messianic Idea in Judaism and Other Essays on Jewish Spirituality* (New York, 1971).

Schorsch, Ismar. *From Text to Context: The Turn to History in Modern Judaism*, Tauber Institute for the Study of European Jewry Series (Hanover, NH, 1994).

Schorsch, Ismar. 'Zacharias Frankel and the European Origins of Conservative Judaism', *Judaism* 30 (1981), 344–54.

Schulte, Christoph. *Die jüdische Aufklärung. Philosophie, Religion, Geschichte* (Munich, 2002).

Schutjer, Kari. *Goethe and Judaism: The Troubled Inheritance of Modern Literature* (Evanston, 2015).

Schwartz, Daniel B. *The First Modern Jew: Spinoza and the History of an Image* (Princeton, 2012).

Schwartz, Dov. *Religion or Halakha: The Philosophy of Rabbi Joseph B. Soloveitchik* (Leiden, 2007).

Scott, H. M. 'Whatever Happened to the Enlightened Despots?', *History* 68 (1983), 245–57.

Senyal, Lopa. *English Literature in Eighteenth Century* (New Delhi, 2006).

Seyferth, Peter, ed. *Den Staat zerschlagen! Anarchistische Staatsverständnisse* (Baden-Baden, 2015).

Sharvit, Gilad, and Karen S. Feldman, eds. *Freud and Monotheism: Moses and the Violent Origins of Religion*, Berkeley Forum in the Humanities (New York, 2018).

Shavit, Yaacov, and Mordechai Eran. *The Hebrew Bible Reborn: From Holy Scripture to the Book of Books. A History of Biblical Culture and the Battles over the Bible in Modern Judaism*, trans. Chaya Naor, Studia Judaica 38 (Berlin, 2007).

Sheehan, Jonathan. *The Enlightenment Bible: Translation, Scholarship, Culture* (Princeton, 2005).

Sieg, Ulrich. *Deutschlands Prophet. Paul de Lagarde und die Ursprünge des modernen Antisemitismus* (Munich, 2007).

Sikka, Sonia. *Herder on Humanity and Cultural Difference: Enlightened Relativism* (Cambridge, 2011).

Smend, Rudolf. 'Lowth in Deutschland', in idem, *Epochen der Bibelkritik*, Gesammelte Studien 3, Beiträge zur evangelischen Theologie 109 (Munich, 1991), 43–62.

Smith, George. *The Chaldean Account of Genesis: Containing the Description of the Creation, the Fall of Man, the Deluge, the Tower of Babel, the Times of the Patriarchs, and Nimrod. Babylonian Fables, and Legends of the Gods. From the Cuneiform Inscriptions* (London, 1876).

Smith, Helmut Walser. *The Continuities of German History: Nation, Religion, and Race across the Long Nineteenth Century* (Cambridge, 2008).

Soloveitchik, Joseph B. *Halakhic Man*, trans. Lawrence J. Kaplan (Philadelphia, 1983).

Sorkin, David. *The Berlin Haskalah and German Religious Thought: Orphans of Knowledge* (London, 2000).

Sorkin, David. *The Religious Enlightenment: Protestants, Jews, and Catholics from London to Vienna* (Princeton, 2008).

Sorkin, David. *The Transformation of German Jewry, 1780–1840* (New York, 1987).

Sparks, George Downing. 'The Law of Moses Historically Considered', *The Sewanee Review* 14, no. 3 (1906): 281–87.

Spendel, Günter. *Josef Kohler. Bild eines Universaljuristen*, Heidelberger Forum 17 (Heidelberg, 1983).

Spitzer, Samuel. *Die jüdische Ehe nach mosaisch talmudischen und den in Oesterreich bestehenden, besonders neuesten Ehegesetzen* (Essek, 1869).

Stanyan, Temple. *The Grecian History*, vol. 1, *Containing the Space of about 1684 Years* (London, 1707), vol. 2, *From the End of the Peloponnesian War to the Death of Philip of Macedon, Containing the Space of Sixty-eight Years* (London, 1739).

Steinthal, Heymann. *Allgemeine Ethik* (Berlin, 1885).

Steinthal, Heymann. 'Die ursprüngliche Form der Sage von Prometheus (Mit Bezug auf: Kuhn, *Die Herabkunft des Feuers und des Göttertranks*)', *Zeitschrift für Völkerpsychologie und Sprachwissenschaft* 2 (1862), 1–29.

Steinthal, Heymann. *Mythos und Religion* (Berlin, 1870).

Stern, Eliyahu. 'Catholic Judaism: The Political Theology of the Nineteenth-Century Russian Jewish Enlightenment', *Harvard Theological Review* 109/4 (2016), 483–511.

Strauch, Dieter. 'Quellen, Aufbau und Inhalt des Gesetzbuches', in *Das schwedische Reichsgesetzbuch (Sveriges Rikes Lag) von 1734. Beiträge zur Entstehungs- und Entwicklungsgeschichte einer vollständigen Kodifikation*, ed. Wolfgang Wagner, Ius Commune: Veröffentlichungen des Max-Planck-Instituts für Europäische Rechtsgeschichte, Sonderhefte: Studien zur europäischen Rechtsgeschichte 29 (Frankfurt, 1986), 61–106.

Strodtmann, Johann Christoph. *Übereinstimmung der deutschen Alterthümer mit den biblischen, sonderlich hebräischen* (Wolfenbüttel, 1755).

Sutcliffe, Adam. *Judaism and Enlightenment*, Ideas in Context (Cambridge, 2005).

Sutcliffe, Adam. *What are Jews For? History, Peoplehood, and Purpose* (Princeton, 2020).

Tasch, Roland. *Samson Raphael Hirsch. Jüdische Erfahrungswelten im historischen Kontext* (Berlin, 2011).

Taylor, Charles. *Hegel* (Cambridge, 1975).

Theimer, Walter. *Geschichte des Sozialismus* (Tübingen, 1988).

Thibaut, Anthon Friedrich Justus. *Ueber die Nothwendigkeit eines allgemeinen bürgerlichen Rechts für Deutschland* (Heidelberg, 1814).

Thirlwall, Connop. 'Speech of the late Bishop of St David's', in *Tracts Issued by the Marriage Law Defence Union,* vol. 1, *Scriptural* (London, 1889), tract 24, pp. 173–76.

Thompson, Thomas L. *The Mythic Past: Biblical Archaeology and the Myth of Israel* (New York, 1999).

Toller, Ernst. *Die Wandlung* (Potsdam, 1919).

Tombo, Rudolph. *Ossian in Germany: Bibliography, General Survey, Ossian's Influence upon Klopstock and the Bards* (New York, 1901).

Totzeck, Markus M. *Die politischen Gesetze des Mose. Entstehung und Einflüsse der politica-judaica Literatur in der Frühen Neuzeit,* Refo500 Academic Studies 49 (Göttingen, 2019).

Trevor, George. *The Scriptural Argument against Marriage with a Deceased Wife's Sister* (London, 1884).

Ussishkin, David. 'The "Lachish Reliefs" and the City of Lachish', *Israel Exploration Journal* 30 (1980), 174–95.

v. O. Review of *Das Sklavenrecht des alten Testaments. Eine rechtsgeschichtliche Studie* by Max Mandl, *Vierteljahrschrift für Volkswirtschaft, Politik und Kulturgeschichte* 25/1 (1888), 103–06.

van Mieroop, Marc. *King Hammurabi of Babylon: A Biography* (Malden, MA, 2005).

Van Rahden, Till. 'Von der Eintracht zur Vielfalt. Juden in der Geschichte des deutschen Bürgertums', in *Juden, Bürger, Deutsche. Zur Geschichte von Vielfalt und Differenz, 1800–1933,* eds Andreas Gotzmann, Rainer Liedtke, and Till van Rahden (Tübingen, 2001), 9–32.

Velthusen, Johann Caspar. *Einfluss frommer Juden und ihrer Harfe auf den Geist roher Nationen, insonderheit auf Ossians Bardenlieder ...* (Leipzig, 1807).

Viesel, Hansjörg, ed. *Literaten an der Wand. Die Münchner Räterepublik und die Schriftsteller* (Frankfurt am Main, 1980).

Voegelin, Eric. *Die politischen Religionen,* 3rd ed., ed. Peter J. Opitz (Munich 2007 [1938]).

Voltaire, *Essai sur les mœurs et l'esprit des nations* (Geneva, 1756).

von der Krone, Kerstin. *Wissenschaft in Öffentlichkei. Die Wissenschaft des Judentums und ihre Zeitschriften,* Studia Judaica 65 (Berlin, 2012).

von der Schulenburg, Sigrid. *Leibniz als Sprachforscher* (Frankfurt, 1973).

Voß, Johann Heinrich. 'Deutschland. An Friedrich Leopold, Graf zu Stolberg [1772]', in idem, *vermischte Gedichte und prosaische Aufsätze* (Leipzig, 1784), 19–23.

Walicki, Andrzej. *A History of Russian Thought: From the Enlightenment to Marxism* (Stanford, 1979).

Wallace, Anne D. *Sisters and the English Household: Domesticity and Women's Autonomy in Nineteenth-century English Literature* (London, 2018).

Wallace, Valerie, and Colin Kidd. 'Biblical Criticism and Scots Presbyterian Dissent in the Age of Robertson Smith', in *Dissent and the Bible in Britain, c. 1650–1950,* eds Scott Mandelbrote and Michael Ledger-Lomas (Oxford, 2013), 233–55.

Warburton, William. *The Divine Legation of Moses Demonstrated, On the Principles of a Religious Deist, From the Omission of the Doctrine of a Future State of Reward and Punishment in the Jewish Dispensation,* 2 parts (London, 1738, 1742).

Waubke, Hans-Günther. *Die Pharisäer in der protestantischen Bibelwissenschaft des 19. Jahrhunderts,* Beiträge zur historischen Theologie 107 (Tübingen, 1998).

Weber, Ferdinand Wilhelm. *Der Profet Jesaja, in Bibelstunden ausgelegt,* 2 vols (Nördlingen, 1875, 1876).

Weber, Ferdinand Wilhelm. *System der altsynagogalen palästinischen Theologie aus Targum, Midrasch und Talmud dargestellt,* eds Franz Delitzsch and Georg Schnedermann, 1st ed. (Leipzig, 1880). Reprinted with a new title as *Die Lehren des Talmud, quellenmässig, systematisch und gemeinverständlich dargestellt,* Schriften des Institutum Judaicum 2/1 (Leipzig, 1886) and in its 2nd ed. as *Jüdische Theologie auf Grund des Talmud und verwandter Schriften gemeinfasslich dargestellt* (Leipzig, 1897).

Wehr, Gerhard. *Europäische Mystik zur Einführung* (Hamburg, 1995).

Weichenhan, Michael. *Der Panbabylonismus. Die Faszination des himmlischen Buches im Zeitalter der Zivilisation* (Berlin, 2016).

Weidner, Daniel. 'Politik und Ästhetik: Lektüre der Bibel bei Michaelis, Herder und de Wette', in *Hebräsiche Poesie und jüdischer Volksgeist. Die Wirkungsgeschichte Johann Gottfried Herders im Judentum Mittel- und Osteuropas,* eds C. Schulte et al. (Hildesheim, 2003), 57–63.

Weidner, Daniel. 'Säkularisierung', in *Enzyklopädie jüdischer Geschichte und Kultur,* vol. 5, ed. Dan Diner (Stuttgart, 2014), 295–301.

Weinbrot, Howard D. *Britannia's Issue: The Rise of British Literature from Dryden to Ossian* (Cambridge, 2007).

Wendel, Saskia. 'Religionsphilosophie nach Kant', *Colloquia Theologica* 2 (2001), 203–14.

Wiedemann, Felix. '"Apologie der Semiten". Der Münchner Semitist und Assyriologe Fritz Hommel zwischen Philo- und Antisemitismus,' *Zeitschrift für Religions-und Geistesgeschichte* 75 (2023).

Wiener, Harold M. *Essays in Pentateuchal Criticism* (Oberlin, 1909).

Wiener, Harold M. *Studies in Biblical Law* (London, 1904).

Wiener, Harold M. *The Origin of the Pentateuch* (London, 1910).

Wiese, Christian. *Challenging Colonial Discourse: Jewish Studies and Protestant Theology in Wilhelmine Germany,* trans. Barbara Harshav and Christian Wiese, Studies in European Judaism 10 (Leiden, 2005).

Wiese, Christian. 'Ein "aufrichtiger Freund des Judentums"? "Judenmission", christliche Judaistik, und Wissenschaft des Judentums im deutschen Kaiserreich am Beispiel Hermann L. Stracks', in *Gottes Sprache in der philologischen Werkstatt. Hebraistik von 15. bis zum 19. Jahrhundert*, eds Giuseppe Veltri and Gerold Necker, Studies in European Judaism 11 (Leiden, 2004), 277–316.

Wiese, Christian, Walter Homolka, and Thomas Brechenmacher, eds. *Jüdische Existenz in der Moderne. Abraham Geiger und die Wissenschaft des Judentums* (Berlin, 2016).

Wilhelm II. *Das Königtum im alten Mesopotamien* (Berlin, 1938).

Wilhelm II. 'Kaiser Wilhelm on "Babel and Bible". (Letter from His Majesty Emperor William II. To Admiral Hollman, President of the Oriental Society)', *The Open Court* 7 (1903), 432–36. Originally published as 'Babel und Bibel. Ein Handschreiben Seiner Majestät Kaiser Wilhelms des Zweiten an das Vorstandsmitglied der Deutschen Orientgesellschaft, Admiral Hollmann', *Die Grenzboten. Zeitschrift für Politik, Literatur und Kunst* 62/8 (1903), 493–96.

Wilke, Carsten, ed. *Biographisches Handbuch der Rabbiner*, vol. 1, *Die Rabbiner der Emanzipationszeit in den deutschen, böhmischen und großpolnischen Ländern, 1781–1871* (Munich, 2004).

Willems, Joachim. *Religiöser Gehalt des Anarchismus und anarchistischer Gehalt der Religion? Die jüdisch-christlich-atheistische Mystik Gustav Landauers zwischen Meister Eckhart und Martin Buber* (Albeck, 2001).

Williamson, George S. *The Longing for Myth in Germany: Religion and Aesthetic Culture from Romanticism to Nietzsche* (Chicago, 2004).

Winckelmann, Johann Joachim. *History of the Art of Antiquity*, trans. Harry Francis Mallgrave (Los Angeles, 2006). First published as *Geschichte der Kunst des Alterthums* (Dresden, 1764).

Winckler, Hugo. *Die Gesetze Hammurabis in Umschrift und Übersetzung* (Leipzig, 1904).

Winckler, Hugo. *Die Gesetze Hammurabis, Königs von Babylon um 2250 v. Chr. Das älteste Gesetzbuch der Welt,* Der Alte Orient 4/4 (Leipzig, 1902).

Wishnitzer, M. 'Moses Mendelssohn', *Jewish Quarterly Review* 25 (1935), 307–10.

Wismann, Heinz. *Penser entre les langues* (Paris, 2012).

Witte, Bernd. *Moses und Homer. Griechen, Juden Deutsche: Eine andere Geschichte der deutschen Kultur* (Berlin, 2018).

Wolf, Immanuel. 'Über den Begriff einer Wissenschaft des Judenthums', *Zeitschrift für die Wissenschaft des Judentums* 1 (1822), 1–24.

Wordsworth, Charles. 'Lev. XVIII. v. 18. "A Wife to her Sister." Explained', in *Tracts Issued by the Marriage Law Defence Union*, vol. 1, *Scriptural* (1889), tract 22, pp. 155–56.

Wordsworth, Christopher. *On Marriage with a Deceased Wife's Sister*, new ed. (London, 1883 [1876]).

Wordsworth, Christopher. 'What the Bishop of Lincoln says; As addressed to the Clergy and Laity of his Diocese at his Triennial Visitation, Oct. 1882', in *Tracts Issued by the Marriage Law Defence Union,* vol. 1, *Scriptural* (1889), tract 1, pp. 7–14.

Wright, Melanie Jane. *Moses in America: The Cultural Uses of Biblical Narrative* (Oxford, 2002).

Wundt, Wilhelm. *Völkerpsychologie,* 10 vols (Leipzig, 1900–20).

Yerushalmi, Yosef Hayim. *Freud's Moses: Judaism Terminable and Interminable,* Franz Rosenzweig Lecture Series (New Haven, 1991).

Ziolkowski, Theodore. *Uses and Abuses of Moses: Literary Representations since the Enlightenment* (Notre Dame, 2016).

Zschackwitz, Johann Ehrenfried. *Erläuterte Teutsche Alterthümer, Worinnen ...* (Frankfurt, 1743).

Zwiep, Irene E. 'Nation and Translation: Steinschneider's *Hebräische Übersetzungen* and the End of Jewish Cultural Nationalism', in *Latin-into-Hebrew,* vol. 1, *Studies,* eds Resianne Fontaine and Gad Freudenthal, Studies in Jewish History and Culture 39/1 (Leiden, 2013), 421–45.

# Index of Persons